Target Organ Toxicology Series

Neurotoxicology

Target Organ Toxicology Series

Series Editors
A. Wallace Hayes, John A. Thomas, and Donald E. Gardner

Neurotoxicology
Hugh A. Tilson and Clifford L. Mitchell, editors, 416 pp., 1992

Toxicology of the Lung
Donald E. Gardner, James D. Crapo, and Edward J. Massaro, editors, 540 pp., 1988

Endocrine Toxicology
John A. Thomas, Kenneth S. Korach and John A. McLachlan, editors, 414 pp., 1985

Immunotoxicology and Immunopharmacology
Jack H. Dean, Michael I. Luster, Albert E. Munson, and Harry Amos, editors, 528 pp., 1985

*Reproductive Toxicology
Robert L. Dixon, editor, 368 pp., 1985

Toxicology of the Blood and Bone Marrow
Richard D. Irons, editor, 192 pp., 1985

Toxicology of the Eye, Ear, and Other Special Senses
A. Wallace Hayes, editor, 264 pp., 1985

*Toxicology of the Immune System
Jack H. Dean and Albert E. Munson, editors, 528 pp., 1985

Cutaneous Toxicity
Victor A. Drill and Paul Lazar, editors, 288 pp., 1984

Intestinal Toxicology
Carol M. Schiller, editor, 252 pp., 1984

Cardiovascular Toxicology
Ethard W. Van Stee, editor, 400 pp., 1982

*Nervous System Toxicology
Clifford L. Mitchell, editor, 392 pp., 1982

*Toxicology of the Liver
Gabriel L. Plaa and William R. Hewitt, editors, 350 pp., 1982

*Developmental Toxicology
Carole A. Kimmel and Judy Buelke-Sam, editors, 354 pp., 1981

*Toxicology of the Kidney
Jerry B. Hook, editor, 288 pp., 1981

---

*Out of print.

Target Organ Toxicology Series

Neurotoxicology

Editors

Hugh A. Tilson, PH.D.
Director, Neurotoxicology Division
Health Effects Research Laboratory
United States Environmental Protection Agency
Research Triangle Park, North Carolina

Clifford L. Mitchell, PH.D.
Molecular and Integrative Neuroscience
National Institute of Environmental Health Sciences
Research Triangle Park, North Carolina

Raven Press   New York

Raven Press, Ltd., 1185 Avenue of the Americas, New York, New York 10036

Made in the United States of America

Library of Congress Cataloging-in-Publication Data

Neurotoxicology / editors, Hugh A. Tilson, Clifford L. Mitchell.
    p.   cm.—(Target organ toxicology series)
  Includes bibliographical references and index.
  ISBN 0-88167-849-X
  1. Neurotoxicology. I. Tilson, Hugh A. II. Mitchell, Clifford
L. III. Series.
  [DNLM: 1. Nervous System—drug effects. 2. Nervous System
Diseases—chemically induced. 3. Neurobiology. WL 100 N49673]
  RC347.5.N483 1991
  616.8—dc20
  DNLM/DLC
  for Library of Congress   91-35188
                                        CIP

9   8   7   6   5   4   3   2   1

Dedication

This volume of the *Target Organ Toxicology* series is dedicated to the late Dr. Robert L. Dixon. Dr. Dixon served meritoriously as the Editor-in-Chief of this highly successful series of monographs. It was through Dr. Dixon's early recognition of the need for *Target Organ Toxicology* that this successful multivolume series became established. From its beginning by Dr. Dixon, this series will continue to review the morphology, physiology, biochemistry, cellular biology, and developmental aspects of various target organs. A description of the tests routinely used to determine toxicity, evaluation of the feasibility of assays used in the assessment of hazards, and the need for applying recent advances in the basic sciences to the development and validation of new testing procedures will be included in this series again. Likewise, a description of the incidence of chemically induced human disease, the reliability of laboratory data extrapolation to humans, and the methods currently used to estimate human risks, are to be contained in this volume.

Dr. Dixon's early investigations into toxicological events affecting the reproductive systems are well known, and were recognized by the Society of Toxicology in 1972, when he received its prestigious Achievement Award. Later, Dr. Dixon went on to serve as President of the Society of Toxicology from 1982–83. Bob was also noted for his role in relating basic toxicological findings to regulatory affairs and risk assessment. His untimely death also created a professional need for the continuance of *Target Organ Toxicology.* Accordingly, and with the respect of a long personal and professional relationship that all the present Editors enjoyed with the late Dr. Dixon, it is with renewed anticipation that this series will continue to be a valuable reference to the next generation of toxicologists. We are indeed indebted to the late Dr. Robert L. Dixon for his foresight in having established this important series in toxicology.

The Editors

Contents

Contributors

W. Kent Anger   *CROET L-474, Oregon Health Sciences University, 3181 Sam Jackson Parkway, Portland, Oregon 97201*

W. K. Boyes   *US EPA, Health Effects Research Laboratory, Neurotoxicology Division, Research Triangle Park, North Carolina 27711*

Philip J. Bushnell   *Neurotoxicology Division, Health Effects Research Laboratory, US Environmental Protection Agency, Research Triangle Park, North Carolina 27711*

Deborah A. Cory-Slechta   *Environmental Health Sciences Center, University of Rochester School of Medicine and Dentistry, P.O. Box BPHYS, Rochester, New York 14642*

Lucio G. Costa   *Department of Environmental Health, University of Washington, SC-34, Seattle, Washington 98195*

K. M. Crofton   *Neurotoxicology Division, Health Effects Research Laboratory, US Environmental Protection Agency, Research Triangle Park, North Carolina 27711*

David A. Eckerman   *Department of Psychology, University of North Carolina, Chapel Hill, North Carolina 27514*

David Gaylor   *National Center for Toxological Research, US Food and Drug Administration, Jefferson, Arkansas 72079*

John W. Griffin   *Department of Neurology, Johns Hopkins University School of Medicine, Meyer Building, Room 5-119, 600 North Wolfe Street, Baltimore, Maryland 21205*

Beverly M. Kulig   *Department of Pharmacology, TNO Medical Biological Laboratory, P.O. Box 45, Rijswijk, 2280 AA The Netherlands*

J. H. C. M. Lammers   *Department of Pharmacology, TNO Medical Biological Laboratory, P.O. Box 45, Rijswijk, 2280 The Netherlands*

Herbert E. Lowndes   *Neurotoxicology Laboratories, College of Pharmacy, Rutgers University, Piscataway, New Jersey 08855-0789*

Robert C. MacPhail   *Neurotoxicology Division, Health Effects Research Laboratory, Office of Research and Development, US Environmental Protection Agency, Research Triangle Park, North Carolina 27751*

Joel L. Mattsson *The Dow Chemical Company, Health and Environmental Sciences, Midland, Michigan 48674*

James P. O'Callaghan *Neurotoxicology Division, US Environmental Protection Agency, Research Triangle Park, North Carolina 27711*

Kenneth R. Reuhl *Neurotoxicology Laboratories, College of Pharmacy, Rutgers University, P.O. Box 789, Piscataway, New Jersey 08855-0789*

J. F. Ross *The Procter and Gamble Company, Miami Valley Laboratories, Cincinnati, Ohio 45239-8707*

William F. Sette *Health Effects Division, Office of Pesticide Programs, US Environmental Protection Agency, 401 M Street, S.W., Washington, D.C. 20460*

William Slikker, Jr. *National Center for Toxicological Research, US Food and Drug Administration, Jefferson, Arkansas 72079*

M. Anthony Verity *Department of Pathology and Laboratory Medicine, Division of Neuropathology, University of California at Los Angeles School of Medicine, Los Angeles, California 90024-1732*

Bellina Veronesi *Neurotoxicology Division, US Environmental Protection Agency, Research Triangle Park, North Carolina 27711*

Charles V. Vorhees *Institute for Developmental Research, Children's Hospital Research Foundation and Departments of Pediatrics and Environmental Health, University of Cincinnati, Elland and Bethesda Avenues, Cincinnatti, Ohio 45229*

Preface

Neurotoxicology studies the adverse effects of chemical, biological, and physical factors on the function and/or structure of the nervous system. As a discipline within toxicology, neurotoxicology has only recently emerged. The large majority of publications, symposia, and conferences dealing with neurotoxicology has occurred in the last ten years. The nervous system has a number of special features not found in other organ systems. Once damaged, it has a modest capacity for repair and regeneration, especially in the central nervous system. This limited ability to regenerate means that subtle damage to the nervous system can have serious, long-lasting effects. Another unique feature of the nervous system is that it is comprised of a number of different cell types interconnected by a complex communication mechanism dependent upon chemical transmission. Agent-induced alterations in neurotransmitter synthesis, release, uptake, metabolism, or signal transduction processes can adversely affect the functioning of the nervous system.

Human exposure to neurotoxic agents may be significant. The U.S. Government Office of Technology Assessment has estimated that 25% of all chemicals may be neurotoxic. The possibility of human exposure to neurotoxicants has led to the promulgation of testing guidelines by regulatory agencies in the United States. Neurotoxicity is also assuming more importance in risk assessment and management decisions. This volume reviews the latest information in basic neurobiology as it applies to neurotoxicology, as well as covers important new developments in risk assessment and regulatory neurotoxicology. This volume includes chapters on the rapidly developing field of molecular biology as it relates to neurotoxicology, the potential advantages and disadvantages of using tissue culture methodologies in neurotoxicology, and the significance of mechanistic neuropathology to neurotoxicology. Recent advances in identifying broadly applicable markers of neuronal damage, as well as site(s) at which neurotoxicants interfere with neurotransmission, are also discussed. Chapters on cellular and molecular neurotoxicology are then followed by other chapters concerning functional indicators of neurotoxicity, including sensory-evoked potentials, schedule-controlled behavior, and neurobehavioral measures of sensory, motor, and cognitive function. These chapters are followed by discussions concerning the developing organism and the expression of developmental neurotoxicity, and recent advances in detecting and quantifying neurotoxicity in humans. Later chapters of this volume introduce the basic concepts of risk assessment as it pertains to neurotoxicity and controversial issues confronting neurotoxicologists in the regulatory domain.

One central theme of this volume is the need to study neurotoxicity in an integrated fashion. The integrated study of neurobehavioral, neurochemical, neurophysiological, and neuroanatomical endpoints will lead to a more precise understanding of the mechanism(s) of neurotoxicity and reduce uncertainties in the risk assessment process.

The editors hope that this volume will promote a greater understanding of neurotoxicity and stimulate those interested in the field to think about and initiate research in this area. We also believe that this volume will be useful to those scientists in government and industry who are responsible for public health and the safe and efficient use of chemicals in our daily lives.

Hugh A. Tilson
Clifford L. Mitchell

Target Organ Toxicology Series

Neurotoxicology

Neurotoxicology, edited by Hugh Tilson
and Clifford Mitchell. Raven Press, Ltd.,
New York © 1992.

1

Determination of Neurotoxicity Using Molecular Biological Approaches

M. Anthony Verity

*Department of Pathology and Laboratory Medicine, Division of Neuropathology,
UCLA School of Medicine, Los Angeles, California 90024-1732*

For the purpose of this discussion, "neurotoxicity" is the capacity of chemical, biological, or physical agents to cause adverse functional or structural change in the nervous system. The definition further includes the concept that an ultimate expression of the toxicity shall be irreversible neuroinjury culminating in a loss of the capacity to metabolize. This latter component does not define a single metabolic event and is still unfixed, but is a well-recognized endpoint. Although neuronal cell lysis is not a mandatory aspect of this concept, a perturbation of normal ionic homeostasis secondary to derangement of membrane permeability and ion transport is a central feature.

The advances and concepts deriving from the advent of molecular biology for biological thinking represent a modern Darwinian-like evolution. Recognition of the cellular mechanisms of inheritance and the genetic readout of structural (e.g., protein) parameters and their related functional parameters is rapidly expanding. Molecular neurobiology is a natural extension of this DNA-based technology and is a prime moving force in neuroscience. In contrast and unfortunately, the incorporation of molecular biological technology in toxicology is developing slowly, and meaningful experiments or concepts which embrace the twin disciplines of molecular neurobiology and neurotoxicology are virtually nonexistent. It is recognized that further advances in the discipline of neurotoxicology must stem from our ability to exploit molecular studies of the nervous system perturbed by endogenous or environmental toxicant-induced injury.

Conventional neurobiology has identified numerous macromolecules specific for nervous system differentiation and interaction likely to represent the molecular basis of memory, learning, and consciousness, etc. Neurotoxicology must identify the structure and interrelations of neurofunctional proteins on which toxicant-induced perturbation may be analyzed. Examples of such proteins include the neurotransmitter receptors, ion channel proteins, neurotransmitter release/ reuptake carrier translocases, neuron- or glial-specific trophic factors, determi-

1

nants of synaptogenesis, and molecules associated with neuron-neuron or neuronal-glial adhesion and communication. Site-directed toxicant interaction with these specific substrates will add specificity and thereby identify neurotoxicant mediation of disturbed cellular function perhaps leading to irreversible injury of defined nerve cell systems.

In an attempt to identify some strategies, concepts, and speculations for future direction, this chapter will identify selected molecular and cellular biological concepts which may provide approaches to an understanding of the mechanisms of neurotoxicity. The choice is representative, not comprehensive, and it carries the stamp of personal bias. Therefore representative examples of molecular neurobiological approaches to understanding neurotoxicant interaction come from studies of neurotransmitter (ligand) receptor-mediated interaction, myelin biogenesis, cell adhesion molecular biology, and disturbed phosphorylation.

MOLECULAR BIOLOGY OF *DROSOPHILA* AS A PARADIGM FOR NEUROTOXICOLOGY STUDIES

The molecular neurobiology of *Drosophila* illustrates principles of conceptual value in neurotoxicology. The advantages of this model are summarized in Table 1. The model allows for a combination of classic (electrophysiological) approaches with molecular genetic approaches for studying macromolecules involved in signaling. An advantage for neurotoxicological studies is the variety of strategies available to clone genes of interest. For example, a protein of interest (e.g., the Na^+ channel) (68) modified during pesticide neurotoxicity (16,59) may exemplify the use of gene cloning as schematized in Fig. 1. However, the most useful feature of *Drosophila* lies in the fact that a gene identified by its mutant phenotype may be cloned also by a variety of methods (Fig. 2). The use of *Drosophila* mutants has allowed the identification of specific genes that are known to encode proteins of interest. Acetylcholine is a major excitatory neurotransmitter and is accompanied by appropriate activities of the biosynthetic enzymes choline acetyltransferase (ChAT) and acetylcholinesterase (AChE) (31). Mutations in the

TABLE 1. Drosophila *studies and potential role in molecular neurotoxicology*

Simple neural organization.
Strong conservation of nervous system proteins.
Developmental studies through embryo, larvae, and adult.
Electrophysiological methods well developed.
Variety of strategies available for gene cloning (see also Figs. 1 and 2).
Genes mapped to precise physical location and banded chromosomes.
Various preparations:
  1. Thoraco-ganglion neurons
  2. Dorsal-longitudinal muscle
  3. Compound eye (e.g., ERG)
  4. Primary cultures of dissociated neurons
  5. Patch clamp

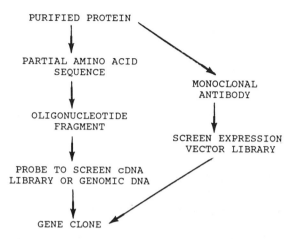

FIG. 1. Flow chart for the experimental methods used for genetic and molecular analysis of proteins. A partial amino acid sequence is obtained from a purified protein, allowing synthesis of an oligonucleotide fragment which can be used to probe a cDNA library to isolate a specific gene clone.

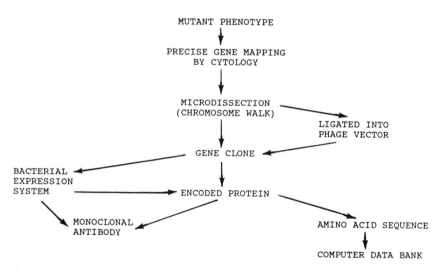

FIG. 2. Flow chart identifying the molecular studies needed to identify a specific protein associated with a mutant phenotype. Mutations may be induced in neuronal cell lines allowing for the identification of phenotypes with specific developmental or biochemical abnormalities of value in neurotoxicant analysis, neurotransmitter receptor deletion, ion channel defect, etc.

genes encoding ChAT and AChE have been isolated and their genes cloned (36,43). These mutants allow for an analysis of the role of these components in development, behavior, electrophysiology, and neurotoxicology of toxicants directed against the cholinergic system. A different genetic approach is used when the identity of the proteins essential to the process of interest and the location of the relevant genes are unknown. For instance, mutations can be induced by irradiation or other mutagens and mutants can be recovered which reveal aberrant developmental, electrophysiological, and/or behavioral activity (29).

The electrophysiology of the Na^+ channel is well documented and molecular analysis of Na^+ channels is being studied in several different organisms including *Drosophila*. The mechanisms which regulate expression, posttranslational modification, membrane distribution, etc. remain to be elucidated. The *Drosophila* homologue of the vertebrate Na^+ channel specifies a polypeptide very similar to that found with the vertebrate channel (28). The central subunit, α-subunit, is formed in all phyla studied and is regarded as the channel proper (16), with as many as six major binding sites for venoms, toxins, and drugs. The pyrethroids and DDT are potent insecticides acting at the Na^+ channel to cause prolonged opening (59,86,91). Costa et al. (24) showed selective phosphorylation of the α-subunit by a cyclic AMP (cAMP)-dependent protein kinase during depolarization; subsequently, DDT and deltamethrin were shown to inhibit Na^+ channel phosphorylation (42). The speculation points to excessive Na^+ influx as a result of inhibition of α-subunit phosphorylation by the pesticides.

This single example suggests a molecular biological approach in which similar phosphorylation studies would be performed in *Drosophila* or neuroblastoma mutants selected for defects in Na^+ channel activity. For instance, a Na^+ channel mutant nap[ts] (no action potential, temperature sensitive) mutation was recovered following screening for temperature sensitive paralytic mutants (90). If homozygous nap[ts] larvae or adults are shifted from 25° to 37°C, they become paralyzed. Normal mobility is regained upon return to 25°C. Extracellular recordings of action potentials reveal a failure in nerve conduction at 37°C (29). Of considerable interest to neurotoxicological investigations is the observation that dissociated neurons from nap[ts] mutants cultured *in vitro* are resistant to the cytotoxic effects of veratridine, a drug causing persistent activation of Na^+ channels (74). This observation provides an excellent example of the possible interrelated studies between a known genetic mutant and a Na^+ channel agonist. The use of such mutants in investigating the toxicant interaction of other presumed ionophoric-like agents may prove valuable. In fact, a variety of mammalian neuroblastoma cell lines have been shown to possess Na^+ channel variant properties (87). These studies were able to couple the variation in cytotoxicity, expressed in terms of neuroblastoma plating efficiency as a function of the activity/presence of the Na^+ channel as measured by $^{22}Na^+$ uptake. As summarized in Tables 2 and 3, the effect of veratridine on the plating efficiency of the wild phenotype neuroblastoma strain, N18, demonstrated a moderate cytopathic effect with veratridine but virtual extinction of colony plating efficiency when co-incubated with ve-

TABLE 2. *Effect of Na$^+$ channel neurotoxins on plating efficiency (viability) of wild neuroblastoma phenotype*

| | Plating efficiency, % | | |
| | | | Veratridine + |
Clone	Control	Veratridine, 40 μM	scorpion toxin, 50 nM
N.18	100	40	0.07

Plating efficiency determined after 10 days. Modified from West and Catterall, ref. 87.

ratridine and scorpion toxin (ScTx). However, following growth in the presence of veratridine with or without ScTx, numerous toxin-resistant neuroblastoma clones were isolated. When compared for their relative plating efficiency and ^{22}Na$^+$ uptake, the clones were found to demonstrate remarkable degrees of cytotoxic resistance correlated with the absence of ScTx-mediated ^{22}Na$^+$ uptake. These studies clearly reveal the value of using neuroblastoma variants to analyze the role of Na$^+$ channel activity in the analysis of unknown neurotoxic function which may have a component of Na$^+$ channel activity as well as (for example) neurotransmitter receptor involvement.

An extension of these studies stems from recent knowledge of the developmental regulation of Na$^+$ channel expression in the rat forebrain (69). These observations examine the changes in gene transcription and Na$^+$ channel subunit association which accompany the developmental increase in the Na$^+$ channel number. Specifically, they identified that channel expression was regulated at the level of gene transcription and that the mRNA level was temporarily correlated with axonal growth and synapse formation. Identification of the processes modulating Na$^+$ channel gene transcription during this period of neuronal development is of importance, but more so for proposed studies in the area of developmental neurotoxicology in which presumed Na$^+$ channel-linked neurotoxicants may be examined for cytotoxic endpoints. A primary culture paradigm in which a similar time course of Na$^+$ channel accumulation has been identified (7) and the formation of the high affinity saxitoxin binding site (representing the fully formed Na$^+$ channel) occurs only following linkage between

TABLE 3. *Na$^+$ channel and cytopathic correlation of toxin-resistant neuroblastoma clones*

| | | ^{22}Na$^+$ uptake + ScTX | |
Clone	Plating efficiency, %	−	+ 200 nM
N.18	0.02	31	116
LV9	71	<2	3
LV26	11	<2	7
LB1	36	7	35

^{22}Na$^+$ uptake with 40 μM veratridine, nmol·min^{-1}.
ScTx, scorpion venom toxin.
Modified from West and Catterall, ref. 87.

the various subunits and their ultimate transport to the cell surface. These closely linked phenomena occur late in channel biogenesis, all or any of which may be the site of neurotoxicant interaction.

MOLECULAR-CELLULAR ASPECTS OF LIGAND BINDING NEURONAL INJURY

The known neurotransmitter and neuromodulator receptors fall broadly into three classes. Class 1 representing the ligand-gated ion channels includes the nicotinic acetylcholine, $GABA_a$, glycine, and the glutamate receptor complex. The class 2 receptors, exemplified by dopamine, histamine, beta-adrenergic, muscarinic acetylcholine, and numerous neuropeptides, are coupled to their effector system by a guanine nucleotide-binding protein (G-protein). The effector systems are variable, including activation of adenylate cyclase, phosphoinositide hydrolysis, formation of arachidonic acid, or possibly direct activation of an ion channel. Class 3 represents a broad group of growth factor type neuropeptide receptors (e.g., nerve growth factor).

In vivo animal model studies are complex and therefore have proven unsuitable for dissecting and analyzing the cascade of events which may result from toxin-membrane ligand or transmembrane interaction in nerve tissue. Significant data on the early aspects of toxin-ligand interaction and their sequelae that may cause neurotoxicity have been obtained with *in vitro* models, especially primary neuronal cultures, glial cultures, and tissue slices. It is likely that further progress in the elucidation of the molecular mechanisms underlying a variety of neurotoxic interactions, and especially the introduction of classic molecular biologic approaches, will include these preparations. The concepts derived from an analysis of a class 1 ligand-gated receptor-neurotoxicity may be used as a prototype.

The toxicity produced by glutamate and related excitatory amino acids (Table 4) is initially associated with acute swelling of neuronal dendrites and cell bodies, followed by a more slowly evolving, Ca^{2+}-dependent neuronal degeneration (22). Distinctive patterns of neuronal loss are produced by different excitatory amino acids, likely reflecting differences in postsynaptic receptor density and/or species (32). At least three classes of glutamate receptor are now recognized, the best defined being the *N*-methyl-D-aspartate (NMDA) receptor characterized by high conductance, voltage-dependent Mg^{2+} blockade and permeability to both Ca^{2+} and Na^+ (61). Two other non-NMDA receptors activated preferentially by quisqualate or kainate have been identified. Subgroups of quisqualate receptor may be present, one linked directly to membrane Na^+ channels while a second produces a delayed second messenger response mediated by receptor interaction with GTP-binding proteins leading to the generation of diacylglycerol (DAG), and inositol 1,4,5-triphosphate (IP_3) with appropriate mobilization of intracellular Ca^{2+} and activation of protein kinase C (60,73). Kainate receptor activation leads to a Na^+ ionophore-mediated response very similar to the quisqualate

TABLE 4. *Neuronotoxins: ligand and nonligand examples of primary dendro-neuronal sites of injury*

Ligand-gated:
Glutamate (also note cystine uptake hypothesis, ref. 57)
   a) NMDA
   b) non-NMDA (kainate, quisqualate)
N-acetyl aspartyl glutamate
S-containing acidic amino acids (cysteic acid, homocysteic acid)
Quinolinic acid
   β-*N*-methylamino-L-alanine (BMAA) (cycad plant derivative)
   β-*N*-oxalylamino-L-alanine (BOAA) (cycad plant derivative)

Non-ligand gated:
MPP^+
gp120 (also HIV envelope protein)
Free radical-oxidant injury
Ionophoric (A23187, triethyllead and organotins)

---

class. Hence, protracted or "sensitized" glutamate receptor stimulation will provide a neurotoxic signal determined in part by the type of effector (i.e., cation channel, enzyme, or second messenger) coupled to the receptor recognition site (89).

Neuronal death associated with sustained increase in intracytoplasmic Ca^{2+} (21,49,64) is central to an analysis of the mechanism and sequelae of pathological glutamate receptor stimulation. Under normal functioning, homeostatic conditions, numerous proteins, and enzymes operating in the neuronal membrane maintain intracytoplasmic Ca^{2+} concentration below 0.1 μM. Such systems include the Ca^{2+}-binding proteins, voltage-sensitive plasma membrane Ca^{2+} channels, plasmalemma Ca^{2+}-ATPases which pump Ca^{2+} across the neuronal membrane, intracellular Ca^{2+} storage organelles including mitochondria and endoplasmic reticulum, and a plasma membrane Na^+/Ca^{2+} exchanger (5,10,15,33,52). Pathologic, prolonged glutamate receptor activation may lead to an increase of intracytoplasmic Ca^{2+} via (a) decreased activity of plasma membrane Ca^{2+}-ATPase (see below for phosphorylation), (b) protracted increase in Ca^{2+} influx via sustained opening of voltage-sensitive Ca^{2+} channels or directly through the NMDA receptor operated Ca^{2+} channel, or (c) a stimulation of the Na^+/Ca^{2+} exchanger whereby elevated cytosolic $[Na^+]$ drives the internal uptake of extracellular Ca^{2+}. In the case of prolonged glutamate receptor activation, little evidence is provided for activation of the voltage-dependent Ca^{2+} channels but likely increase $[Ca^{2+}]$ is due to persistent activation of the receptor-coupled cation channel.

Mattson et al. (51) have provided interesting observations for the role of Na^+-dependent Ca^{2+} extrusion in protecting neurons against a variety of toxic insults including glutamate excitotoxicity, K^+ depolarization, or A23187-mediated Ca^{2+} influx. In co-cultures of rat hippocampal neurons and mouse neuroblastoma hybrid (NCV-20) cells, depolarizing levels of extracellular K^+, A23187, or glutamate killed hippocampal neurons selectively, leaving NCV-20 cells unaffected. Both cells showed similar sustained rises in intracellular Ca^{2+}, and the ability of the NCV-20 cell to reduce its Ca^{2+} load and survive was dependent on $[Na^+]_e$. These findings suggest that differences in cell Ca^{2+}-regulating systems, especially Na^+/Ca^{2+} translocation at the plasmalemma, may determine whether a neuron degenerates in the face of a toxic/excitotoxic challenge associated with cytoplasmic Ca^{2+} elevation (46). The neuronal plasmalemmal Ca^{2+}-ATPase pump may be activated by phosphorylation, limited proteolysis, or direct interaction with calmodulin (15). Calmodulin binding to Ca^{2+}-ATPase can be modulated by protein kinase C-mediated phosphorylation (4).

A second manifestation of pathological glutamate receptor activation results in the "toxic" amplification of several second and third messenger responses. Ca^{2+}-mediated activation of phospholipase C (PLC) or directly coupled via G-protein activation leads to the formation of DAG and the inositol phosphates IP3 and IP4; the former driving the activation of protein kinase C (PKC), the latter capable of receptor-mediated release of Ca^{2+} from intraneuronal stores. Hence, under abusive receptor stimulation, neuronal proteins may be modified via PKC-mediated protein phosphorylation (78). Manev et al. (50) extended these findings, drawing attention to the translocation of PKC from the cytoplasm to the neuronal membrane with activation, possibly associated with the generation of membrane alkylglycerols. Such prolonged translocation of PKC leads to aberrant phosphorylation (see also below) and prolonged toxic loss of intracellular Ca^{2+} homeostasis. As indicated above, the regulation by calmodulin of the Ca^{2+}-ATPase pump strongly suggests a role for PKC activation in neuronal irreversible injury. Of interest in this respect is the protective action afforded by inhibitors of PKC translocation (e.g., glycosphingolipids) (26,50) or its down regulation.

Prolonged stimulation of glutamate receptors activates guanylate cyclase with subsequent elevation of cyclic GMP (cGMP) (63). Guanylate cyclase inhibition is associated with neurotoxicity, whereas addition of cyclic GMP prevents the toxicity mediated via the NMDA receptor (30). The results indicate that guanylate cyclase/GMP does not mediate neurotoxicity but inhibition potentiates while activation appears protective. The pattern of toxicity mediated by guanylate cyclase inhibition was mimicked by oxygen radical generation, suggesting a coupling between free radical-mediated cytotoxicity and the cGMP axis. The free radical-mediated damage associated with guanylate cyclase inhibition likely arises from a Ca^{2+}-activated process, one candidate being the phospholipase A_2-arachidonic acid pathway since phospholipase A_2 appears to be part of the pathway leading to activation of guanylate cyclase (72). Oxygen radicals are potent activators of guanylate cyclase and it has been hypothesized that the product of this activation, GMP, may serve as a feedback inhibitor of oxygen radical formation.

Although this discussion has concentrated on events associated with glutamate receptor-linked signal transduction, other sequelae may be triggered. It is now recognized that receptor-linked stimuli are associated with transcription coupling in neurons or the pheochromocytoma cell line, PC12, with the activation of cellular immediate-early response genes (55,56). An increase in the transcription of inducible genes encoding specific nuclear proteins (c-fos, c-jun) may act as third nuclear messengers by activating specific gene expression programs. The occupation of specific oligonucleotide sequences located in the 5' regulatory region may modulate gene expression possibly linked to the synthesis of proteins involved in long-term potentiation, the formation of synaptic (dystrophic) spines, or the maintenance of appropriate synaptic connectivity during development (6,75). While the functional role of these proto-oncogenes is still unclear, care must be expressed in analyzing the consequence of physiological versus pathological receptor activation; nontoxic glutamate addition induces c-fos mRNA compared to a relative failure to increase c-fos mRNA following long-term, abusive stimulation. Morgan and Curran (55) were able to demonstrate c-fos induction in PC12 cells either by ligand-gated receptor interaction (e.g., veratridine) or by agents or conditions that affected voltage-dependent Ca^{2+} channels. Moreover, the induction was dependent on $[Ca^{2+}]_e$ leading to elevated $[Ca^{2+}]_i$ and a calmodulin-dependent step. Events following the calmodulin-dependent step are unknown but the concept recognizes that c-fos and allied genes are candidates for coupling neuron membrane events to long-term adaptive changes in transcription. Notwithstanding that this may be a model for long-term potentiation, of immediate relevance is the modification of neurotoxicity likely to occur in the newly "adapted" neuron, a condition analogous to the heat stress/oxidant injury protection of early stress events. A further example of c-fos and ornithine decarboxylase gene expression as an early marker of neurotoxicity was observed by Vendrell et al. (80) in studies of the organochlorine pesticide, lindane.

This discussion has concentrated on the central role played by an alteration in intracellular Ca^{2+} homeostasis in mediating irreversible neural injury, but the precise mechanism of cytotoxicity remains unclear. A sustained increase in cytosolic Ca^{2+} can stimulate endogenous proteases and phospholipases and activate an endonuclease that cleaves host chromatin into oligonucleosome length fragments. McConkey et al. (54) coupled the stimulation of DNA fragmentation by the Ca^{2+} ionophore A23187 with thymocyte killing that could be blocked by the endonuclease inhibitor aurintricarboxylic acid, whereas inhibitors of other Ca^{2+}-dependent degradative processes did not block cell death. Glucocorticoid-induced thymocyte DNA fragmentation and 2,3,7,8-tetrachlorodibenzo-*p*-dioxin (TCDD) cell killing were also dependent on a sustained increase in cytosolic Ca^{2+} concentration (54,55). These results strongly suggest that chromatin cleavage is the lesion responsible for apoptosis, a form of cell injury morphologically similar to that seen in the prolonged phase of glutamate, organolead, and organotin neurotoxicity (11). A further observation of interest was the demonstration that concentrations of A23187 greater than 1 μM which caused more extensive elevation of cytosolic Ca^{2+} concentration did not cause DNA fragmentation. It appears that a cytosolic Ca^{2+} threshold exists for the activation of endogenous DNA fragmentation; above this level, cells die by mechanisms prior to and unrelated to the onset of endonuclease activation and DNA strand breakage. Similar dose-response mechanisms have been observed for methyl mercury toxicity (85) and cystamine (62). Cystamine binds to membrane protein-SH with the formation of mixed disulfides and inhibits Ca^{2+} efflux via the plasmalemma Ca^{2+}-translocase. The resulting $[Ca^{2+}]_i$ elevation is sufficient to activate PLA_2 with phospholipid hydrolysis and proteolysis. Only pretreatment of cells with inhibitors of Ca^{2+}-activated proteases protected cells from cytotoxicity. Analogous observations have been made with diphtheria toxin, a potent toxin due to inhibition of ribosomal activity by ADP-ribosylation of eEF-2 (41) and a cause of segmental demyelination. The cytotoxicity was thought to be explained in terms of inhibition of translation. Methyl mercury also blocks elongation (19), but the inhibition of translation can be dissociated from the onset of neurotoxicity. It has now been shown that diphtheria toxin-induced cytolysis is preceded by internucleosomal DNA fragmentation typical of apoptosis and it is likely, signifies endonuclease activation (18). These observations received critical confirmation from Cantoni et al. (14) who demonstrated the key role of $[Ca^{2+}]_i$. The calcium chelator Quin 2 prevented H_2O_2-induced DNA breakage, was unaffected by $[Ca^{2+}]_e$, was cytoprotective, and did not influence the generation of the $\cdot OH$ radical.

The mitogenic lectins phytohemagglutinin and concanavalin A stimulate a dose-dependent sustained increase in $[Ca^{2+}]_i$ without endonuclease activation. Both these agents stimulate PKC. Of interest was that co-incubation with the Ca^{2+} ionophore A23187 and phorbol ester *prevented* DNA fragmentation, suggesting that PKC activation suppressed endonuclease activation somewhat analogous to that found in the glutamate model.

A further dimension to elucidating the mechanism of glutamate toxicity stems from the observation of Murphy et al. (57) who found that addition of glutamate to a neuronal cell line (N18-RE-105) or immature cortical neurons inhibited the transport of cystine leading to a time-dependent reduction in glutathione, a major intracellular thiol reductant. The hybrid cells do not express NMDA receptors but glutamate competes with a Cl^--dependent membrane transport system normally exchanging extracellular cystine for intracellular glutamate. These observations, now extended to immature cortical neurons, couple the Ca^{2+}-dependent quisqualate type of glutamate neuronotoxicity to inhibition of high affinity cystine uptake and depletion of intracellular reduced glutathione (GSH). These interactions provide a fertile model for genetic mutant studies, neuronal specificity and toxin sensitivity, glial-neuronal interactions, Cl^--dependent glutamate uptake, and oxidative stress secondary to intracellular thiol depletion.

MOLECULAR ASPECTS OF MYELIN NEUROTOXICOLOGY

Abnormalities in peripheral and central myelin have resulted from toxicant interaction. Myelination is a major event in development of the central and peripheral nervous system and recent molecular studies of myelin biogenesis

(13) have proven useful in examining mechanisms underlying this nervous system-specific process and provide the basis for an understanding of some toxicant-mediated dysmyelination states.

In the CNS, myelin represents an extrusion of the oligodendroglial plasma membrane. The cytoplasm associates with axons at varying distances from the cell body in contrast to the close proximity of the Schwann cell and myelin sheath in the peripheral nervous system. Two major and two minor proteins comprise approximately 95% of the protein content of the membrane (Table 5). The major proteins, myelin basic protein (MPB) and the proteolipid proteins (PLP), contribute approximately 75%. These proteins assume a structural role with specific physical properties. The MBPs are hydrophilic, while the proteolipids are extremely hydrophobic (47). At least six isoforms of MBPs have been identified varying in molecular mass from 14 to 21.5 kD. The MBPs are believed to be located at the major dense line of myelin formed by apposition of the cytoplasmic surface of the oligodendroglial plasma membrane. The MPBs may undergo posttranslational modification (e.g., phosphorylation, methylation), and the latter may be important for membrane compaction (71). Myelin PLPs exist as two isoforms and the gene has been assigned to the human and mouse X-chromosome (88). MAG is the major myelin glycoprotein, constituting approximately 1% of total myelin protein (67). Approximately 30% of MAG is carbohydrate and the protein exists in two polypeptide isoforms with apparent masses of 72 and 67 kD. A significant portion of MAG is exposed to the extracellular surface of the unit bilayer and is proposed to mediate cell-cell interactions in the nervous system, analogous to other adhesion molecules, e.g., N-CAM (40). Although the physiological role is unclear, immunohistochemical data suggest a preferential localization to the periaxonal region (77). The enzyme CNP (2′,3′-cyclic nucleotide 3′-phosphodiesterase) catalyzes the hydrolysis of several cyclic nucleoside monophosphates and is a component of the "Wolfgram protein" fraction, a heterogeneous group of high MW proteins.

TABLE 5. *CNS myelin proteins*

Type	Isoforms	Synthesis site	Posttranslation modification	Location
MBP	4–6 (14–21.5 kD)	F	Phosphorylation, methylation, N-acylation	Major dense line
PLP	2 (PLP, DM20)	MB	Fatty acylation	
MAG	2 (72, 67 kD)	MB	Glycoprotein Phosphorylation	Regulation of adhesion, periaxonal
CNP	2 polypeptide chains	F	Wolfgram protein	Myelin assoc. enzyme, cytoplasmic face of plasmalemma

F, free ribosomes; MB, membrane-based ribosomes.

The biological relevance of posttranslational modification of myelin proteins is unclear, but examples of disturbance in posttranslational acetylation, phosphorylation, or methylation have resulted in myelin structural abnormality (27,71). For instance, there is evidence that the methylation of MBPs is important for the maintenance and assembly of the multilamellar structure of myelin. Cycloleucine and sinefungin are inhibitors of protein methylation acting at different steps in the transmethylation process. Amur et al. (3) demonstrated that sinefungin inhibited MBP-methylase activity in cultured cerebral cells and ultrastructural examination revealed myelin-like figures devoid of typical multilamellar periodicity and compactness reminiscent of the vacuolated myelin observed in nitrous oxide-treated animals (70). Nitrous oxide toxicity may block transmethylation via oxidation of vitamin B12 that can no longer function in the methylation of homocysteine to methionine. Chickens injected with cycloleucine developed myelin vacuolation similar to that seen in animals administered N_2O. Cycloleucine inhibits the production of S-adenosyl methionine, the substrate of methyl transferase. These studies identify a posttranslational defect in methylation associated with myelin vacuolization likely secondary to a disorder in compacting.

In contrast, toxicant-induced defects in myelin protein phosphorylation have not been identified and the physiological role of MBP phosphorylation is unclear. Vartanian et al. (79) demonstrated PKC-dependent phosphorylation of MBP following oligodendroglial adherence. MBP phosphorylation may occur prior to its incorporation into compact myelin, thereby reducing the positive charge on the monomer, promoting dimer formation, and allowing for myelin compaction at the major dense line. Hence, toxicant-induced defects at the dense line may reflect either genetically expressed abnormalities in MBP and/or defects in phosphorylation coupled to abnormal PKC activity. Myelin PLP and DM20 proteolipids may be acylated *in vivo* (9) and *in vitro* (8,9) at or near the myelin membrane, in contrast to the delay in incorporation of newly synthesized PLP into myelin, a reflection of the noncytoplasmic transport of PLP mRNA. The molecular significance or evidence of toxicant-mediated defect in acylation is unclear but Townsend and Benjamins (76) revealed continued acylation of PLP following the addition of monensin known to block the transport of newly formed PLP to the membrane.

The myelin proteins have different subcellular sites of transcription and synthesis and follow different paths of assembly into the complete myelin membrane. MBPs and CNP are synthesized on free ribosomes (12,23), while PLP and MAG are synthesized on membrane-bound ribosomes (23). Of equal or greater significance, especially in view of possible mechanisms of toxicant interaction, are the studies indicating different routes of assembly of PLP and MBP into the myelin membrane. The intracellular route followed by PLP involves passage through the Golgi zone resulting in delayed incorporation into terminal myelin. However, MBP is likely synthesized close to the site of assembly into the membrane. These observations have received strong support from *in situ* hybridization studies (82–84) in which PLP mRNA labeling remained localized to the oligo-dendroglial cell body at all stages of myelinogenesis. A schema summarizing these observations is presented in Fig. 3.

Studies using the dysmyelinating mouse mutants have identified factors thought to regulate the expression of myelin-associated genes and the assembly of proteins into the myelin membrane (13). These studies have identified points in myelin biogenesis and/or maintenance at which the molecular biology of toxicant-induced injury may be understood. For instance, dysregulation of the gene at the level of splicing to produce mRNAs will disturb the proportional distribution known to occur with development. Excess steroids modulate the level of translation of the MBP mRNAs (81). Toxicant dysregulation may exist at the level of translocation of the MBP mRNA from cell soma to the oligoden-droglial process as in the Golgi-vesiculation of protein prior to transport to site of incorporation. Finally, as previously indicated, dysregulation of ultimate myelin expression may occur after synthesis but prior to incorporation into the membrane.

Toxicant-induced dysmyelination may be loosely classified: A, defects in myelin assembly; B, vacuolar demyelination; and C, primary demyelination. It is likely that both central and peripheral myelin may be involved in these broad descriptive categories. Defective myelin protein biogenesis, lipogenesis, assembly, or translocation will be reflected predominantly in a morphological pattern of hypomyelination. It is not expected that evidence of myelin catabolism will be

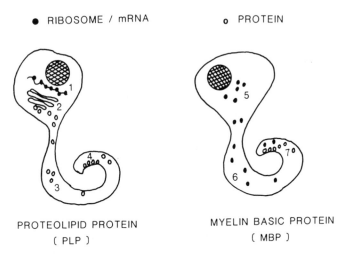

FIG. 3. Comparison of sites of proteolipid protein and myelin basic protein synthesis and incorporation within developing myelin membrane. *1*, Synthesis of protein on membrane-bound ribosomes; *2*, translocation through Golgi cisternae; *3*, cytoplasmic movement to myelin biogenesis site; *4*, incorporation into forming myelin; *5*, synthesis on free ribosomes in perinuclear and cytoplasmic projection; *6*, active translocation of mRNA to site of myelin synthesis; *7*, translation and posttranslational modification of protein at site of myelin incorporation.

TABLE 6. *Classification of genetic and toxicant-induced myelin injury*

Class	Site of lesion	Example
Defects in myelin assembly		
1. Gene expression	PLP	Jimpy, Pelizeus-Merzbacher
	MBP	Shiverer
2. Translocation	PLP	Jimpy, Monensin
3. Posttranslation, assembly	MBP	Quaker, trimethyltin, cycloleucine, N_2O, p-chlorophenylalanine
Vacuolar demyelination		Organotin (TET), hexachlorophene, cuprizone
Primary demyelination	eEF-2	Diphtheria toxin Pb, Te
	DNA	

PLP, proteolipid protein; MBP, myelin basic protein.

a feature. Examples of genetic and toxicant-induced demyelination are summarized in Table 6.

NEUROTOXICOLOGY OF CELL ADHESION

Cell-cell interaction requires contact between cell surfaces or between cells and the extracellular matrix. Immunological methods and quantitative assays of cell binding have allowed for the isolation of specific molecules involved in neuron-neuron or neuron-glial adhesion. The identification, biochemical characterization, role, and dynamic experiments pertaining to these molecules depends heavily on the use of polyspecific and monoclonal antibodies.

NCAM, possibly the first identified and characterized neural cell adhesion molecule, comprises a family of related glycoproteins with molecular weights 180, 140, and 120 kD which result from the translation of alternatively spliced mRNAs (25,58). The larger NCAM forms are transmembrane glycoproteins differing in the length of their cytoplasmic domains. The smallest (120 kD) form lacks a transmembrane domain but is attached via a phosphatidylinositol linkage (39). All NCAMs contain a high content of polysialic acid comprising as much as 30% by weight, which decreases during development and maturation. NCAM produces cell-cell adhesion by homophilic binding (NCAM molecules on apposing cell surfaces bind directly to one another). A variety of structure-function relationships may be of importance in neurotoxicological studies. The expression of the larger NCAM form appears to accompany neuronal differentiation and the 180 kD-specific NCAM epitopes are abundant in differentiating neurons but absent during proliferation (65). Secondly, variations in polysialic acid content occur between cell soma and the neuritic process, reflecting the stage of neuron development. Variations in the intracellular domain of NCAM will likely alter cellular responses. For instance, the 180 kD form has greatly reduced mobility within the cell membrane compared to the 140 kD form, likely a result of the binding of submembranous cytoskeletal fodrin to the 180 kD polypeptide but not to the 140 or 120 kD forms (86). Of possibly greater significance to neurotoxicological studies, especially in relation to toxicant-induced changes in cell transduction signaling, is the lack of cytoplasmic or transmembrane domain for the smallest NCAM polypeptide and its attachment to the cell surface via a phospholipase-C sensitive phosphatidylinositol linkage.

While NCAM may provide a polyspecific role in developing neuronal interaction, the identification of L1 and NgCAM as Ca^{2+}-independent neural adhesion molecules confers some specificity for neuron-glial cell adhesion (34,35,44). Little can be gained at this time by cataloging in detail the plethora of molecules associated with cell-cell interaction. However, the ability of neurons-astrocytes-oligodendroglial cells to modulate the amount and topographic pattern of expression of adhesion molecules provides a mechanism for morphoregulation in the developing CNS and a fertile field for studies in developmental neurotoxicology.

A prototypical example may be given in studies of cerebellar granule cell migration, especially in view of the tantalizing but still speculative role of adhesion molecules and/or cytoskeleton in the genesis of fetal methyl mercury neurotoxicity (17,20). The genesis of granule cell migration from the external granule layer via its movement along a Bergmann-glial cell process, translocation through the molecular layer, and final migration into the internal granule layer is well established. Immunohistochemical and *in vitro* studies have revealed granule cell-granule cell interactions mediated by NCAM and L1/NgCAM, while AMOG and cytotactin may be involved in granule cell adhesion along the Bergmann-glial fibers. An *in vitro* assay developed by Lindner et al. (48) allows for studies of the migration of [^{3}H]thymidine-labeled granule cells from the external granule layer through the molecular layer into the internal granule layer using cerebellar explants. Small fragments of cerebellar cortex are pulse labeled with [^{3}H]thymidine. Labeled nuclei are found in the external granule layer. Following explant culture for 3 to 5 days, the majority of labeled nuclei are found in the internal granule layer, reflecting granule cell migration. Using this system, studies have suggested that anti-L1/NgCAM interferes with the initial postmitotic granule cell process outgrowth, while the astrocyte antigens AMOG and cytotactin are involved in elongation along the Bergmann-glial fiber. This study paradigm will likely prove of great value in developmental neurotoxicological studies (viz. methyl mercury) as semiquantitative measure of [^{3}H]thymidine-labeled neuronal movement may be expressed in terms of both toxicant addition and immunohistochemistry of key adhesion molecules.

Although toxicant interaction may impair neuron-glia or axon-Schwann cell contact by modulating biosynthesis or membrane incorporation of the adhesion molecules, loss of such contact will have far-reaching effects on morphology, cell movement, and viability. Secondly, NCAM-mediated membrane adhesion is associated with an induction of ChAT activity (1) or tyrosine hydroxylase (2). Hatten has added a further dimension to these studies in explaining the change in morphology occurring in astrocytes grown in the absence of neurons to their morphological differentiation when co-cultured with purified granule neurons (37). Cell binding between neurons and astroglia was rapid, and following neuronal attachment, astroglial DNA synthesis decreased dramatically. These studies revealed specificity toward granule cell membrane–astroglial membrane interaction, a morphological glial transformation from a flat epithelial shape to a highly differentiated, elongated, Bergmann-glial type, and the possibility that the membrane elements involved in glial growth regulation included specific neuron-glial interaction molecules (38). These *in vitro* experiments identify a key neurotoxicologic concept: a mechanism for glial proliferation (gliosis) as a response to neuronal cytotoxic injury and loss.

CONCLUSION AND SUMMARY

In this chapter we have identified the present and exciting role of molecular neurobiology in the advancement of neurotoxicology. The chapter has included highly selective molecular concepts which may provide approaches to an understanding of mechanisms in neurotoxicity, especially developmental neurotoxicology. It is hoped that conceptual ideas derived from rapid advances in the molecular biology of *Drosophila,* gene cloning, and gene mapping will allow for identification of biological systems through which molecular analyses of drug and toxicant interaction may be studied. Of equal importance is the opportunity now provided through a variety of *in vitro* and tissue culture paradigms for the time- and dose-dependent molecular analysis of neuronal and/or glial irreversible injury and the accompanying cytomorphology. The coupling of morphological change to second and third signal messenger cascade events is an exciting thrust for the future. Doubtless, ideas and concepts will be derived that will impact directly on cytoprotection and therapeutic reversibility. For instance, the use of glycosphingolipids, PKC down regulation, intracellular Ca^{2+} chelation, modulation of GMP, and phospholipase A2 activation with subsequent arachidonic acid release may all be considered sites for therapeutic modulation.

ACKNOWLEDGMENTS

I would like to acknowledge the numerous discussions held with H. Tilson, K. Reuhl, H. Lowndes, and my colleague T. Sarafian in the authorship of this chapter. I would like to acknowledge and thank the Burroughs-Wellcome Foundation for a Senior Investigator Travel Award which allowed completion of this study.

I am indebted to Scott D. Brooks and Stephen Kaufman for manuscript preparation.

REFERENCES

1. Acheson AL, Rutishauser U. Neural cell adhesion molecule regulates cell contact-mediated changes in choline acetyltransferase activity of embryonic chick sympathetic neurons. *J Cell Biol* 1988;106: 479–486.
2. Acheson AL, Thoenen H. Cell contact-mediated regulation of tyrosine hydroxylase synthesis in cultured bovine adrenal chromaffin cells. *J Cell Biol* 1983;97:925–928.
3. Amur SG, Shanker G, Cochran JM, Ved HS, Pieringer RA. Correlation between inhibition of myelin basic protein (arginine) methyltransferase by sinefungin and lack of compact myelin formation in cultures of cerebral cells from embryonic mice. *J Neurosci Res* 1986;16:367–376.
4. Apel ED, Byford MB, Au D, Walsh KA, Storm DR. Identification of the protein kinase C phosphorylation site in neuromodulin. *Biochemistry* 1990;29:2330–2335.
5. Baker PF. The sodium-calcium exchange system. In: *Calcium and the cell.* New York: Wiley 1986;73–92.
6. Balazs R, Jorgensen OS, Hack N. *N*-methyl-D-aspartate promotes the survival of cerebellar granule cells in culture. *Neuroscience* 1988;27:437–451.
7. Berwald-Netter Y, Martin-Moutot N, Koulakoff A, Couraud F. Na$^+$ channel-associated scorpion toxin receptor sites as probes for neuronal evolution in vivo and in vitro. *Proc Natl Acad Sci USA* 1981;78:1245–1249.
8. Bizzozero OA, Soto EF, Pasquini JM. Myelin proteolipid protein. *Neurochem Int* 1983;5:729–736.
9. Bizzozero OA, McGarry JF, Lees MB. Acylation of rat brain myelin proteolipid protein with different fatty acids. *J Neurochem* 1986;47:772–778.
10. Blaustein MP. Calcium transport and buffering in neurons. *Trends Neurosci* 1988;11:438–443.
11. Brown AW, Aldridge WN, Street BW, Verschoyle RD. The behavioral and neuropathologic sequelae of intoxication by trimethyltin compounds in the rat. *Am J Pathol* 1979;97:59–82.
12. Campagnoni AT, Carey GD, Yu Y-T. In vitro synthesis of the myelin basic proteins: subcellular site of synthesis. *J Neurochem* 1980;34:677–686.
13. Campagnoni AT, Macklin WB. Cellular and molecular aspects of myelin protein gene expression. *Mol Neurobiol* 1988;2:41–89.
14. Cantoni O, Sestili P, Cattabeni F, Bellomo G, Pou S, Cohen M, Cerutti P. Calcium chelator Quin 2 prevents hydrogen-peroxide-induced DNA breakage and cytotoxicity. *Eur J Biochem* 1989;182:209–212.
15. Carafoli E. Intracellular calcium homeostasis. *Ann Rev Biochem* 1987;56:395–433.
16. Catterall WA. Neurotoxins that act on voltage-sensitive Na$^+$ channels in excitable membranes. *Ann Rev Pharmacol Toxicol* 1980;20:15–43.
17. Chang LW, Reuhl KR. Ultrastructural study of the latent effects of methyl mercury on the nervous system after prenatal exposure. *Environ Res* 1977;13:171–185.
18. Chang MP, Bramhall J, Graves S, Bonavida B, Wisnieski BJ. Internucleosomal DNA cleavage precedes diphtheria-toxin induced cytolysis. Evidence that cell lysis is not a simple consequence of translation inhibition. *J Biol Chem* 1989;264:15261–15267.
19. Cheung MK, Verity MA. Experimental methyl mercury neurotoxicity: locus of mercurial inhibition of brain protein synthesis in vivo and in vitro. *J Neurochem* 1985;44:1799–1808.
20. Choi BH. Effects of methyl mercury on neuroepithelial germinal cells in the developing telencephalic vesicles of mice. *Acta Neuropathol* 1991;81:359–365.
21. Choi DW. Ionic dependence of glutamate neurotoxicity in cortical cell culture. *J Neurosci* 1987;7: 369–379.
22. Choi DW. Glutamate neurotoxicity and diseases of the nervous system. *Neuron* 1988;1:623–634.
23. Colman DR, Kreibich G, Frey AB, Sabatini DD. Synthesis and incorporation of myelin polypeptides into CNS myelin. *J Cell Biol* 1982;95:598–608.
24. Costa MRC, Casnellie JE, Caterall WA. Selective phosphorylation of the alpha-subunit of the sodium channel by cyclic AMP-dependent protein kinase. *J Biol Chem* 1982;257:7918–7921.
25. Cunningham BA, Hemperly JJ, Murray BA, Prediger EA, Brackenbury R, Edelman GM. Neural cell adhesion molecule: structure, immunoglobulin-like domains, cell surface modulation, and alternative RNA splicing. *Science* 1987;236:799.

26. Favaron M, Manev H, Siman R, Bertolino M, Szekely AM, Deerausquin G, Guidotti A, Costa E. Down-regulation of protein kinase C protects cerebellar granule neurons in primary culture from glutamate-induced neuronal death. *Proc Natl Acad Sci USA* 1990;87:1983–1987.
27. Gandy G, Jacobsen W, Sidman R. Inhibition of a transmethylation reaction in the central nervous system—an experimental model for subacute combined degeneration of the cord. *J Physiol* (Lond) 1973;233:1P–3P.
28. Ganetzky B, Loughney K, Wu C-F. Analysis of mutations effecting sodium channels in Drosophila. In: Kao, CY Levinson SR, eds. *Tetrodotoxin, saxitoxin and the molecular biology of the sodium channel.* New York: NY Academy of Science, 1986;325–337.
29. Ganetzky B, Wu C-F. Neurogenetics of membrane excitability in Drosophila. *Ann Rev Genet* 1986;20:13–44.
30. Garthwaite G, Garthwaite J. Cyclic GMP and cell death in rat cerebellar slices. *Neuroscience* 1988;26:321–326.
31. Gerschenfeld HM. Chemical transmission in invertebrate central nervous systems and neuromuscular junctions. *Physiol Rev* 1973;53:1–119.
32. Greenamyre JT, Olson JM, Penney JB, Young AB. Autoradiographic characterization of *N*-methyl-D-aspartate, quisqualate, and kainate-sensitive glutamate binding sites. *J Pharmacol Exp Ther* 1985;233:254–263.
33. Greenberg DA. Calcium channels and calcium channel antagonists. *Ann Neurol* 1987;21:317–330.
34. Grumet M, Edelman GM. Heterotypic binding between neuronal membrane vesicles and glial cells is mediated by a specific cell adhesion molecule. *J Cell Biol* 1984;98:1746.
35. Grumet M, Hoffman S, Chuong CM, Edelman GM. Polypeptide components and binding functions and neuron-glial cell adhesion molecules. *Proc Natl Acad Sci USA* 1984;81:7989.
36. Hall LMC, Spierer P. The *ace* locus of Drosophila melanogester: structural gene for acetyl cholinesterase with an unusual 5' leader. *EMBO J* 1986;5:2949–2954.
37. Hatten ME. Neuronal regulation of astroglial morphology and proliferation in vitro. *J Cell Biol* 1985;100:384–396.
38. Hatten ME. Neuronal inhibition of astroglial cell proliferation is membrane mediated. *J Cell Biol* 1987;104:1353–1360.
39. He HT, Barbet J, Chaix JC, Goridis C. Phosphatidyl inositol is involved in the membrane attachment of N-CAM-120, the smallest component of the neural cell adhesion molecule. *EMBO J* 1986;5:2489.
40. Holley JA, Yu RK. Localization of glycoconjugates recognized by the HNK-1 antibody in mouse and chick embryos during early neuron development. *Dev Neurosci* 1987;9:107–119.
41. Honjo T, Nishizuka Y, Hayaishi O, Kato I. Diphtheria toxin-dependent adenosine diphosphate ribosylation of aminoacyl transferase II and inhibition of protein synthesis. *J Biol Chem* 1968;243:3553–3555.
42. Ishikawa Y, Charalambous P, Matsumura F. Modification by pyrethroids and DDT of phosphorylation activities of rat brain Na^+-channel. *Biochem Pharmacol* 1989;38:2449–2457.
43. Itoh N, Slemmon JR, Hawke DH, Williamson R, Morita E, Itakura K, Roberts E, Shively JE, Crawford GD, Salvaterra PM. Cloning of Drosophila choline acetyltransferase cDNA. *Proc Natl Acad Sci USA* 1986;83:4081–4085.
44. Keilhauer G, Faissner A, Schachner M. Differential inhibition of neuron-neuron, neuron-astrocyte and astrocyte-astrocyte adhesion by L1, L2, and N-CAM antibodies. *Nature* 1985;316:728.
45. Konat G, Gantt G, Hogan EL. Acylation of myelin proteolipid protein in subcellular fractions of rat brainstem. *Neurochem Int* 1986;9:545–549.
46. Kudo Y. Neuronal death in vitro: parallelism between survivability of hippocampal neurons and sustained elevation of cytosolic calcium after exposure to glutamate receptor agonists. *Exp Brain Res* 1988;73:447–458.
47. Lees MB. Proteolipids. *Scand J Immunol* 1982;15:147–166.
48. Lindner J, Rathjen FG, Schachner M. L1 mono- and polyclonal antibodies modify cell migration in early postnatal mouse cerebellum. *Nature* 1983;305:427.
49. Manev H, Favaron M, Guidotti A, Costa E. Delayed increase of calcium influx elicited by glutamate: role in neuronal death. *Mol Pharmacol* 1989;36:106–112.
50. Manev H, Favaron M, Vicini S, Guidotti A, Costa E. Glutamate-induced neuronal death in primary cultures of cerebellar granule cells: protection by synthetic derivatives of endogenous sphingolipids. *J Pharmacol Exp Ther* 1990;252:419–427.

51. Mattson MP, Guthrie PB, Kater SB. A role for sodium-dependent calcium extrusion in protection against neuronal excitotoxicity. *FASEB J* 1989;3:2519–2526.
52. McBurney RN, Neering IR. Neuronal calcium homeostasis. *Trends Neurosci* 1987;10:164–169.
53. McConkey DJ, Hartzell P, Duddy SK, Hakansson H, Orrenius S. 2,3,7,8-tetrachlorodibenzo-p-dioxin (TCDD) kills immature thymocytes by Ca^{2+}-mediated endonuclease activation. *Science* 1988;242:256–259.
54. McConkey DJ, Hartzell P, Nicotera P, Orrenius S. Calcium-activated DNA fragmentation kills immature thymocytes. *FASEB J* 1989;3:1843–1849.
55. Morgan JI, Curran T. Role of ion flux in the control of c-fos expression. *Nature* 1986;322:552–555.
56. Morgan JI, Curran T. Stimulus-transcription coupling in neurons: role of cellular immediate-early genes. *Trends Neurosci* 1989;12:459–462.
57. Murphy TH, Miyamoto M, Sastre A, Schnaar RL, Coyle JT. Glutamate toxicity in a neuronal cell line involves inhibition of cystine transport leading to oxidative stress. *Neuron* 1989;2:1547–1558.
58. Murray BA, Hemperly JJ, Prediger EA, Edelman GM, Cunningham BA. Alternatively spliced mRNA's code for different polypeptide chains of the chicken neural cell adhesion molecule (N-CAM). *J Cell Biol* 1986;102:189.
59. Narahashi T. Modulation of nerve membrane sodium channels by chemicals. *J Physiol (Paris)* 1985;77:1093–1101.
60. Nicoletti F, Wroblewski JT, Novelli A, Alho H, Guidotti A, Costa E. The activation of inositol phospholipid metabolism as a signal-transducing system for excitatory amino acids in primary cultures of cerebellar granule cells. *J Neurosci* 1986;6:1905–1911.
61. Nicoll RA, Kauer JA, Malenka RC. The current excitement in long term potentiation. *Neuron* 1988;1:97–103.
62. Nicotera P, Hartzell P, Baldi C, Svensson S-A, Bellomo G, Orrenius S. Cystamine induces toxicity in hepatocytes through the elevation of cytosolic Ca^{2+} and the stimulation of a non-lysosomal proteolytic system. *J Biol Chem* 1986;261:14628–14635.
63. Novelli A, Nicoletti F, Wroblewski JT, Alho H, Costa E, Guidotti A. Excitatory amino acid receptors coupled with guanylate cyclase in primary culture of cerebellar granule cells. *J Neurosci* 1987;7:40–47.
64. Ogura A, Miyamoto M, Kudo Y. Neuronal death in vitro: Parall Olney JW. Insighting exitotoxic cytocide among central neurons. *Adv Exp Med Biol* 1986;203:631–645.
65. Pollerberg GE, Sadoul R, Goridis C, Schachner M. Selective expression of the 180-kD component of the neural cell adhesion molecule N-CAM during development. *J Cell Biol* 1985;101:1921.
66. Pollerberg GE, Schachner M, Davoust J. Differentiation state-dependent surface mobilities of two forms of the neural cell adhesion molecule. *Nature* 1986;324:462.
67. Quarles RH. Glycoproteins in myelin and myelin related membranes. In: Hashim G, ed. *Myelin chemistry and biology*. New York: Liss, 1979;55–77.
68. Salkoff L, Butler A, Wei A, Scavarda N, Giffen K, Ifune C, Goodman R, Mandel G. Genomic organization and deduced amino acid sequence of a putative Na^+ channel gene in Drosophila. *Science* 1987;237:744–749.
69. Scheinman RI, Auld VJ, Goldin AL, Davidson N, Dunn RJ, Catterall WA. Developmental regulation of sodium channel expression in the rat forebrain. *J Biol Chem* 1989;264:10660–10666.
70. Scott JM, Wilson P, Dinn JJ, Weir DJ. Pathogenesis of subacute combined degeneration: a result of methyl group deficiency. *Lancet* 1981;2:334–337.
71. Small DH, Carnegie PR, Anderson R. Cycloleucine-induced vacuolation of myelin is associated with inhibition of protein methylation. *Neurosci Lett* 1981;21:287–292.
72. Snider RM, McKinney M, Forray C, Richelson E. Neurotransmitter receptors mediate cyclic GMP formation by involvement of arachadonic acid and lipoxygenase. *Proc Natl Acad Sci USA* 1984;81:3905–3909.
73. Sugiyama H, Ito I, Hirono C. A new type of glutamate receptor linked to inositol phospholipid metabolism. *Nature* 1987;325:531–533.
74. Suzuki N, Wu C-F. Voltage sensitivity to Na^+ channel-specific neurotoxins in cultured neurons from temperature-sensitive paralytic mutants of Drosophila. *J Neurogenet* 1984;1:225–238.
75. Szekely AM, Barbaccia ML, Alho H, Costa E. In primary cultures of cerebellar granule cells the

activation of N-methyl-D-aspartate-sensitive glutamate receptors induces c-fos mRNA expression. *Mol Pharmacol* 1989;35:401–408.

76. Townsend LE, Benjamins JA. Effects of monensin on post translational processing of myelin proteins. *J Neurochem* 1983;40:1333–1339.

77. Trapp BD, Quarles RH. Presence of the myelin-associated glycoprotein correlates with alterations in the periodicity of peripheral myelin. *J Cell Biol* 1982;92:877–882.

78. Vaccarino F, Guidotti A, Costa D. Ganglioside inhibition of glutamate-mediated protein kinase C translocation in primary cultures of cerebellar neurons. *Proc Natl Acad Sci USA* 1987;84: 8707–8711.

79. Vartanian T, Szuchet S, Dawson G, Campagnoni AT. Oligodendrocyte adhesion actives protein kinase C-mediated phosphorylation of myelin basic protein. *Science* 1986;234:1395–1398.

80. Vendrell M, Zawia NH, Serratosa J, Bondy SC. C-fos and ornithine decarboxylase gene expression in brain as early markers of neurotoxicity. *Brain Res* 1991;544:291–296.

81. Verdi JN, Kampf K, Campagnoni AT. Translational regulation of myelin protein synthesis by steroids. *J Neurochem* 1989;52:321–324.

82. Verity AN, Campagnoni AT. Myelin basic protein gene expression in the quaking mouse. *Brain Disfunct* 1988;1:272–284.

83. Verity AN, Campagnoni AT. Regional expression of myelin protein genes in the developing mouse brain: in situ hybridization studies. *J Neurosci Res* 1988;21:238–248.

84. Verity AN, Levine MS, Campagnoni AT. Gene expression in the gimpy mutant: evidence for fewer oligodendrocytes expressing myelin protein genes and impaired translocation of myelin basic protein mRNA. *Dev Neurosci* 1990;12:359–372.

85. Verity MA, Sarafian T. Role of oxidative injury in the pathogenesis of methyl mercury neurotoxicity. Symposium on Advances in Mercury Toxicology. Tokyo, Japan; 1990;19.

86. Vijverberg HPM, DeWeille JR. The interaction of pyrethroids with voltage-dependent sodium channels. *Neurotoxicology* 1985;6:23–34.

87. West GJ, Caterall WA. Selection of variant neuroblastoma clones with missing or altered sodium channels. *Proc Natl Acad Sci USA* 1979;76:4136–4140.

88. Willard HF, Riordin JR. Assignment of the gene for myelin proteolipid protein to the X-chromosome: implications for X-linked myelin disorders. *Science* 1985;230:940–942.

89. Wroblewski JT, Danysz W. Modulation of glutamate receptors: molecular mechanisms and functional implications. *Ann Rev Pharmacol Toxicol* 1989;29:441–474.

90. Wu C-F, Ganetzky B, Jan YN, Jan LY, Benzer S. A Drosophila mutant with a temperature-sensitive block in nerve conduction. *Proc Natl Acad Sci USA* 1978;75:4047–4051.

91. Yamamoto K, Quandt FN, Narahashi T. Modification of single sodium channels by the insecticide tetramethrin. *Brain Res* 1983;274:344–349.

Neurotoxicology, edited by Hugh Tilson
and Clifford Mitchell. Published by
Raven Press. Ltd., New York 1992.

2

The Use of Cell Culture
for Evaluating Neurotoxicity

Bellina Veronesi

*Neurotoxicology Division, US Environmental Protection Agency,
Research Triangle Park, North Carolina 27711*

The need to develop acceptable alternatives to conventional animal toxicity testing is generally recognized by toxicologists in industrial, academic, and regulatory sectors. Reasons for this range from the increasing number of chemicals being developed and introduced into the environment and workplace without adequate human-risk evaluation, the escalating costs and time required for routine toxicity assessment, the need to address the concerns of animal-welfare activists, and recent challenges to the validity of extrapolating data derived from animals to the "at risk" human population (37,50,61,65,106,122,124,150,173). These reasons have incited the need to develop faster and more cost-effective tests to evaluate chemicals for toxic potential. The number of chemicals already present in the environment is staggering. More than 60,000 chemicals are listed in the Environmental Protection Agency's (EPA) inventory of toxic substances and approximately 1,500 "notices of intent" to manufacture new chemicals are received by the Agency annually (108). Given these numbers, conventional neurotoxicity testing for EPA regulatory guidelines (45,159) follows a tiered approach, starting with general testing procedures and moving to the more specific. First-tier tests involve acute exposure in which LD50 levels, morbidity, and gross measures of functional status (i.e., functional observational battery) are examined. A second tier level involves assessing morphology, delayed toxicity, tolerance, and reversibility of effects. Finally, subtle changes in sensory function are measured and toxicity mechanisms are characterized. The length of time and the cost needed to complete the EPA's neurotoxicity guideline studies are other considerations (135). Acute toxicity tests and observations last up to 14 days, subchronic testing using repeated exposures can span 90 days, and chronic toxicity tests using repeated exposure can last for 2 years. The cost of testing a new chemical can be between $2,500 and $1,300,000 (124).

Although concern about the use of laboratory animals in biomedical research is not a new issue, public anxiety about the ethics, necessity, and validity of

The research described in this article has been reviewed by the Health Effects Research Laboratory, U.S. Environmental Protection Agency, and approved for publication. Approval does not signify that the contents necessarily reflect the views and policies of the Agency nor does mention of trade names or commercial products constitute endorsement or recommendation for use.

animal experiments is rising with unprecedented vigor (174). Animal-welfare activists demand that scientists reduce the number of animals by refining their methodologies and replacing animal experiments with alternative test systems. If left unaddressed or bluntly ignored by regulatory agencies, more radical voices will come to dominate public opinion. The validity of human toxicity reports based on animal data is becoming increasingly challenged (50,150,172,173). The sex, age, health, nutritional status, and genetic makeup greatly affect an organism's response to toxic substances. Ambiguities and uncertainties inherent in using other species to evaluate human health risk are compounded by the strong possibility that data obtained on young, healthy animals may not apply to a human population that is old, diseased, malnourished, or diverse in its genetic makeup. Since the human population is more heterogeneous than most conventional experimental animal strains, the range of doses producing an effect on humans may be larger than that for animals (102). Furthermore, many consequences of toxicity, especially neurotoxicity, may not manifest until senescence.

Concerns over the efficiency and validity of whole animal testing for chemical evaluation have led scientists to examine the potential of cell and tissue culture for Hazard Identification, a major component of the Risk Assessment process that is concerned with evaluating chemicals for their potential risks to human health. Hazard Identification defines the adverse effect resulting from chemical exposure and determines the dose-response relationship. One classic test for Hazard Identification is the LD_{50}, which defines the chemical dose required to produce lethality in 50% of an exposed population. Although no longer required by US regulatory agencies, the LD_{50} toxicity test is one of the most familiar to the public, the one targeted most frequently by animal-welfare activists, and the one test that *in vitro* toxicity models, in the form of toxicity "screens," can most effectively complement or replace (9,25,61,107).

IN VITRO TESTING: RATIONALE/ADVANTAGES

As our understanding of neurotoxic mechanisms becomes more sophisticated, expressions of toxicity more subtle than animal morbidity or lethality must be used as endpoints. Current thinking recognizes that conventional neurotoxic endpoints measured in the whole animal are subsequent to those initiating in the cell. This has, in recent years, turned the focus of toxicology from observational and descriptive to mechanistic. Technical advances in cell and tissue culture techniques have paralleled this new thinking in toxicology. Analytical instruments now monitor and measure toxicity at the cellular rather than at the organismal level. Development of a wider range of transformed neural and glia cell lines, the availability of human cell lines, and standardization of commercially available nutrient and other support media have been major factors in stimulating the use of cell and tissue culture models for neurotoxicity evaluation.

The use of cell cultures in the analysis of chemical and drug toxicity has been presented in numerous reviews (10,31,63,70,106,118,120,122,125,149). *In vitro*

TABLE 1. *Advantages of* in vitro *toxicity testing*

Uniform chemical and physical environment
Toxic exposure continuous or intermittent
Exposure parameters strictly controlled
Small amounts of test chemical needed
Systemic (e.g., hepatic) effects bypassed
Range of donor species available
Human materials available

systems offer unique advantages over whole animals for toxicity testing (Table 1). Test chemicals, dissolved in the feeding media, come into direct tissue contact without filtration by the blood-brain barrier and with minimal lipid-aqueous partitioning. Because the chemicals are dissolved in the nutrient fluids, toxic exposure is continuous and easily quantitated. Experimental parameters can be rigorously standardized, thus reducing variability between experiments. The dose of the test chemical can be controlled in terms of the amount of parent compound or active metabolite being delivered to the cell population or an individual target cell. Cell cultures can be sampled intermittently throughout exposure and the media can be sampled for high-performance liquid chromatography (HPLC) or mass spectrometry-gas–liquid chromatography (MS-GLC) analysis to detect metabolites. Toxicity testing can be conducted on specially synthesized, novel compounds since only small quantities of test chemicals are needed for culture systems. This feature also reduces the problem of toxic waste disposal. *In vitro* systems offer a uniform chemical and physical environment for toxic exposure. Systemic effects imposed by other organs (e.g., immune, endocrine, digestive, vascular, renal), which can interact with and compound a chemical's primary effect, are bypassed. Systems can be chosen which employ chemically defined nutrient media, thereby eliminating hepatic influences (i.e., mixed-function oxidation, phase II conjugation) to provide a means to assess a chemical's "intrinsic" neurotoxicity. Alternatively, microsomal additives in the form of embryo extract or S9 fractions can be added to the feeding media, imparting a limited metabolic competence to the system (16). The response of human cells can be compared with that of other species (e.g., avian, amphibian, rodent) to address questions of interspecies selective toxicity. For these reasons and others, cell and tissue culture systems must be considered valuable experimental adjuncts to whole animals in examining the toxic properties of chemical agents.

IN VITRO TESTING: SKEPTICISMS AND OBSTACLES

Skepticism surrounding the use of *in vitro* systems models to assess toxicity are based on the repeated objection that cells which are no longer part of normal tissue may develop an altered appearance and metabolism, and consequently, an altered response to test chemicals. Bureaucratic obstacles also exist to adopting *in vitro* techniques for toxicity testing. In many cases, regulations require specific tests (i.e., guidelines) for toxicity testing (45). These guidelines are highly standardized experimental protocols whose design features only laboratory animals. A legitimate fear of litigation arises if a product is tested with a nonstandardized methodology (61). To date, most regulatory agencies rely solely on animal testing to predict human chemical risk. However, the Food and Drug Administration, the EPA, the National Institutes of Health, and the Consumer Product Safety Commission are examining ways to integrate *in vitro* testing approaches (170). Although no test guidelines or framework for *in vitro* procedures have been established for the regulatory process, the strategies needed for their validation and subsequent incorporation into regulatory guidelines are being established (11,53,93).

IN VITRO MODELS OF NEUROTOXICITY

The nervous system consists of a highly specialized, heterogeneous, but integrated population of cells. Neurotoxicity is one of the most serious toxicological events since damage to even a small number of localized neurons can have profound consequences for the overall performance of the organism (121). It was recently proposed that 3 to 5% of existing industrial chemicals (i.e., 1,800–3,000) may be neurotoxic (112), with other estimates running much higher (159). In view of this, neurotoxicity assessment is increasingly emphasized by our legislators (121,159).

Some of the unique features of the nervous system (e.g., lack of regeneration; spatial separation of the cell body and axon; interdependence of the glia, nerve, and myelin sheath; neurotransmission) are specifically targeted by neurotoxicants. Many of the above features can be isolated in culture and experimentally exploited for toxicity testing. The use of various cultured cells and tissues models to evaluate neurotoxicity has been reviewed (8,23,33,60,75,86,110,129,163,171). Systems ranging from transformed cell lines (i.e., neuroblastomas, gliomas, and pheochromocytomas) to organotypic explants provide rapid, inexpensive models for either screening putative neurotoxicants or addressing their mechanism(s) of action. Six major categories of nervous system cultures are amenable to assessing neurotoxicity: whole embryo, organ, organotypic, reaggregates, primary cell, and cell lines. The major distinction among these culture systems is their complexity and resemblance *in situ* to the tissue from which they were derived. Historically, most of these culture systems have been used to address neurobiological rather than toxicological questions.

Certain technical fundamentals are necessary for culturing nerve cells and tissues and are relatively generic to other *in vitro* systems. Technical parameters are devised that optimize structural, biochemical, and physiological differentiation in cultured nerve cells in spite of their isolation from the body. Sterile techniques are required for all procedures since cultured cells and tissues can no longer rely on the immune system. In some of the less differentiated *in vitro* systems (e.g., cell lines), antibiotics are routinely incorporated into the feeding media, whereas antibiotics are deleterious to the dividing and differentiating cells of embryonic tissue explants. Careful control of the bathing and feeding solutions is vital to both the viability of the system and the valid interpretation of the results. The bathing solutions are composed of salts, amino acids, and ions in proportions devised to correct for osmotic pressure and to maintain an optimal pH. The housing chambers used for different *in vitro* systems vary in shape and material. For short-term cell cultures, plastic petri dishes or flasks are used; for suspension cultures, which promote large-scale cell growth, roller bottles are used; for long-term organotypic cultures, alcar dishes or Maximow glass chamber-assemblies, free of silicon impurities, are necessary. Dialyzed rat-tail collagen (97) or commercial products such as Poly-lysine (161) provide a suitable substrate for neuritic outgrowth of explants, whereas dissociated primary cells and cell lines can grow directly on the surface of plastic flasks.

Numerous technical pitfalls surround the culturing of nerve tissues (54,129) (Table 2). The more intact explant systems require nutrient media supplemented with animal (i.e., bovine, fetal calf, horse) serum or human placental serum. Sera, however, contain a variety of hormones, unknown drugs, and metabolites, and present distinct problems to cultured cells and tissues which are no longer protected by a blood-brain barrier. The housing chambers of cultured cells and tissues allow the metabolic production of CO_2 to accumulate in the nutrient fluid, causing pH shifts in the microenvironment which could affect the response of the nerve cultures to toxic challenge. Alterations in the osmolarity of the culture medium, resulting from high concentrations of the dissolved test chemical, may also occur. Reactions (e.g., protein denaturation) between the test chemical and components in the nutrient medium may affect the availability of essential nutrients to the cultured cells. The protein binding properties of test chemicals may also affect the dose-response curve. Many neurotoxic solvents are relatively insoluble in aqueous media (e.g., *n*-hexane), form insoluble particulates, evaporate over time due to their volatility, or precipitate when added to tissue culture medium (e.g., lead). These physiochemical considerations raise problems with obtaining adequate concentrations of the test compound or physically precipitating critical nutrients. Defined nutrient media supplemented with hormones and other nutrients, or undefined media enriched with animal or placental sera, may not reflect the actual *in vivo* physiological milieu of cells and protect the cells from neurotoxic damage. If neurotoxic potency of a test chemical *in vivo*

TABLE 2. *Technical caveats of tissue culture*

Animal sera in nutrient media
pH shifts due to housing chamber design
Altered osmolarity
Reactions between test chemical/media components
Test compound insolubility
Contamination imitating cytotoxicity

is closely controlled by metabolic parameters such as biotransformation, lipid-aqueous partitioning, and distribution, then the expression may be masked in a chemically defined-nutrient media system, and data gleaned from such a culture system will be meaningless when related to whole animal studies. On the other hand, a chemical's intrinsic neurotoxicity can be assessed by using chemically defined media. Contamination can perturb the culture system and produce effects resembling neurotoxicity. Therefore, the viability of the culture itself must be monitored microscopically and biochemically since an unhealthy culture, stressed by bacterial, fungal, or mycoplasmic contaminants, will respond differently to neurotoxic insult. In spite of these concerns, the many advantages presented by *in vitro* models for neurotoxicity testing present a strong incentive for continued development and increased utilization. In the following section various cell and tissue culture systems that have been used in neurotoxicology will be described (Table 3).

Whole Embryo and Organ Cultures

A great deal of morphological, physiological, and biochemical detail is present in the cultured embryo, and in which normal development of diverse cellular and tissue interactions continue. Therefore its usefulness for teratological studies is widely accepted (5). Whole embryo cultures parallel the growth and development of rodent embryos throughout much of the period of organogenesis and the critical stage of neural tube closure. For this technique, rat or mouse embryos are removed on day 8 to 10 of gestation, dissected free of decidua, and cultured

TABLE 3. In vitro *models used in neurotoxicity studies*

In vitro model	Neurotoxicant tested
Organotypic	hexacarbons
	methyl ethyl ketone
	thallium
	triethyltin
	acrylamide
	pesticides
Hippocampal slice	trimethyltin
	triethyltin
Reaggregate	pesticides
	methyl mercury
Primary cell	pesticides
	lead acetate
	methyl bromide
	methyl mercury
	cadmium
	zinc
	acrylamide
Cell lines	pesticides
	hexacarbons
	acrylamide

in medium containing 50 to 100% serum in a roller bottle system for 48 to 72 hours. In this period the neural tube closes, and the brain, heart, and other organs develop. Embryos can be maintained for 3 to 4 days, during which time they accumulate protein and nucleic acids at rates comparable to those observed *in vivo.* The embryo may be exposed via the nutrient media to test chemicals throughout development, using a wide range of development and differentiation endpoints.

Organ cultures obtained from pre- and postnatal animals are less intact than the whole embryo preparations. In these preparations, a whole ganglion, pineal gland, hypophysis, etc., is cultured and many neuron-neuron, neuron-glia, and glia-glia spatial and neurochemical relationships are maintained. Drawbacks to using tissues cultured as a solid mass involve problems with gaseous diffusion and nutrient exchange which results in anoxic damage to the preparation. In addition, it is difficult to visualize neurotoxic damage morphologically with whole embryos and organs unless histology is done on the preparation.

Organotypic Explants

A less intact system than organ culture uses explants of brain and spinal cord tissue. Such organotypic cultures are derived from relatively undifferentiated embryonic tissue and develop into integrated neuronal and glial populations. In these preparations there is rapid neuronal development characterized by dendritic and axonal growth, synaptogenesis, and myelination of axons, and the synaptic connections and spatial relationships of nerve cells *in situ* are preserved. Although technically demanding, organotypic explants can remain viable for up to a year, making them unique *in vitro* paradigms for chronic neurotoxicity studies. This technique, pioneered by Peterson et al. (115), offers several advantages over the less intact culture systems described below. Cytoarchitecture is optimally preserved and morphological changes induced by neurotoxicants can be monitored by bright-field microscopy with little or no disturbance to the cultures. Myelination, in the presence of sera-supplemented media, occurs reliably and is easily detected by light microscopy. A variety of neurochemical, electrophysiological, and morphological endpoints can be applied to the organotypic preparation. The most successful application of organotypic explants in neurotoxicology has used spinal cord co-cultured with muscle to provide facsimiles of sensory-motor targets (163,171). Such models have been used to examine the neurotoxicity of hexacarbon solvents (164,165), triethyltin (68), thallium (78,144), acrylamide (139), and pesticides (158,169). In addition to spinal cord co-cultures, other levels of the neuraxis (e.g., embryonic inner ear) have been tested with ototoxicants (143). As with organ cultures, hypoxia and necrosis occur at the explant center due to its thickness. In spite of the explant flattening at the periphery, microscopic examination of individual cells is limited by the thickness of the explant.

BRAIN SLICES

Brain slices, especially the hippocampal slice, have been used extensively for neurotoxicity studies (51,52,94). These preparations retain many of their *in situ* properties, such as cytoarchitecture and neural circuitry (35,151). Cultures can be maintained for several months by suspending the tissue slice in nondefined media (57). Several notable studies have used the hippocampal slice preparation to study the effects of organoalkyl metals such as trimethyltin and triethyltin (3,51,52).

SUSPENSION AND REAGGREGATE CULTURES

Suspension cultures are used to produce the large cell populations needed for biochemical endpoints (69,154). Tissues are enzymatically digested, gently agitated into a suspension of individual cells, and decanted into bottles, flasks, or dishes for incubation. With time, the individual cells form a confluent monocellular layer and further growth stops. The related brain reaggregate technique is characterized by three-dimensional cytoarchitecture, high cell mass yields for neurochemical analysis, and long survival time using chemically defined media (7,85,99,131,155). The basic methodology was introduced by Moscona (103) who showed that dissociated, undifferentiated fetal cells could reassemble spontaneously *in vitro* to form histotypic, three-dimensional assemblies. Subsequent investigations have demonstrated that fetal brain aggregates form patterns of cell alignment similar to those *in vivo* and undergo extensive morphological differentiation, including synaptogenesis and myelination (156,157). Biochemically, neural-reaggregate cultures develop expressions of neuronal and glial proteins that closely resemble *in vivo* ontogeny (83,84,99). The donor age, brain site, and dissociation procedures are very critical to this technique and influence the degree of cytoarchitectural restoration. Suspension and reaggregate cultures represent valuable experimental models for neurotoxicity studies. Rotation-mediated aggregating cultures of fetal rat brain cells have been used to examine the acute and subchronic effects on cholinesterase activity of organophosphorus and carbamate pesticides (132). Methyl mercury neurotoxicity has also been examined in suspensions of cerebellar granule neurons (127).

PRIMARY CELL CULTURES

One of the most widely used *in vitro* methodologies for basic neurobiological research is the primary cell culture technique. Outlined in a recent technical manual by Shahar and colleagues (137), the dissection techniques and nutrient media are unique to each level of the neuraxis and are, for the most part, empirically derived. For these preparations, tissues are removed from the embryonic or fetal animal, dispersed mechanically or enzymatically, and cultured in plastic flasks to form monolayers. Sensory neurons of the sympathetic and dorsal root ganglia (18,32,141), Schwann cells (6,96,97), oligodendroglia (26,101), and astrocytes (19,79) as well as a vast array of sites throughout the neuraxis, have been successfully established as primary cultures. Retinal neurons and photoreceptors have also been cultured from neonatal mouse in serum-free, chemically defined media and serve as models for assessing the survival and differentiation of photoreceptor cells critical to visual processes (117). The availability of human donor materials for primary nerve cultures is very limited. Recently, however, a self-replicating culture was developed (123) from human cerebral cortical neurons obtained from a patient with unilateral megalencephaly, a developmental neuropathy in which immature neurons proliferate continually (67). This self-replicating, primary nerve culture, known as HCN-1A, has many morphological and biochemical characteristics of human neurons *in situ,* including positive staining for various neurotransmitters (i.e., GABA). Lymphocytes, derived from blood or whole spleen, are also valuable primary cells for neurotoxic assays since they contain muscarinic receptors (28) and house acetylcholinesterase (AChE) and neurotoxic esterase (NTE) (36,27), the presumed target of the delayed neuropathy (OPIDN) associated with some pesticide exposure (88). Human lymphocytes have been proposed for biomonitoring pesticide exposure in high-risk populations (95) and automated techniques have recently been developed to assay AChE and NTE activity in cultured lymphocytes exposed to test pesticides (27).

Primary nerve cultures can be maintained for several months by using supplemented nutrient media and a fibronectin or collagen substratum which encourages neuronal adhesion and neuritic extension. Primary cell cultures retain many normal properties and can be prepared from cells at various developmental stages. However, substantial physiological differences may occur among primary cultures derived from different embryological or adult stages (48). Primary neural cells, especially astrocytes (90), show a pronounced plasticity. Minor alterations in the technical parameters (i.e., donor age, dissociation procedures, seeding density, amount of serum, trophic factors) will influence the characteristic of the culture (22,152). The technique used for dispersing the tissue or the organ may also affect culture properties (56). Cells isolated by physical or nonenzymatic chemical methods are less viable than those isolated by treating the tissue with proteolytic enzymes (e.g., collagenase, hyaluronidase). It is generally difficult to prepare large quantities or primary cultures and they contain many different cell types. However, a variety of manipulations ranging from mechanical agitation, adjustments in the nutrient media, and addition of antimitotic agents will promote the survival of one neural cell type over another (e.g., glia versus neuronal). For example, cultures can be treated with cytosine arabinose, an antimitotic agent that inhibits the growth of nonneuronal cells (i.e., fibroblasts) but permits the growth of neurons and glia. This treatment allows one to compare the differential response of mixed (neuron and glia) and purified (neuronal) cultures to neurotoxic exposure.

Until recently, primary cultures have not been rigorously applied in neurotoxicity studies, with some notable exceptions. Spinal cord cultures containing both neuronal and nonneuronal cells have been exposed to organophosphate esters and neurotoxic amino acids and indexed with several biochemical markers of cholinergic function (60). Primary cerebral cell cultures have been exposed to acrylamide using viability endpoints (e.g., cell outgrowth, attachment, and diameter) (86,87). Primary cultures of fetal astroglia, oligodendroglia, and human neuroblastoma SK-N-SH, SY5Y cell lines have been tested for viability after exposure to lead acetate using Trypan blue dye exclusion (153). Other studies have examined lead acetate toxicity in primary cultures of cerebral astrocytes and cerebellar granular neurons with both electron microscopic and metabolic endpoints (82) and aluminum (133) and ethanol (147) neurotoxicity in dorsal root ganglia neurons. Davenport and colleagues (30) have used primary cerebellar cultures and cytotoxic endpoints to examine methyl bromide and methyl chloride neurotoxicity. Primary nerve cells from various brain regions have been used to screen neurotoxicants and establish structure-activity relationships among heavy metals (i.e., methylmercury, cadmium zinc, organotins) using the neutral red cytotoxicity assay (14). Rat retinal cells fused with mouse neuroblastoma cells have been used to evaluate the neurotoxic effects of glutamate and related compounds (104).

Cell Lines and Cell Strains

The longevity of a primary nerve culture varies from a few days to several months according to donor age, species, tissue source, and nutrient conditions. However, primary cells can become self-replicating cell lines either spontaneously (126) or by exposure to viral (91), chemical (77), or physical (13) agents. The extent and nature of this transformation depend on both the cell and the transforming agent. Continuous cell lines can also be established from tumors of neural and glial origin or by fusion of neurons and glioma cells (20,74). The life span of a cell line is approximately 4 to 50 divisions (47) and is divided into log phases which are characterized by progressively longer population-doubling times, degeneration, and death (76). Cell lines of limited life span often undergo "crisis" after which their growth potential changes and their life span becomes unlimited ("immortal") (12). Cell lines can also be cloned to form cell strains, which are homogeneous populations with the potential for continuous propagation (see below).

Many criteria exist for identifying established cell lines. According to Diamond and Baird (34), these features include the capacity to undergo an unlimited number of cell divisions, altered cell and colony morphology, lack of locomotion, lack of contact inhibition, lack of density-dependent inhibition of cell multiplication, loss of anchorage dependence, and high fibrinolytic activity. Freshney (55) and Nelson (109) have reviewed various methodologies and endpoints commonly used for cell line cultures. Cell lines can grow as monolayers, in suspension, or can be kept continuously rotating in roller tubes and bottles. Although many neural cell lines have been characterized (74) and are available from several species, including human, the variety of neural cell types is rather limited, including adrenal chromaffin cells (PC12), sympathetic ganglion neurons, gliomas, neuroblastomas, astrocytomas, and some unspecified brain tumors. Most of these have retained key neurochemical features (20), making them suitable for neurotoxicity studies. With recent developments in cell fusion techniques, neuronal primary cells of defined origin can be "immortalized" by fusion to a transformed cell line. The NG108-15 hybridoma, for example, was developed by fusion of the mouse neuroblastoma clone N18TG2 with the rat glioma clone C6BU1 using Sendai virus (72). Its characteristics include nerve fiber extension, synapse formation, choline acetyltransferase activity, and receptors for various neurotransmitters (29). Although hybridoma technology makes it possible to immortalize various cell types, many neuronal properties are lost from such preparations after a certain number of passages. To prolong the lifetime of cell lines while maintaining their phenotype, transfection of the primary culture with oncogenes has been attempted (105). Although long-term survival of these transfected cell lines has not been established, this technology may provide a means to immortalize cells that have proven difficult to maintain in culture.

Probably the most utilized cell line for neurotoxicity studies is the neuroblastoma. These cells, derived from a neoplasm over 40 years ago, form homogenous populations, proliferate rapidly in chemically defined media, extend neuritic-like processes, and contain several neurotransmitter precursor molecules (i.e., choline acetyltransferase and tyrosine hydroxylase). Neuroblastomas of mouse (Neuro-2a, NB41A3) and human (IMR-32, SK-N-MC) origin are commercially available (American Type Culture Collection, Rockville, MD). With neuroblastoma cells it is particularly important, and sometimes very difficult, to distinguish between morphologic "differentiation" and cytotoxicity (41). Several studies have used mouse neuroblastomas for specific neurotoxicity questions (21,44,113, 134,168), whereas others have used the cell line for screening a number of neuro- and nonneurotoxicants (160,167). Widely used is the NIE115 neuroblastoma, cloned from the murine neuroblastoma C1300 (4). This cell line has been proposed (24) as a substitute for the "Hen Test" which the EPA pesticide registry (46) uses to identify OPIDN-causing chemicals. Since these cells contain NTE activity, several investigators have used the NIE115 line to assess OPIDN (21,44) and have demonstrated a parallel ranking in NTE threshold inhibition levels in both hen brain and the neuroblastoma for various organophosphates. Further characterization of NIE115 and additional testing with putative OPIDN-causing chemicals are necessary to evaluate further its use as an adjunct for the hen test.

The pheochromocytoma and its clone, the PC12, exhibit many of the properties of mature, terminally differentiated sympathetic neurons and have become increasingly popular in neurobiological research (71). The PC12 tumor cell line grows readily under standard culture conditions, and in the presence of nanogram quantities of nerve growth factor, the cells hypertrophy and produce short neuritic process within 24 hours of plating. The PC12 has become the premiere *in vitro* model to study nerve growth action on catecholamine biosynthesis and secretion, as well as neuronal differentiation. The PC12 contains dense-core granules similar to those housed in both adrenal chromaffin cells and sympathetic neurons. In addition to containing catecholamines and catecholamine-synthesizing enzymes, the PC12 also synthesizes and releases acetylcholine and contains choline acetyltransferase, the enzyme catalyzing the synthesis of acetylcholine.

Cell lines (neuroblastomas, PC12 cells) have been co-cultured with muscle cells (L6) to examine the effects of neurotoxicants on neuromuscular junctions. Such preparations provide structurally and functionally coupled targets of pesticides and are amenable to both histochemical and electrophysiological endpoints (49,89,119,130). Co-cultures of primary culture hepatocytes with a C6 glioma or the NIE115 neuroblastoma have also been established to provide an *in vitro* preparation with limited hepatic capability (40).

A subclass of the cell line is the *cell strain,* which is obtained by selecting from an established line a small number of cells having a common biological characteristic (e.g., a specific enzyme) that persists throughout passages and can be used as a marker (47). Although derived from a genetically and phenotypically identical stock, the individual cells in the cell strain population may not remain homogeneous indefinitely and may require subcloning to maintain the desired phenotypic characteristics. Because of their uniformity, however, cell strains can be expected to respond more uniformly to neurotoxicant exposure than cell lines. Cell strains are normally grown as monolayers in closed bottles or flasks, and require less frequent renewal of their nutrient medium than cell lines or primary cells because of their limited metabolism.

The advantages of using cell lines and strains in neurotoxicity assessment is that they have a wide literature base, are available through commercial repositories, are available from a wide range of donor species including human, can survive in chemically defined media, and are relatively inexpensive to maintain and easily monitored with cytotoxic and neurotoxic endpoints. Skepticism underlies their use for neurotoxicology testing, however, since transformed cells may not accurately reflect their *in situ* counterparts in response to chemical exposure (129). However, these fears can be minimized by strategic experimental design and use of limited, well-characterized endpoints.

MECHANISTIC *IN VITRO* MODELS OF NEUROTOXICITY

In view of the variety of nerve cell and tissue culture models available, and in keeping with the complexity of the nervous system, it is apparent that no single *in vitro* preparation can be relied on to detect all possible neurotoxic endpoints. Rather, *in vitro* neurotoxicity might best be addressed using a tiered approach which progresses from the simple, inexpensive, and less specific models (e.g., cell lines) to more technically demanding, complex, and definitive paradigms (e.g., organotypic explants or slices). The first tier acts as the preliminary screen, differentiating neurotoxicants from cytotoxicants by using nonspecific and neural-specific endpoints. Subsequent screens are used to differentiate classes (e.g., metals, cholinesterase inhibitors, solvents, etc.) of neurotoxicants using specific targeted endpoints. An additional level of inquiry uses the *in vitro* model to ask specific mechanistic questions that may be difficult to obtain in the whole animal. Discussions of *in vitro* systems screens have focused on how well they imitate the animal's response to toxic insult. The mechanistic *in vitro* model should instead be conceptualized as providing an isolated system that can be manipulated to yield unique toxicity data, elusive in the whole animal. Such models try to preserve a unique structural, electrophysiological, or biochemical property of nervous tissue that is targeted by the toxic chemical and can be exploited as an endpoint. Generally speaking, the mechanistic model system is neither complete nor faithful to the animal model in all aspects, but it tends to provide unique information if an isolated single endpoint is examined and the data are interpreted within the context of that endpoint. The initial design of mechanistic models is largely empirical and focuses on correlating *in vivo* and *in vitro* endpoints by incorporating metabolic, neurochemical, or neuropathic targets into the model system. An example of an *in vitro* mechanistic study is the work of Veronesi and colleagues who used organotypic cultures of spinal cord explants with attached dorsal root ganglia, co-cultured with muscle tissue, to investigate the neurotoxic effects of subchronic exposure to a number of n-hexane metabolites and the synergistic effect of methyl ethyl ketone on hexacarbon neuropathy (162,164,165,166). Using ultrastructural and metabolic endpoints, these studies lead to a more definitive identification of the primary neurotoxic diketone metabolite and to a more lucid interpretation of the sequence of axonal degeneration in the toxic "dying-back" process that had only been surmised from animal studies. Kasuya (92) examined the cytotoxic responses of neurons, glia cells, and fibroblasts to inorganic and organic mercury compounds in experiments which led to a better understanding of their structure-activity relationship and mode of neurotoxic action. With the same system, interaction of selenium, vitamin E, and methylcobalamin on mercury toxicity was examined.

NEUROTOXICITY SCREENS

The emphasis on animal replacement instead of reduction has limited toxicologists from recognizing the full potential of *in vitro* systems for neurotoxic screening (33). Screens promise to provide a quick and inexpensive way for assessing potential neurotoxicity. Currently, cell and tissue models are available which can identify those chemicals most likely to present toxicological hazards, prioritize those requiring further toxicity evaluation, differentiate neurotoxicants

TABLE 4. *Criteria for* in vitro *screens*

Low incidence of false negatives and false positives
High correlation with *in vivo* toxicity data
Sensitive, simple, rapid, economical, versatile

from cytotoxicants, and rank neurotoxicants for toxic potency. Screening implies a preliminary or "first tier" evaluation that will be followed by more stringent testing. The main criterion for a screen is that it gives an accurate correlation with *in vivo* neurotoxicity and has a low incidence of false negatives and false positives (Table 4). Knowledge of mechanistic relationships is not a high priority, nor is the interpretability of the test. The idealized screen is sensitive, reliable, simple, rapid, economical, and versatile (53).

Risk Assessment is by far the most challenging application for *in vitro* screens. For this, cell culture modeling must be based on toxicokinetic considerations (i.e., solubility, protein binding, kinetics, metabolic activation, octanol/water partitioning coefficient data etc.) as well as mechanistic endpoints to predict target and dose-response effects. Exposure conditions and repair mechanisms are other considerations. Such models must employ a statistically sufficient population of cells and be amenable to pharmacokinetics and multidose testing. A Risk Assessment screen must be able to identify those chemicals which act through the mechanism being measured. Although the economic profile of an *in vitro* screen is always important, the higher costs and longer time commitments of Risk Assessment models are acceptable if regulatory agencies will consider the data for chemical classification and regulatory decisions. In reality, the various routes of "real world" exposure, the diverse metabolic handling of xenobiotics, and the heterogeneity of "at-risk" human populations exaggerate the difficulty of designing these models at the present time.

NEUROTOXICITY SCREENS: CHOOSING A TEST BATTERY

Defining a test battery for differentiating cytotoxicants from neurotoxicants or for identifying neurotoxic chemical classes requires considerable strategy and thoughtful design. Since cell culture systems poorly mimic the intact nervous system, it is important to use a variety of *in vitro* models and multidisciplinary endpoints to parallel neurotoxicity in whole animals. The *in vitro* system should not be constructed as an imitation of the animal, but as a self-contained, manipulatable system in which a relevant endpoint can be tested independent of all other influences. Since specific mechanisms of neurotoxic action are rarely known, a variety of cells housing clearly defined neurochemical or morphological targets should be featured in the test battery (Table 5). Such models should be highly predictive for identifying chemicals known to produce neurotoxicity via that endpoint in animals. The actual *in vitro* models chosen for the test battery should include neural cells and tissues from various species (e.g., avian, mouse,

TABLE 5. *Sample test battery for neurotoxicity screening*

Nonneural cells/tissues
  (i.e., nontarget tissues, hepatic, kidney, lung)
Neural cells/tissues
  Cell lines:
    Neuroblastoma (C1300, NIE115, IMR-32, SK-N-MC)
    Astrocytomas U-87 MG; Glioma HS 683
    PC12
  Co-cultures of neuroblastoma/PC12 with L6 muscle/hepatocytes
  Lymphocytes
  Brain reaggregates

rat) which feature unequivocal neurospecific functions. Neuroblastomas and PC12s are favorite cell lines for neurotoxicity screens. Cell lines such as the NIE115, a cell clone of the murine neuroblastoma C1300 (4), and lymphocytes containing AChE and NTE activity are also promising candidates for pesticide testing. Co-cultures of PC12 or neuroblastoma and L6 myoblasts yield functionally and structurally coupled "neuromuscular" preparations often targeted by pesticides and are amenable to morphological, histochemical, and electrophysiological endpoints. Although not fully characterized, use of human cells gives some feeling for the neurotoxic response in the relevant species. Among those that should be routinely incorporated in the test battery are the IMR-32 and SK-N-MC neuroblastomas, the U-87 MG and U-373 astrocytomas, and the HS 683 glioma cell lines (American Type Culture Collection, Rockville, MD). Data derived from human nerve cells are more extrapolatable to the human condition and supply insight into species differences in neurotoxic susceptibilities. Nonneuronal cells with readily indexed basal level functions should be included to differentiate between neurotoxicants and other xenobiotics.

CYTOTOXIC AND NEUROTOXIC ENDPOINTS

Two levels of toxicity undermine the structure and function of all cells, and may be indexed in assembling a screen for neurotoxicants. Basal level endpoints reflect generic functions required by the cell for viability, whereas tissue- or organ-specific endpoints are those unique to the function or morphology of a particular tissue or organ (37,107,145). In general, basal level cell functions always support the tissue-specific functions, and the degree of such dependence determines a toxic chemical's primary target tissue. The expression of toxic insult can range from cell death (i.e., sudden lysis) to mild and transient alterations in synthetic or metabolic function (Table 6). Cell lethality can be measured by quantitating the total cellular protein, vital dye uptake assays, or mitochondrial viability tests. The concentration of a test chemical that produces 50% depression of a cell's protein content (i.e., IC50), can be compared with the *in vivo* IC50s of other neurotoxic chemicals to rank its relative neurotoxicity (146). This assay

TABLE 6. Cytotoxic/viability endpoints

Cell protein measurements
Vital dye uptake
Mitochondrial viability
Cell growth and reproduction (i.e., cell number,
    DNA content, plating efficiency, etc.,)
Energy regulation
Macromolecular synthesis
Cell-cell communication
Transport processes
Lipid peroxidation
Microscopic endpoints
In situ hybridization

can be performed in combination with microsomal additives so that the effects of chemical intermediates can be tested. Vital dye assays are based on the ability of cells with damaged membranes to allow the entry of stains such as Trypan blue (116) or neutral red (15,111). Although vital dye assays are used to index cell lethality, permeability of the cell membrane does not necessarily indicate cell death since some cells are able to repair the underlying damage to their membranes and reestablish volume control (2). Cell growth and reproduction are common measurements of basal level functions and can be quantitatively evaluated through changes in cell number, total protein and DNA content, colony formation, and plating efficiency.

In contrast to cytotoxic endpoints which measure lethality, viability endpoints are used to assess the degree of damage or monitor the progression of events during intoxication (122). Such endpoints are designed to evaluate membrane integrity, loss of ions or cofactors, energy regulation (oxidative-reduction status), macromolecular synthesis, uridine uptake (140), cell-cell communication and recognition via gap junction integrity (39), biosynthetic reactions, transport processes, or specific enzyme changes. More refined viability tests that border on neurotoxicity endpoints (106) include production of a specific biochemical end-product (e.g., antibody and isozyme expression), tests for cell surface effects (receptor binding, ionic influx, and efflux tests), and lipid peroxidation as an index of membrane (i.e., myelin sheath) stability. Morphological endpoints to monitor viability include multinuclearity, cellular giantism, vacuolization, chromosome breaks, and mitochondrial swelling. Scanning electron microscopy can reveal abnormalities in the cell surface (e.g., blebbing), and transmission electron microscopy can describe a wealth of organelle changes. A variety of techniques using conventional scanning and electron microscopy (42,98,114,148) have been modified for tissue culture preparations and are routinely used. More sophisticated and technically demanding morphological endpoints, including in situ hybridization (80,81) and neural-specific histochemical and immunocytochemical stains (58,59), can also be applied to tissue culture preparations.

Although they have limited ability to predict neural-specific effects, cytotoxicity

TABLE 7. *Neural-specific endpoints*

Neuronal enzymes: Glutamic acid decarboxylase, neuron-specific enolase,
  dopamine hydroxylase, NTE, AChE, choline acetyltransferase, tyrosine hydroxylase)
Neuronal receptors
Electrophysiological endpoints
Morphological (i.e., neuritic extension) endpoints

NTE, neurotoxic esterase; AChE, acetylcholinesterase.

and viability endpoints provide essential information on the intrinsic toxicity of pure chemicals, mixtures, and formulations and must be included in the neurotoxicity test battery to differentiate cytotoxicants from neurotoxicants. If the cytotoxic or basal level cellular endpoints are affected while the neuronal ones are unaffected, the test chemical is judged cytotoxic, not neurotoxic. Conversely, if neurotoxic endpoints are affected at a dose that has no affect on nonneural cells, the test chemical is considered neurotoxic. Further testing is required to establish the degree of change when both neuronal and nonneuronal endpoints are affected.

For the screening of putative neurotoxicants, neural-specific endpoints representing neurochemical, neuromorphological, and neurotransmission functions are necessary (Table 7). No single endpoint can encompass the range of neurotoxic targets, so a variety should be incorporated in the neurotoxicity screen. Biochemical determinations of specific enzymes or receptors offer extremely rapid, accurate, and quantifiable endpoints that might be suitably adapted for neurotoxicity screens. Since neurotoxicants can target the neurotransmitter systems at the presynaptic release and postsynaptic receptor levels, assays for neural enzymes such as glutamic acid decarboxylase, dopamine β-hydroxylase, neuron-specific enolase activity, NTE and AChE activity, choline acetyltransferase, tyrosine hydroxylase, and other relevant enzymes are reliable assays and relatively easy to perform. Electrophysiological measurements have been routinely applied to organotypic cultures, brain slices, and reaggregate cultures. In hippocampal slice preparations, altered neurotransmission can be detected as an extracellular change in the electrical response, i.e., the excitatory postsynaptic potential, the paired-pulse methods for the inhibitory basket cells, and long-term potentiation recordings (151).

EXPERIMENTAL DESIGN AND TECHNICAL CONCERNS FOR SCREENING NEUROTOXICANTS

Important considerations in designing a meaningful *in vitro* neurotoxicity study apply to toxicity testing in general, but some are peculiar to those using cultured nerve cells (Table 8). For screening of neurotoxic chemicals, the source and passage of the neural cell lines and the cell density are critical parameters since they may influence the expression of toxicity. The duration of exposure (i.e.,

TABLE 8. *Technical considerations for screening toxicants*

Duration of exposure
Solubility of test agent
Solvent effect
Interaction between media/test agent
Cell density
Chemically defined media/microsomal additives
Source of cell lines
Source of test chemicals

acute versus subchronic treatments), the developmental status (i.e., log phase of cell lines; embryonic age of tissue) of the test system, and the concept of reversibility or recovery are also important. Other technical concerns that have been shown to influence the interpretability and reliability of test results include the solubility of test agents, solvent effects, and interactions between components of complex test samples (146). The influence of the vehicle or solvent system may contribute a cytotoxic component to the cells or may structurally and physically alter the test agent. The dependence of the cell culture system on physical factors such as the temperature at which cells are cultured and incubated, osmolarity, cell density, etc., should also be evaluated. Technical procedures for preparing the biological components, definition of culture media and conditions, and establishing criteria for determining normal states of growth and differentiation should be clearly reported. The test protocol must list the various parameters such as duration of exposure to test toxicant, the endpoint assays, presence of serum during exposure to test chemicals, and other specific factors related to dosing. As a rule, general cytotoxic responses should be reported for cells in log-phase growth in contrast to lag or stationary phase cells. However, the selection of growth phases for neurotoxicity depends on the parameters being measured. For example, cellular proliferation requires log phase cells, whereas biochemical modifications can be measured in either log or stationary phase cells. Other variables, such as the use of exogenous agents (i.e., nerve growth factor) in the nutrient media, need also to be considered since they may alter the neurotoxic sensitivities of some cell lines (153). Preferably, systems should be chosen which use chemically defined nutrient media to facilitate standardization. To be technically and scientifically acceptable, the cell lines, the culture conditions (media, serum, subculturing regimens), and the test conditions must be standardized. The greater the reliance on commercially available materials and supplies such as cell lines, culture plates, and defined media, the more easily the test technology can be transferred to other laboratories. All validation activities (see below) must be conducted blind, and the test chemicals must be quality assured, coded, and if possible obtained from repository laboratories (11,25,146). Microsomal (S9) additives can be added to the feeding media to impart a limited metabolic competence to the cell culture system (16). If secondary phenomena such as lipid-aqueous partitioning are unique controlling mechanisms in the bioavailability of a test chemical, models of more differentiated systems than cell lines should be included in the battery.

On a purely technical level, the test chemicals used to calibrate the test battery should be of the highest purity with established physical, chemical, and biological properties, since impurities, decomposition, or adulteration of test chemicals may result in a shifting of the dose-response curve which would affect the interpretation of the test results. Prototype neurotoxicants should be chosen with well-defined target sites (142). A series of chemical analogs, which vary in their acute neurotoxicity, should compose the set with related nonneurotoxicants serving as negative controls. Chemicals which are nontoxic and chemicals which are active toxicants in organs other than nerve tissue are included to assess the system's ability to differentiate neural and nonneural target organ toxicants. When choosing test chemicals, large numbers of those representing each neurotoxic class (e.g., cholinesterase inhibitors, solvents, metals) should be selected in order to obtain statistically accurate estimates of the false positive and false negative rates. These rates are used to compute the predictive power of the test battery in the context of practical toxicity testing. Every test system will have its own characteristic false positive and false negative rates and it is critical to define these rates since a high percentage of false positives can lower the confidence of using a particular screen (53). In summary, selection of test chemicals should include those with an established toxicological database, sufficient numbers of toxic and nontoxic test chemicals to obtain accurate estimates of false positives and false negatives, sufficient numbers of chemically related test chemicals to establish structure-activity relationship, a selection of mixtures to investigate interactive toxicology, and test chemicals which are nonneurotoxic to establish selective toxicity. Subgrouping test chemicals by mechanism of action or target site can be used to identify and confine specific test systems which are predictive for particular neural mechanisms. A test system may be highly predictive for chemicals which produce their effects by a certain mechanism, yet fail to identify neurotoxic chemicals which act through others. Validation of test systems to deal with complex mixtures is also critical since commercial products are in fact formulas of various substances. However, testing of mixtures should be introduced into the validation process only after their purified components are tested individually.

THE VALIDATION PROCESS

After a test battery consisting of a variety of neural and nonneural cell models and appropriate test chemicals and endpoints is assembled, it must be validated. In brief, validation of a test battery establishes its credibility for a specific purpose. Validation establishes *relativity* (i.e., provides information that contributes to chemical safety evaluation) and *reliability* (i.e., gives consistent results in different laboratories and at different times in the same laboratory) (53). Relativity or interpretability implies that the data have meaning for some scientific decision-making process. Obviously a test which produces consistent data is useless if it is unclear how that data impact on the safety evaluation process. Reliability means that when the test is conducted in a technically competent secondary laboratory or in the same laboratory at different times, the results are reproduced accurately. This aspect of the validation process determines whether or not the methodology can be transferred from one laboratory to another. To facilitate this critical aspect of the validation process, the cell lines, the culture conditions (i.e., media, serum, etc.), and the exposure parameters (e.g., dose, duration of exposure, vehicle, etc.) require standardization, a process facilitated by using repositories of reference cell lines, support media, and test chemicals. The confidence one has in a test battery is ultimately determined by the correlation between *in vitro* and *in vivo* neurotoxicity. Ideally, the test battery correlates toxicity results from cell culture, whole animal, and human clinical findings, and epidemiology studies. Although, one strives for 100% correlation between the *in vitro* and *in vivo* endpoints when testing a series of chemicals, it is unlikely since even data from animal tests are never 100% predictive of human risk.

Much debate and experimental effort have been directed toward outlining the validation process (9,11,17,53,62,64,106,128,136,146). The concept of validation must be firmly defined if regulatory acceptance of *in vitro* screens is anticipated. Goldberg (62,64) and others have suggested that development of *in vitro* test models should progress from the empirical to the more defined. In this scheme of validation, the questions are initially unfocused and the methodologies heuristic. Subsequent *in vitro* models are designed that mimic specific *in vivo* targets and incorporate mechanistically based endpoints. Scala (128) has proposed that the validation of an *in vitro* screen should proceed in three phases, with each stage addressing a different degree of interpretability and reliability. In this approach, *in vitro* models are initially established with mechanistic endpoints incorporated. A limited number ($N = 10 - 100$) of chemicals that are known to target that mechanism are first tested. Chemical selection for this initial stage of the validation process is used to define the relevant endpoints and as quality control standards. If one were attempting to develop an *in vitro* test battery to screen pesticides, the systems chosen might consist of cell lines (e.g., neuroblastoma, PC12) and primary cultures (i.e., lymphocytes) which house the vulnerable target enzymes of pesticides (i.e., AChE, NTE). Co-cultures of neural cells and muscle would also be included to provide neuromuscular targets of the agents. Initially, this battery would be tested with a group of known AChE and NTE inhibitors to assess endpoint (e.g., enzyme depression, histochemical changes) sensitivity. Using these endpoints, the rates of false positive and negatives would be established. The next stage of test development would use the same battery but now test with a larger group ($N = 100 - 1,000$) of AChE and NTE inhibitors, nonesterase inhibitors, nontoxic structural analogs, and nonneurotoxicants. This subsequent set of test chemicals would more fully evaluate the battery's test potential. At this stage of testing, the battery would be tested in secondary laboratories and the methodological quirks identified and resolved. Again, the rates of false positives and negatives would be calculated. Ultimately, a test battery for pesticide screening would be established with broad laboratory performance, defined interpretability, and reliability rates and would have undergone extensive chemical testing in a variety of laboratories. The third chemical test set ($N \geq$ 1,000) would more fully define the performance characteristics of the battery. When brought to this stage of validation, the test battery and its methodology should be firmly standardized and broadly accepted. The regulatory status of the test battery at this stage would, hypothetically, be considered equivalent to the corresponding whole animal test.

FUTURE DIRECTION

When used appropriately, and in conjunction with animal studies, cell culture systems can improve the sensitivity of neurotoxic prediction and provide a more detailed understanding of the cellular mechanisms responsible for neurotoxic reactions. If mammalian cell culture systems are to reduce or significantly supplant animals for neurotoxicity testing, a broader spectrum of cell types derived from varied species and a more comprehensive range of specific, sensitive, and automated endpoints must be developed. Further efforts must also be directed toward obtaining and characterizing additional nerve cell lines and primary cells derived from human tissues (73) to further our understanding of the extrapolation process. Although many of the technical parameters for maintaining animal cell lines have been detailed, they are less established for human cell lines. Development of such techniques represents an exciting challenge for *in vitro* neurobiologists. In identifying human cells and tissues, technological, ethical, and legal obstacles must be appreciated and surmounted (1,43,138). Although the supply of normal human cells for toxicological testing activities is limited, interactions with organizations such as the National Disease Research Interchange, which manages the largest collection of human tissues preserved for transplantation and research, and NIGMS, a human mutant cell repository (Coriell Institute Medical Research, Camden, NJ), should be encouraged. In general, the use of cell and tissue culture for toxicity screening requires stringent standardization of methodologies to maximize specificity, sensitivity, and reproducibility. Major impediments to standardization include the use of poorly characterized cells and the use of culture media of uncertain composition. Totally defined media that are able to support the differentiated functions and sustained growth of the nerve cell must be developed (100). Standardization of culture conditions must be extended to the exposure parameters so that factors such as the solvent vehicle, the length of exposure, the relationship of dose to cell density, etc., are clearly defined.

Probably the most apt and efficacious use of cell culture in the toxicity testing scheme will be in replacement of the LD_{50} test (107,124). Animal-welfare groups and many toxicologists want to see the LD_{50} replaced by a rough estimate of acute toxicity, an effort so far resisted by regulatory authorities. In recent years, several key laboratories throughout the world have focused on generating data and developing *in vitro* test procedures that parallel LD_{50} (9,11,25,61,93). A recent study conducted by the Multicenter Evaluation of *In Vitro* Cytotoxicologists (i.e., MEIC) in Scandinavia has reported very encouraging results in relating cytotoxic endpoints in human and rodent cell lines as predictive of acute lethal concentrations in humans, and mouse and rat LD_{50} dosages for a range of chemicals (38). The striking feature of this study is that all of the different *in vitro* systems used, in combination with the various cytotoxic endpoints, predicted human toxicity. The similar response of the various systems supports the basal cytotoxicity concept. Other areas in which neurotoxicity test systems will impact heavily is in defining intrinsic neurotoxicity, differentiating cytotoxic from neurotoxic chemicals, defining structure-activity relationships, and providing continuous dose-response data for neurotoxicants. In this context, the Center for Alternatives to Animal Testing (i.e., CAAT), housed within the Johns Hopkins School of Hygiene and Public Health (Baltimore, MD), provides an information network and a research funding program for the development, validation, and dissemination of *in vitro* toxicity test models.

What role can *in vitro* neurotoxicity tests hope to play in the regulatory arena? Although cell and tissue culture screens have been routinely used in industry for product development, such testing strategies have yet to be incorporated into the federal guidelines by regulatory agencies. In its current state, *in vitro* tests are not intended to replace existing *in vivo* test procedures. Rather, their incorporation at the earliest stages of the Risk-Assessment process can complement animal tests and reduce the number of animals being used for routine toxicity testing by identifying chemicals having the lowest probability of neurotoxicity. As *in vitro* models become more firmly established and validated by industrial and academic communities, they may eventually be considered for regulatory use. This goal can only be attained with continued interaction among international governments, industries, and scientific communities.

SUMMARY

This chapter familiarizes the reader with the need to develop, validate, and utilize *in vitro* models to test chemicals for neurotoxic potential. The major advantages and disadvantages of using cell and tissue cultures, factors which have stimulated and hampered the promulgation of *in vitro* models for neurotoxicity testing, and recent improvements in tissue culture methodologies are discussed. The rationale for using tissue culture for evaluating neurotoxicants and the unique aspects of culturing nervous tissue are described. Major considerations for designing screening tests and factors used in selecting a screening battery are discussed. Topics such as parameters in the validation process, choice and quantification of endpoints (i.e., cytotoxic and neurotoxic), cost and technical requirements, choice of test chemicals, and other quality control aspects are discussed.

REFERENCES

1. Adams EM. The ethical responsibilities at issue. In: Schiller CM, ed. *Human tissue for in vitro research: an analysis of the scientist's responsibility in the acquisition and utilization of human tissue for in vitro research.* Tissue Culture Association, 1977;8–24.
2. Aeschbacher M, Reinhardt CA, Zbinden G. A rapid cell membrane permeability test using fluorescent dyes and flow cytometry. *Cell Biol Toxicol* 1986;2:247–255.
3. Allen CN, Fonnun F. Trimethyltin inhibits the activity of hippocampal neurons recorded *in vitro. Neurotoxicology* 1984;5:23–30.
4. Amano T, Richelson E, Nirenberg M. Neurotransmitter synthesis by neuroblastoma clones. *Proc Natl Acad Sci USA* 1972;69:258–263.
5. Anderson D. *In vitro* methods for teratology testing. In: Anderson D, Conning DM, eds. *Experimental toxicology—the basic issues.* Royal Society of Chemistry, 1988;335–347.
6. Assouyline JG, Bosch EP, Lim R. Purification of rat Schwann cells from cultures of peripheral nerve. In: Shahar A, de Villis J, Vernadakis A, Haber B, eds. *A dissection and tissue culture manual of the nervous system.* New York, A.R. Liss, 1989;247–250.
7. Atterwill CK. Brain reaggregate cultures in neurotoxicological investigations. In: Atterwill CK, Steele CE, eds. *In vitro methods in toxicology.* Cambridge, 1987;133–164.
8. Atterwill CK, Walum E. Neurotoxicology *in vitro:* model systems and practical applications. *Toxicol in vitro,* 1989;3:159–161.
9. Balls M, Bridges JW. The FRAME research programme on *in vitro* cytotoxicology. In: Goldberg AM, ed. *Alternative methods in toxicology,* vol. 2. pp. 61. New York: Mary Ann Liebert, 1984;61.
10. Balls M, Clothier R. Differentiated cell and organ culture in toxicity testing. *Acta Pharm Toxicol* 1983;52(Suppl. II):115–137.
11. Balls M, Horner SA. The FRAME interlaboratory programme on *in vitro* cytotoxicity. *Fd Chem Toxic* 1985;23:209–213.
12. Beatty PW, Lembach KJ, Holscher MA, Neal RA. Effects of 2,3,7,8-tetrachlorodibenzo-p-dioxin (TCDD) on mammalian cells in tissue cultures. *Toxicol Appl Pharmacol* 1975;31:309–312.
13. Borek C, Hall EJ. Effect of split doses of x-rays on neoplastic transformation of single cells. *Nature* 1974;252:499–501.
14. Borenfreund E, Babich H. *In vitro* cytotoxicity of heavy metals, acrylamide, and organotin salts to neural cells and fibroblasts. *Cell Biol Toxicol* 1987;3:63–73.
15. Borenfreund E, Puerner JA. Toxicity determined *in vitro* by morphological alterations and neutral red absorption. *Toxicol Lett* 1985;24:119–124.
16. Borenfreund E, Puerner JA. Short-term quantitative *in vitro* cytotoxicity assay involving an S-9 activating system. *Cancer Letters,* 34:243–248.
17. Borenfreund E, Shopsis C. Toxicity monitored with a correlated set of cell culture assays. *Xenobiotica* 1985;15:705–711.
18. Bottenstein, JE. Serum-free cultures of dissociated chick embryo sensory neurons. In: Shahar A, de Villis J, Vernadakis A, Haber B, eds. *A dissection and tissue culture manual of the nervous system.* New York: A.R. Liss, 1989;227–228.
19. Bottenstein JE, Michler-Stuke A. Serum free culture of dissociated neonatal rat cortical astrocytes. In: Shahar A, de Villis J, Vernadakis A, Haber B, eds. *A dissection and tissue culture manual of the nervous system.* New York: A.R. Liss, 1989;109–111.
20. Bottenstein JE, Sato G. *Cell culture and the neurosciences.* New York: Plenum Press, 1985.
21. Carrington CD, Carrington MN, Abou-Donia MB. Neurotoxic esterase in cultured cells: an *in vitro* alternative for the study of organophosphorus compound-induced delayed neurotoxicity. In: Goldberg AM, ed. *Alternative methods in toxicology, Vol. 3: In Vitro Toxicology.* New York: Mary Ann Liebert, 1985;457–465.
22. Casper D, Davies P. Regulation of choline acetyltransferase activity by cell density in a cultured human neuroblastoma cell line. *Dev Neurosci* 1988;10:245–255.

23. Chambers PL. An overview of *in vitro* neurotoxicity assays. In: Thompson GR, ed. *Alternatives to in vivo bioassays for risk assessment.* Hazlet, New Jersey: Forum for Scientific Excellence, 1984;224–240.

24. Claudio L. An analysis of the current status of *in vitro* testing in neurotoxicology. Draft report prepared for U.S. E.P.A. fellowship program/OPP, Washington, DC, 1990;1–25.

25. Clothier RH, Hulme L, Ahmed AB, Reeves HL, Smith M, Balls M. *In vitro* cytotoxicity of 150; chemicals to 3T3-L1 cells, assessed by the FRAME kenacid blue method. *ATLA* 1988;16: 84–95.

26. Cole R, de Villis J. Preparation of astrocyte and oligodendrocyte cultures from primary rat glial cultures. In: Shahar A, de Villis J, Vernadakis A, Haber B, eds. *A dissection and tissue culture manual of the nervous system.* New York: A.R. Liss, 1989;121–133.

27. Correll L, Ehrich M. A microassay method for neurotoxic esterase determinations. *Fundamental Appl Toxicol* 1991;16(1):110–116.

28. Costa LG, Kaylor G, Murphy SD. Muscarinic cholinergic binding sites on rat lymphocytes. *Immunopharmacology* 1988;16:139–149.

29. Dahms NM, Schnaar RL. Ganglioside composition is regulated during differentiation in the neuroblastoma X glioma hybrid cell line NB 108-115. *J Neurosci* 1983;3:806–817.

30. Davenport CJ, Williams DA, Morgan KT. Neurotoxicology using cell culture. *CIIT Activities Chemical Institute of Toxicology,* 1989;9, no. 1.

31. Dawson M, Dryden WF. Tissue culture in the study of the effect of drugs. *J Pharm Sci* 1967;56: 545–561.

32. De Boni U, Goldenberg S. Isolation of pure fractions of viable dorsal root ganglionic neurons from rabbit or mouse using percoll gradients. In: Shahar A, de Villis J, Vernadakis A, Haber B, eds. *A dissection and tissue culture manual of the nervous system.* New York: A.R. Liss, 1989;230–232.

33. Dewar AJ. Neurotoxicity. In: Balls M, Riddell RJ, Worden AN, eds. *Animals and alternatives in toxicity testing.* London: Academic Press, 1983;230–284.

34. Diamond L, Baird WM. Chemical carcinogenesis *in vitro.* In: Rothblat GH, Cristofalo VJ, eds. *Growth, nutrition and metabolism of cells in culture,* vol. 3. New York: Academic Press, 1977;421.

35. DiScenna P. Method and myth in maintaining brain slices. In: Schurr A, Teyler TJ, Tseng MT, eds. *Brain slices.* Basel: Karger, 1987;10–21.

36. Dudek BR, Richardson RJ. Evidence for the existence of neurotoxic esterase in neural and lymphatic tissue of the adult hen. *Biochem Pharmacol* 1982;31:1117–1121.

37. Ekwall B. Screening of toxic compounds in mammalian cell cultures. *Ann NY Acad Sci* 1983;407: 64–77.

38. Ekwall B, Bondesson I, Castell JV, Gomezlechon MJ, Hellberg S, et al. Cytotoxicity evaluation of the 1st 10 MEIC chemicals—acute lethal toxicity in man predicted by cytotoxicity in 5 cellular assays and by oral LD_{50} tests in rodents. *ATLA* 1989;17:83–100.

39. Elmore E, Milman HA, Wyatt GP. Applications of the chinese hamster V79 metabolic cooperation assay in toxicology. In: Milman HA, Elmore E, eds. *Biochemical mechanisms and regulation of intercellular communication. Adv. in Mod. Environ. Tox. XIV.* Princeton Scientific, 1987;265–292.

40. Ericsson AC, Walum E. Cytotoxicity of cyclophosphamide and acrylamide in glioma and neuroblastoma cell lines co-cultured with liver cells. *Toxicol Lett* 1984;20:251–256.

41. Erkell LJ, Walum E. Differentiation of cultured neuroblastoma cells by urea derivatives. *FEBS Lett* 1970;104:401–404.

42. Evan AP, Dail WG, Dammrose D, Palmer C. Scanning electron microscopy of cell surfaces following removal of extracellular material. *Anat Rec* 1976;185:433–466.

43. Falk HL. The toxic substances control act and *in vitro* toxicity testing. In: Schiller CM, ed. *Human tissue for in vitro research: an analysis of the scientist's responsibility in the acquisition and utilization of human tissue for in vitro research.* Tissue Culture Association, 1977;89–108.

44. Fedalei A, Nardone RL. An *in vitro* alternative for testing the effect of organophosphates on neurotoxic esterase activity. In: Goldberg AM, ed. *Alternative methods in toxicology, vol. I: product safety evaluation.* New York: Mary Ann Liebert, 1983;252–267.

45. Federal Register, vol. 46, no. 14. Testing standards and guidelines work group; toxicity testing guidelines, availability. No. 7075. January, 1981.

46. Federal Register, vol. 50, no. 188. Acute delayed neurotoxicity of organophosphorus substances. No. 798.8540. February, 1985.

47. Fedoroff S. Proposed usage of animal tissue culture terms. *In Vitro* 1966;2:155–159.
48. Fedoroff S. Primary cultures, cell lines and cell strains: terminology and characteristics. In: Fedoroff S, Hertz L, eds. *Cell, tissue and organ cultures in neurobiology.* New York: Academic Press, 1977;265.
49. Fischbach GD, Nelson PG. Electrical properties of chick skeletal muscle fibers developing in cell culture. *J Cell Physiol* 1981;78:289–300.
50. Fletcher AP. Drug safety tests and subsequent clinical experience. *J R Soc Med* 1978;71:693–696.
51. Fountain SB, Teyler TJ. Characterizing neurotoxicity using the *in vitro* hippocampal brain slice preparation: heavy metals. In: Shahar A, Goldberg AM, eds. *Model systems in neurotoxicology: alternative approaches to animal testing.* New York: A.R. Liss, 1987;199–231.
52. Fountain SB, Ting YLT, Hennes SK, Teyler TJ. Triethyltin exposure suppresses synaptic transmission in area CA1 of the rat hippocampal slice. *Neurotoxicol Teratol* 1988;10:539–548.
53. Frazier J. Scientific criteria for validation of *in vitro* toxicity tests. *Organization for Economic Cooperation and Development (OECD),* Environmental monographs, no. 36, 1990;3–62.
54. Frazier JM, Bradlaw JA. Technical problems associated with *in vitro* toxicity testing systems. CAAT Technical Report No. 1. The Johns Hopkins Center for Alternatives to Animal Testing. Baltimore, 1989.
55. Freshney RI. Introduction: principles of sterile technique and cell propagation. In: Freshney RI, ed. *Animal cell culture: a practical approach.* New York: A.R. Liss, 1986;1–11.
56. Fry JR, Bridges JW. The metabolism of xenobiotic in cell suspension and cell culture. In: Bridges JW, Chasseaud LF, eds. *Progress in drug metabolism,* vol. 2. London: J Wiley, 1977;71.
57. Gahwiler BH. Organotypic slice cultures: a model for interdisciplinary studies. *Prog Clin Biol Res* 1987;253:13–18.
58. Gahwiler B, Rietschin L. Immunoperoxidase cytochemistry of cultured nerve cells. In: Shahar A, de Villis J, Vernadakis A, Haber B, eds. *A dissection and tissue culture manual of the nervous system.* New York: A.R. Liss, 1989;310–312.
59. Gahwiler B, Simmer J, Robertson R. Staining for acetylcholinesterase. In: Shahar A, de Villis J, Vernadakis A, Haber B, eds. *A dissection and tissue culture manual of the nervous system.* New York: A.R. Liss, 1989;306–308.
60. Goldberg AM. Mechanisms of neurotoxicity as studied in tissue culture systems. *Toxicology* 1980;17:201–208.
61. Goldberg AM, ed. *Alternative methods in toxicology, acute toxicity testing alternative approaches, vol. 2, Product safety evaluation, vol. 5.* New York: Mary Ann Liebert, 1984.
62. Goldberg AM. An approach to the development of *in vitro* toxicological methods. *Food Chem Toxicol* 1985;23:205–208.
63. Goldberg AM, ed. *Alternative methods in toxicology, in vitro toxicology approaches to validation, vol. 5.* New York: Mary Ann Liebert, 1987.
64. Goldberg AM, ed. *Alternative methods in toxicology, progress in in vitro toxicology, vol. 6.* New York: Mary Ann Liebert, 1989.
65. Goldberg AM, Frazier JM. Alternatives to animals in toxicity testing. *Sci Am* 1989;261:24–30.
66. Goldberg AM, Brookes N, Burt DR. The use of spinal cord cell cultures in the study of neurotoxicological agents. *Toxicology* 1980;17:233–235.
67. Goodman R. Hemispherectomy and its alternatives in the treatment of intractable epilepsy in patients with infantile hemiplegia. *Dev Med Child Neurol* 1986;28:251–258.
68. Graham DI, Kim SU, Gonatas NK, Guyotte L. The neurotoxic effects of triethyltin sulfate on myelinating cultures of mouse spinal cord. *J Neuropathol Exp Neurol* 1975;34:401–412.
69. Griffiths B. Scaling-up of animal cell cultures. In: Freshney RI, ed. *Animal cell culture: a practical approach.* New York: A.R. Liss, 1986;33–77.
70. Grisham JW, Smith GJ. Predictive and mechanistic evaluation of toxic responses in mammalian cell culture systems. *Pharmacol Rev* 1984;36:151s–171s.
71. Guroff G. PC12 cells as a model of neuronal differentiation. In: Bottenstein J, Sato G, eds. *Cell culture in the neurosciences.* Plenum Press, 1985;245–272.
72. Hamprecht B. Structural, electrophysiological, biochemical and pharmacological properties of neuroblastoma-glioma cell hybrids in culture. *Int Rev Cytol* 1977;49:99–170.
73. Harris CC, Trump BF, Stroner GD. Normal tissue and cell culture. *Methods Cell Biol* 1980;21A:1–500;21B:1–492.
74. Harvey AL. *The pharmacology of nerve and muscle in tissue culture.* London: Croom Helm, 1984.

75. Harvey AL. Possible developments in neurotoxicity testing *in vitro*. *Xenobiotica* 1988;18:625–632.
76. Hayflick L, Moorhead PS. The serial cultivation of human diploid cell strains. *Exp Cell Res* 1961;25:585–621.
77. Heidelberger C. Chemical oncogenesis in culture. *Adv Can Res* 1973;18:317–366.
78. Hendelman WJ. The effect of thallium on peripheral nervous tissue in culture: a light and electron microscopic study. *Anat Rec* 1969;163:198–199.
79. Hertz LH, Juurlink BHJ, Hertz E, Fosmark H, Schousboe A. Preparation of primary cultures of mouse (rat) astrocytes. In: Shahar A, de Villis J, Vernadakis A, Haber B, eds. *A dissection and tissue culture manual of the nervous system*. New York: A.R. Liss, 1989;105–108.
80. Holmes E, de Vellis J. *In situ* hybridization for detection of mRNAs in cell culture preparations. In: Shahar A, de Villis J, Vernadakis A, Haber B, eds. *A dissection and tissue culture manual of the nervous system*. New York: A.R. Liss, 1989;259–269.
81. Holmes E, Hermanson G, Cole R, de Vellis J. The developmental expression of glial specific mRNAs in primary cultures of rat brain visualized by *in situ* hybridization. *J Neurosci Res* 1988;19:389–396.
82. Holtzman D, Olson JE, De Vries C, Bensch K. Lead toxicity in primary cultured astrocytes and cerebellar granular neurons. *Toxicol Appl Pharmacol* 1987;89:211–225.
83. Honegger P. Biochemical differentiation in serum-free aggregating brain cultures. In: Bottenstein J, Sato G, eds. *Cell culture in the neurosciences*. London: Plenum, 1985;22–33.
84. Honegger P, Richelson E. Neurotransmitter synthesis, storage and release by aggregating cell cultures of rat brain. *Brain Res* 1979;162:89–101.
85. Honegger P, Lenoir D, Favrod P. Growth and differentiation of aggregating fetal cells in a serum-free medium. *Nature* 1979;282:305–308.
86. Hooisma J. Tissue culture and neurotoxicology. *Neurobehavioral toxicology and teratology*. 1982;4:617–622.
87. Hooisma J, DeGroot DM, Magchielse T, Muijser H. Sensitivity of several cell systems to acrylamide. *Toxicology* 1980;17:161–167.
88. Johnson MK. The delayed neuropathy caused by some organophosphorous esters: mechanism and challenge. *CRC Crit Rev Toxicol* 1975;3:289–316.
89. Juang MS. An electrophysiological study of the action of methylmercuric chloride and mercuric chloride on the sciatic nerve-sartorius muscle preparation of the frog. *Toxicol Appl Pharmacol* 1976;37:339–348.
90. Juurlink BHJ, Hertz L. Plasticity of astrocytes in primary cultures: an experimental tool and a reason for methodological caution. *Dev Neurosci* 1985;7:263–277.
91. Kaplan AS, ed. *Virus transformation and endogenous viruses*. New York: Academic Press, 1974.
92. Kasuya M. Effects of inorganic aryl, alkyl and other mercury compounds on the outgrowth of cells and fibers from dorsal root ganglia in tissue culture. *Toxicol Appl Pharm* 1972;23:136–146.
93. Knox P, Uphill PF, Fry JR, Benford J, Balls M. The FRAME multicentre project on *in vitro* cytotoxicity. *Food Chem Toxicol* 1986;24:457–463.
94. Kuroda Y. Brain slices, assay systems for the neurotoxicity of environmental pollutants and drugs on mammalian central nervous system. In: Holmstedt B, Lauwerys R, Mercier M, Roberfroid M, eds. *Mechanisms of toxicity and hazard evaluation*. Amsterdam: Elsevier, 1980;59–62.
95. Lotti M. Organophosphate-induced delayed polyneuropathy in humans: perspectives for biomonitoring. *Trends Pharmacol Sci* 1987;8:175–176.
96. Manthorpe M, Varon S. Purification of neonatal mouse dorsal root ganglion Schwann cells. In: Shahar A, de Villis J, Vernadakis A, Haber B, eds. *A dissection and tissue culture manual of the nervous system*. New York: A.R. Liss, 1989;243–246.
97. Masurovsky EB, Peterson ER. Photo-reconstituted collagen gel for tissue culture substances. *Exp Cell Res* 1973;76:447–448.
98. Matsuda S, Uehara Y. Scanning electron microscopy of neuronal development in ganglia after removal of connective tissue components. In: Shahar A, de Villis J, Vernadakis A, Haber B, eds. *A dissection and tissue culture manual of the nervous system*. New York: A.R. Liss, 1989;272–273.
99. Matthieu JM, Honegger P, Favrod P, Poduslo JK, Kristic R. Aggregating brain cell cultures: a model to study brain development. In: Monset-Chouchard M, Minkowski A, eds. *Physiological and biochemical basis for perinatal medicine*. Basel: Karger, 1981;359–366.

100. Maurer HR. Towards chemically defined, serum free media for mammalian cell culture. In: Freshney RI, ed. *Animal cell culture: a practical approach.* New York: A.R. Liss, 1986;13–30.
101. McCarthy KD, de Vellis J. Preparation of separate astroglial and oligodendroglial cell cultures from rat cerebral tissue. *J Cell Biol* 1980;85:890–902.
102. Menzer RE. Selection of animal models for data interpretation. In: Tardiff R, Rodricks J, eds. *Toxic substances and human risk.* New York: Plenum Press, 1987;133–152.
103. Moscona AA. Rotation-mediated histogenetic aggregation of dissociated cells: a quantifiable approach to cell interactions *in vitro. Exp Cell Res* 1961;22:455–475.
104. Murphy TH, Malouf AT, Sastre A. Calcium-dependent glutamate cytotoxicity in a neuronal cell line. *Brain Res* 1988;444:325–332.
105. Nagata Y, Diamond B, Bloom BR. The generation of human monocyte/macrophage cell lines. *Nature* 1983;306:597–599.
106. Nardone RM. Toxicity testing *in vitro.* In: Rothblatt GH, Cristofalo VJ, eds. *Growth, nutrition, and metabolism of cells in culture, vol. III.* New York: Academic Press, 1977;471–495.
107. Nardone RM. The LD$_{50}$ Test and *in vitro* toxicology strategies. *Acta Pharm Tox* 1981;52(Suppl. II):65–79.
108. National Academy of Sciences. *Toxicity testing: strategies to determine needs and priorities.* Washington, DC: National Academy Press, 1984.
109. Nelson PG. Neuronal cell lines. In: Fedoroff S, Hertz L, eds. *Cell, tissue, and organ cultures in neurobiology.* New York: Academic Press, 1977;347–365.
110. Nelson PG. Neuronal cell cultures as toxicological test systems. *Environ Health Perspect* 1978;26:125–133.
111. Nemes Z, Dietz R, Luth JB, Gomba S, Hackenthal F, Gross F. The pharmacological relevance of vital staining with neutral red. *Experientia* 1979;35:1475–1476.
112. O'Donohue J. Neurotoxicity risk assessment in industry. *American Association for the Advancement of Science Symposium.* May 26, 1986.
113. Oortgiesen M, van Kleef R, Vijverberg HPM. Effects of pyrethroids on neurotransmitter operated ion channels in cultured mouse neuroblastoma cells. *Pesticide biochemistry and physiology.* 1989;34:164–173.
114. Paul J. General morphological techniques. In: *Cell and tissue culture.* New York: Churchill Livingston, 1975;331–353.
115. Peterson ER, Crain SM, Murray MR. Differentiation and prolonged maintenance of biochemically active spinal cord cultures (rat, chick, and human). *Z Zellforsch* 1965;66:130–154.
116. Phillips HJ. Dye exclusion tests for cell viability. In: Kruse PF, Patterson MK, eds. *Tissue culture: methods and applications.* New York: Academic Press, 1973;406–408.
117. Politi LE, Lehar M, Adler R. Development of neonatal mouse retinal neurons and photoreceptors in low density cell culture. *Inv Ophthalmol Visual Sci* 1988;29:534–543.
118. Purchase IFH, Conning DM. International conference on practical *in vitro* toxicology. *Food Chem Toxicol* 1986;24:447–818.
119. Puro DG, Nirenberg M. On the specificity of synapse formation. *Proc Natl Acad Sci USA* 1976;73:3544–3548.
120. Rees KR. Cells in culture in toxicity testing: a review. *J R Soc Med* 1980;73:261–264.
121. Reiter LW. Testimony before the subcommittee on toxic substances, Environmental Oversight, Research and Development: Committee on Environment and Public Works; United States Senate, 1990.
122. Rofe PC. Tissue culture and toxicology. *Food Cosmet Toxicol vol. 9.* Great Britain: Pergamon Press, 1971;683–696.
123. Ronnett GV, Hester LD, Nye JS, Connors K, Snyder SH. Human cortical neuronal cell line: establishment from a patient with unilateral megalencephaly. *Science* 1990;248:603–605.
124. Rowan AN. The LD50—The beginning of the end. *Int J Study Anim Prob* 1983;4:4–7.
125. Rowan AN, Goldberg AM. Perspectives on alternatives to current animal testing techniques in preclinical toxicology. *Ann Rev Pharmacol Toxicol* 1985;25:225–247.
126. Sanford KK. Biologic manifestations of oncogenesis *in vitro:* a critique. *J Natl Cancer Inst* 1974;53:1647–1659.
127. Sarafian T, Hagler J, Vartavarian L, Verity MA. Rapid cell death induced by methyl mercury in suspension of cerebellar granule neurons. *J Neuropathol Exp Neurol* 1989;48:1–10.
128. Scala RA. Theoretical approaches to validation. In: Goldberg AM, ed. *In vitro toxicology—approaches to validation. Alternative methods in toxicology, vol. 5.* New York: Mary Ann Liebert, 1987;1–10.

129. Schrier BK. Nervous system cultures as toxicologic test systems. In: Mitchell CL, ed. *Target organ toxicology series: Nervous system toxicology.* New York: Raven Press, 1982;337–348.
130. Schubert D, Heinemann S, Kidokoro Y. Cholinergic metabolism and synapse formation by a rat nerve cell line. *Proc Natl Acad Sci USA* 1977;74:2579–2583.
131. Seeds NW. Neural cell reaggregation culture. In: Shahar A, de Villis J, Vernadakis A, Haber B, eds. *A dissection and tissue culture manual of the nervous system.* New York: A.R. Liss, 1989;251–256.
132. Segal LM, Federoff S. The acute and subchronic effects of organophosphorous and carbamate pesticides on cholinesterase activity in aggregate cultures of neural cells from the fetal rat brain. *Toxicol in vitro.* 1989;3:111–122.
133. Seil FJ, Lampert PW. Neurofibrillary spheroids induced by aluminum phosphate in dorsal root ganglia neurons *in vitro. J Neuropathol Exp Neurol* 1969;28:74–85.
134. Selkoe DJ, Lukenbill-Edds L, Shelanski ML. Effects of neurotoxic industrial solvents on cultured neuroblastoma cells: methyl n-butyl ketone, n-hexane and derivatives. *J Neuropathol Exp Neurol* 1978;37:768–789.
135. Sette WF. Briefing paper on proposed OPP response to CSPI petition to adopt new guidelines and require more extensive neurotoxicity testing under FIFRA. Toxicology Branch, Hazard Evaluation Division, Office of Pesticide (OPP) Program, Washington, DC, 1987.
136. Shadduck JP, Render J, Everitt J, Meccoli RA, Essex-Sorlie D. An approach to validation: comparison of six materials in three tests. In: Goldberg AM, ed. *In vitro toxicology—approaches to validation. Alternative methods in toxicology.* vol. 5. New York: Mary Ann Liebert, 1987;75–78.
137. Shahar A, de Villis J, Vernadakis A, Haber B. *A dissection and tissue culture manual of the nervous system.* New York: A.R. Liss, 1989.
138. Shapo M. The legal responsibilities at issue—emphasis on informed consent. In: Schiller CM, ed. *Human tissue for in vitro research: an analysis of the scientist's responsibility in the acquisition and utilization of human tissue for in vitro research.* Tissue Culture Association, 1977;24–44.
139. Sharma RP, Obersteiner EJ. Acrylamide cytotoxicity in chick ganglion cultures. *Toxicol Appl Pharmacol* 1977;42:149–156.
140. Shopsis C, Sathe S. Uridine uptake as a cytotoxicity test: correlations with the draize test. *Toxicology* 1984;29:195–206.
141. Smith RA, Orr DJ, McInnes IB. A refined primary culture system of adult mouse sensory neurons for neurotoxicity testing. *Prog Clin Biol Res* 1987;253:69–75.
142. Spencer PS, Schaumburg HH. Classification of neurotoxic disease: a morphological approach. In: Spencer PS, Schaumburg HH, eds. *Experimental and clinical neurotoxicology.* Baltimore: Williams and Wilkins Press, 1980;92–101.
143. Spencer PS, Crain SM, Bornstein MR, Peterson ER, Van de Water T. Chemical neurotoxicity and analysis in organotypic cultures of sensory and motor neurons. *Food Chem Toxicol* 1986;24:539–545.
144. Spencer PS, Peterson ER, Madrid R, Raine CS. Effects of thallium on neuronal mitochondria in organotypic cord-ganglion-muscle combination cultures. *J Cell Biol* 1973;58:79–95.
145. Stammatti AP, Silano V, Zucco F. Toxicology investigations with cell cultures systems. *Toxicology* 1981;20:91–153.
146. Stark DM, Shopsis C, Borenfreund E, Babich H. Progress and problems in evaluating and validating alternative assays in toxicology. *Food Chem Toxicol* 1986;24:449–455.
147. Storey ND, Smith RA, Orr DJ. Effects of ethanol and acetaldehyde on primary cultured adult mouse DRG neurons. *J Anat* 1987;152:43–55.
148. Straussman Y, Shahar A. *In situ* embedding of cultures in 32-mm plastic dishes and processing for transmission electron microscopy. In: Shahar A, de Villis J, Vernadakis A, Haber B, eds. *A dissection and tissue culture manual of the nervous system.* New York: A.R. Liss, 1989;274–276.
149. Tardiff RG. *In vitro* methods of toxicity evaluation. *Ann Rev Pharmacol Toxicol* 1978;18:357–369.
150. Task Force of Past Presidents. Animal data hazard evaluation: paths and pitfalls. *Fundam Appl Toxicol* 1982;2:101–107.
151. Teyler TJ. The brain slice preparation: hippocampus. *Brain Res Bull* 1980;5:391–403.
152. Tiffany-Castiglioni E, Perez-Polo JR. The role of nerve growth factor *in vitro* in cell resistance to 6-hydroxydopamine toxicity. *Exp Cell Res* 1978;121:179–189.
153. Tiffany-Castiglioni E, Zmudzki J, Bratton GR. Cellular targets of lead neurotoxicity: *in vitro* models. *Toxicology* 1986;42:303–315.

154. Tolbert WR, Feder J. Large-scale cell culture technology. *Ann Rep Ferment Proc* 1983;6:35–74.
155. Trapp BD, Richelson E. Usefulness for rotation-mediated aggregating cell cultures. In: Spencer PS, Schaumburg HH, eds. *Experimental and clinical neurotoxicology.* Baltimore: Williams and Wilkins Press, 1980;803–820.
156. Trapp BD, Honegger P, Richelson E, Webster HD. Morphological differentiation of mechanically dissociated fetal rat brain in aggregating cell cultures. *Brain Res* 1979;160:117–130.
157. Trapp BD, Webster H, Johnson D, Quarles RH, Cohen SR, Murray MR. Myelin formation in rotation-mediated aggregating cell cultures: immunocytochemical, electron microscopic, and biochemical observations. *J Neurosci* 1982;2:986–993.
158. Tuler SM, Bowen JM. Toxic effects of organophosphates on nerve cell growth and ultrastructure in culture. *J Toxicol Envir Health* 1989;27:209–223.
159. US Congress, Office of Technology Assessment, Neurotoxicity. Identifying and controlling poisons of the nervous system. Washington, DC, US Government Printing Office; 1990; OTA-Ba-436.
160. Varnbo I, Peterson A, Walum E. Effects of toxic chemicals on the respiratory activity of cultured mouse neuroblastoma cells. *Xenobiotica* 1985;15:727–733.
161. Varon S. The culture of chick embryo dorsal root ganglion cells on polylysine-coated plastic. *Neurochem Res* 1979;4:155–173.
162. Veronesi B. An ultrastructural study of methyl ethyl ketone's effect on cultured nerve tissue. *Neurotoxicology* 1984;5:31–44.
163. Veronesi B. *In vitro* modeling in neurotoxicology. In: Abou-Donia MB, ed. *Neurotoxicology—a textbook.* In press.
164. Veronesi B, Lington A, Spencer PS. A tissue-culture model of methyl ethyl ketone's potentiation of n-hexane neurotoxicity. *Neurotoxicology* 1984;5:43–52.
165. Veronesi B, Peterson ER, Spencer PS. Reproduction and analysis of methyl n-butyl ketone neuropathy in organotypic tissue culture. In: Spencer PS, Schaumburg HH, eds. *Experimental and clinical neurotoxicology.* Baltimore: Williams and Wilkins Press, 1980;863–873.
166. Veronesi B, Peterson ER, Spencer PS. Ultrastructural studies of the dying-back process VI. Examination of nerve fibers undergoing giant axonal degeneration in organotypic culture. *J Neuropathol Exp Neurol* 1983;42:153–165.
167. Walum E, Peterson A. Acute toxicity testing in cultures of mouse neuroblastoma cells. *Acta Pharmacol Toxicol* 1983;52(Suppl. II):100–114.
168. Walum E, Ekblad-Sekund G, Nybverg E, Gustafsson L. Cultured neuroblastoma cells in neurotoxicological models: acrylamide induced neurite disintegration. In: Shahar A, Goldberg AM, eds. *Model systems in neurotoxicology: alternative approaches to animal testing.* New York: A.R. Liss, 1987;121–136.
169. Watanabe PG, Sharma RP. Neurotoxicity of organophosphates: effects of tri-o-tolyl phosphate in chick ganglia cell cultures. *J Comp Pathol* 1975;85:373–381.
170. Weiss R. Test tube toxicology—new tests may reduce the need for animals in product safety testing. *Science News* 1988;133:42–45.
171. Yonezawa T, Murray B, Borstein B, Peterson ER. Organotypic cultures of nerve tissue as a model system for neurotoxicity investigation and screening. In: Spencer PS, Schaumburg HH, eds. *Experimental and clinical neurotoxicology.* Baltimore: Williams and Wilkins Press, 1980;788–802.
172. Zbinden G. *Progress in Toxicology: Special Topics, Vol. 1.* Springer-Verlag, Berlin.
173. Zbinden G. Predictive value of animal studies in toxicology. CMR Annual Lecture, July 1987. Center for Medicines Research, Carshalton, Great Britain, 1987.
174. Zbinden G. Reduction and replacement of laboratory animals in toxicological testing and research: Interim Report 1984–1987. *Biomed Envir Sci* 1988;1:90–100.

Neurotoxicology, edited by Hugh Tilson
and Clifford Mitchell. Raven Press, Ltd.,
New York © 1992.

3

Neurotoxicant-Induced Axonal Degeneration

John W. Griffin

*Department of Neurology, Johns Hopkins University School of Medicine,
Baltimore, Maryland 21205*

This chapter will summarize the types of axonal alterations induced by neurotoxicants, and the known and suspected mechanisms of that injury. The current understanding of axonal organization will be reviewed first, underlining unresolved issues in the field of axonal biology. The answers to some of these outstanding questions will provide necessary bases for further understanding of disease pathogenesis. The converse is equally intriguing: development of models of neurotoxic injury and pathogenetic mechanisms will provide new tools for neurobiologic study.

AXONAL ORGANIZATION

In most neurons the axon is an easily identifiable process that conveys electrical impulses from the nerve cell body toward the target. Among the exceptions to this organization, the primary sensory neuron is the most prominent. In this cell a single axon arises from the cell body as a stem process which then bifurcates into a peripheral and a central branch. Action potentials are initiated by sensory transducers in the periphery and conducted back along the peripheral process toward the cell body, then up the central process to the central nervous system.

In general, large neurons give rise to large axons and small neurons to small axons. However, the relationships between neuronal perikarya and their axons can vary dramatically. In most small neurons of the CNS, the axons represent only a fraction of the cytoplasmic volume of the nerve cell body, but in neurons with long, large axons, such as the lower motor neurons or primary sensory neurons contributing to the sciatic nerve, the axonal volume may dwarf that of the perikaryon by 100 to 1,000-fold.

In view of the spectrum of axonal designs, axons are best defined cytologically. The first of their distinctive characteristics is one that undoubtedly underlies a part of their particular susceptibility to disease: axons lack ribosomes and thus the machinery for synthesis of any substantial amounts of protein. The demarcation of the "ribosome-free zone" that constitutes the axon begins in the axon hillock, the specialized region of the cell body from which the axon arises (1).

Many of the other distinctive features of the axon represent specializations of the cytoskeleton. The axon cytoskeleton appears deceptively simple in routine electron micrographs. It consists of longitudinally oriented microtubules and neurofilaments, the intermediate filaments of neurons. These elements appear to be interlaced by innumerable sidearms or crosslinkers (2,3). However, from a biochemical standpoint, the axonal cytoskeleton is much more complex. One of the microtubule-associated proteins, MAP-1, is prominent in axons, while MAP-2 is not found in the axon, but is prominent in perikarya and dendrites. Similarly, a number of posttranslational modifications of cytoskeletal elements helps to distinguish axons from the remainder of the neuron. For example, phosphorylated neurofilaments are normally prominent in axons but not in perikarya or dendrites (4–6). The neurofilaments are composed of three related intermediate filament proteins, each products of separate genes. The 68 kd core protein (NF-L) and its larger homologues of 145 kD (NF-M) and 200 kD (NF-H) share a 310 amino acid rod domain (7). The difference in molecular weights reflects in large part differences in the carboxy terminal tail region. This tail region contains different numbers of "KSP" repeats—repeated sequences of the three amino acids, lysine, serine, and proline. These potential serine phosphorylation sites number up to 50 for NF-H. The extent of NF-H phosphorylation can be measured biochemically (8,9), or estimated by use of monoclonal antibodies that react with specific epitopes when they are phosphorylated or nonphosphorylated (6). Many of the widely used monoclonal reagents "see" relatively little NF-H phosphorylation in the nerve cell body and dendrites, and more prominent phosphorylation in the axon. The effects of phosphorylation are incompletely understood, but may include regulation of homologous and heterologous protein interactions such as the spacing of neurofilaments and their velocity of axonal transport (10–12).

The axonal microtubules also undergo a variety of posttranslational changes, some of which are associated with increased stability. These changes include acetylation and detyrosination. Taken together, these features—absence of ribosomes and mRNA, the absence of MAP-2, and the presence of "phosphorylated neurofilaments" and markers of microtubule stability—provide cytologic markers of the axon. Many of these same features also appear either to produce vulnerabilities to neurotoxic injury or to provide markers of injury.

The axon is normally maintained by the axonal transport systems. Fast bidirectional transport conveys membrane-associated materials up and down the axon. The mechanism is a form of microtubule-based motility. Anterograde transport from the cell body down the axon uses as the "motor" the protein kinesin. Retrograde transport utilizes a form of brain dynein, microtubule-associated protein 1 (MAP-1) (13). Slow axonal transport carries the cytoskeletal proteins of the axon at rates of .2 to 2 mm/day (14,15). Not only is the mechanism unknown, but the proportion of the cytoskeleton that is moving is controversial. Slow transport has been considered to be an exclusively anterograde process, but recent evidence suggests that at least in special settings, retrograde transport of neurofilaments may occur (16).

AXONAL TRANSECTION: THE "SIMPLEST" AXONAL INJURY

Physical interruption of the axon by transection, crush, or freezing of a site along the nerve results in degeneration of axon distal to the injury. This process of Wallerian degeneration undoubtedly unmasks the dependence of the axon on delivery of materials from the nerve cell body via axonal transport, but the precise sequence of cellular changes that leads to rapid breakdown of the axon is unknown. A brief review of both Wallerian degeneration of the axon in the distal stump and the responses of the nerve cell body and proximal axon are essential to understanding the similarities and differences in neurotoxic injuries.

Changes in the axon of the distal stump. Wallerian degeneration can be divided into an initial phase of axonal breakdown and a subsequent phase of myelin clearance. Axonal breakdown occurs at different rates in the central nervous system (CNS) and the peripheral nervous system (PNS). For example, in the CNS, recognizable axoplasm persists for several days in rodents, whereas in the PNS of young rats and mice, axoplasm is reduced to granular and amorphous debris throughout the distal stump within 24 hours. The rate is also affected by the age of the animal, being slower in aged individuals. Wallerian degeneration reflects active, locally mediated mechanisms of axonal destruction, rather than simple passive starvation of the axon by interruption of axonal transport. It is likely that the process is set in motion by increases in intraaxonal calcium concentrations. In transected rat sciatic nerves, calcium concentrations rise within mitochondria within hours after injury, and by 24 hours are markedly elevated in the axoplasm as well. The axon is known to contain ion-sensitive proteases closely associated with the neurofilaments (17). Schlaepfer and Bunge (18) demonstrated in tissue culture that the distal stump of transected neurites survived longer in media containing calcium chelators than in the presence of calcium. The outstanding question is the mechanism by which axotomy induces a rise in intraaxonal calcium concentration all along the distal stump. The calcium concentration of the endoneurial fluid is higher by orders of magnitude than the normal intraaxonal concentration, so increasing calcium entry across the axolemma is likely to be involved, but the mechanism of breakdown of axolemma barrier function to calcium is unknown.

The possibility that macrophage-derived factors participate in this process, perhaps by opening the axolemma, has been raised by recent studies showing the early appearance of macrophages in the distal stumps of degenerating nerves (19,20) and by the observation by Lunn and co-workers (21) that administration of antibodies against specific macrophage components can delay the onset of axonal breakdown. It should be emphasized that in both the PNS and the CNS, there are populations of resident macrophages, predominantly in the perivascular space. In addition, the CNS has its population of stellate microglial cells. The responses of these cells in Wallerian degeneration have received little attention, but their presence means that attraction of macrophages from the circulation need not necessarily be invoked in order to postulate roles of macrophages in the early phases of axonal degeneration.

Because of the emphasis placed on the spatial-temporal sequence of axonal changes in neurotoxic injury and the prominence of changes in the cytoskeleton, it is instructive to examine the evolution of changes in a mouse model of "slow" Wallerian degeneration. The C57BL/6/Ola strain, closely related to standard C57BL/6 mice, has the capacity for prolonged axonal survival in the distal stump (21–23). For example, in the sciatic nerve the motor amplitude evoked by stimulation of the distal stump falls only by 50% in the first 2 weeks after nerve transection, whereas it has disappeared entirely in standard C57BL/6 mice within 48 hours (21). The basis for this prolongation of axonal survival is uncertain, but it appears to depend on differences intrinsic to the C57BL/6/Ola nerves themselves, rather than on defective macrophages. Chimera studies showed that Wallerian degeneration was delayed even in bone marrow chimeras with normal C57BL/6 macrophages (22,23).

Glass and Griffin examined the spatial-temporal sequence of changes in the distal stumps of transected nerves in the C57BL/6/Ola mice (16). The axons underwent a series of changes in the axonal cytoskeleton, including the formation of neurofilamentous axonal swellings. This change, as described below, is often associated with specific neurotoxins. The prominence of neurofilamentous changes in this model serves as a reminder that these changes need not require neurotoxic injury, but can be seen when Wallerian degeneration occurs in "slow motion."

Responses of nerve cell body to axotomy. Following axotomy, nerve cell bodies in the PNS undergo a stereotypical series of changes in synthesis. Synthesis for some cytoskeletal proteins promptly increases, as reflected in increased abundance of mRNA, increased incorporation of amino acids, and increased proportions of these labeled proteins within the slow component of axonal transport. These proteins include tubulin and actin (14,24–26), proteins of potential use in longitudinal extension of growing axons during regeneration. In contrast, content of mRNAs for the neurofilament proteins decreases dramatically, as does the proportion of labeled neurofilament proteins within the slow component (14,24–27). The extent of phosphorylation of the neurofilaments within the proximal stump is unchanged (12).

The structural correlate of this decrease in synthesis of neurofilament proteins is a decrease in caliber of the axon. This decrease, presumably reflecting the role of neurofilaments in influencing axonal caliber, proceeds from proximal to distal down the axon at the approximate rate of slow axonal transport (14,25). This somatofugal pattern of axonal atrophy appears to reflect the entry into the axon from the cell body of a slow component relatively deficient in neurofilaments, and its subsequent slow transport down the axon.

PATTERNS OF NEUROTOXIC INJURY

A useful classification of axonal pathology in neurotoxic disorders is division into *distal axonal degeneration and disorders of axonal caliber,* including neurofilamentous axonal swelling. These categories are not mutually exclusive; some disorders, such as 2,5-hexanedione intoxication, produce both prominent changes in axonal caliber and also distal axonal degeneration. From a clinical standpoint, the deficits in function are produced by axonal breakdown, but the changes in caliber have been particularly susceptible to analyses of pathogenesis, and appear to correlate in some disorders with subsequent axonal breakdown. A useful means of reviewing these disorders is to compare and contrast the synthesis and axonal transport of neurofilament proteins in neurotoxic models, and then to return to the problem of what initiates frank axonal breakdown.

Neurotoxicant-Induced Disorders of Axonal Caliber: Altered Synthesis and/or Transport of Neurofilaments

In neurotoxic injuries, several agents are associated with early and prominent neurofilamentous swellings and/or with regions of neurofilament depletion and axonal atrophy. Four examples are outlined in Table 1. These disorders are each

TABLE 1. *Examples of neurofilamentous swellings and/or regions of neurofilament depletion and axonal atrophy*

Model	Perikaryon	Proximal axon	Distal axon	Perikaryal synthesis	Axonal transport
IDPN	Normal	Massive accumulations of NFs	Atrophy	Normal	Retarded NF transport
2,5-HD	Normal	Atrophy	Multifocal accumulations of NFs	Normal	Accelerated NF transport
Aluminum	Accumulation of NFs	Some NF swellings	Atrophy	Reduction in synthesis of many proteins	Retarded NF transport
Acrylamide	Axotomy-like changes	Atrophy	Some NF swellings; distal axonal degeneration	Increased tubulin synthesis; reduced NF synthesis	Reduced abundance of NF in the slow component
Axotomy	Chromatolysis	Atrophy	Wallerian degeneration	Increased tubulin synthesis; reduced NF synthesis	Reduced abundance of NF in the slow component

NF, neurofilament.

characterized by a distinctive distribution and time course of neurofilament accumulation. The distribution of neurofilament depletion is distal in some and proximal in others. The pathogenesis of these disorders appears to result from changes in the kinetics of neurofilament transport.

β,β'-iminodipropionitrile

β,β'-iminodipropionitrile (IDPN) administration produces extravagant axonal enlargements in the most proximal regions of large caliber fibers. When single high doses are given to young animals, the neurofilament density in the large axons begins to increase in the region distal to the axon hillock within 24 hours. Within several days the proximal axon undergoes enlargement, with densely packed neurofilaments distending whole internodes (28,29). With time the more distal regions of the axon undergo a decrease in caliber (30). Following a single large dose, the proximal swellings decrease within the first month, but neurofilamentous swellings are seen in more distal regions of the axon (31) (Fig. 1). This pattern suggests that the accumulated neurofilaments migrate from proximal to distal with time. If continuous administration is maintained, the swellings remain concentrated in the proximal axon and this migration does not occur (30).

IDPN impairs the anterograde transport of neurofilaments all along the axon (32,33), yet it has no effect on synthesis of neurofilament proteins, as measured by mRNA levels (34). Thus it appears that the axon receives its normal complement of neurofilaments, but is selectively unable to transport them normally beyond the most proximal regions. The mechanism underlying this defect is not clear. Indeed, it remains uncertain whether IDPN is itself toxic, or whether it requires conversion to a neurotoxic metabolite (35). IDPN administration is associated with increased neurofilament phosphorylation (10,36). *In vitro* IDPN increases the gelation of neurofilaments; this effect is eliminated by salt treatment of the neurofilament preparations, suggesting that neurofilament-associated proteins may be involved (36).

Whatever its molecular mechanism, the agent has the ability to dissociate microtubules from neurofilaments all along the axon (37–39), producing a distinctive segregation. This model provided some of the early evidence supporting the role of microtubules in fast axonal transport; rapidly transported organelles were segregated with the microtubules rather than the surrounding neurofilaments (38,39).

2,5-Hexanedione

2,5-Hexanedione, the neurotoxic metabolite of methyl *n*-butyl ketone and *n*-hexane, produces neurofilamentous axonal swellings that tend to begin in distal regions of long axons (40). It also produces a striking decrease in caliber of the

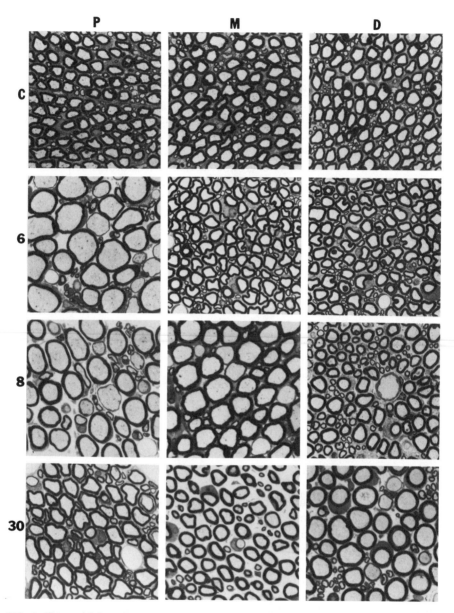

FIG. 1. This panel demonstrates the formation and subsequent proximal-to-distal migration of neurofilamentous anoxal swellings induced by a single injection of β,β^+-iminodipropionitrile (IDPN). The vertical columns labeled P, M, and D represent the most proximal, the middle, and the most distal regions of the L5 rat ventral root; these levels are separated by 8–10 mm. The top row (C) demonstrates that nerve fibers are of comparable sizes at all three levels in normal ventral root. The second row (6) demonstrates that 6 days after administration of IDPN there is marked axonal swelling in the most proximal portion of the ventral root, reflecting the accumulation of newly synthesized neurofilaments entering the axon from the cell body; these neurofilaments are blocked in their proximal-to-distal transport by the action of IDPN. The third row (8) illustrates that by 8 days after administration of a single dose of IDPN the swellings are migrating from proximal to distal, and have reached the mid-portion of the ventral root. By 30 days (30) the neurofilamentous swellings have migrated from the proximal and mid-regions of the root, but are still present within the distal ventral root. (Reprinted with permission from Griffin et al., Schwann cell proliferation and migration during paranodal demyelination. J. Neurosci. 1987;7:682–699.)

most proximal axons (12,41). These changes are accompanied by Wallerian-like degeneration of long axons (40). Like IDPN, this agent produces dissociation of neurofilaments from microtubules (37,42). In contrast to IDPN, 2,5-hexanedione has the capacity to accelerate neurofilament transport (12,41). Also in contrast to IDPN, the neurofilaments in the proximal axon are less heavily phosphorylated than comparable regions of normal nerves (12). The acceleration of neurofilament transport occurs in the absence of changes in content of mRNAs for the neurofilament proteins (12). It is likely that accelerated neurofilament transport underlies the proximal axonal atrophy (41).

The molecular mechanism of 2,5-hexanedione toxicity has received extensive study. The gamma diketones form pyrroyl adducts on lysine residues of proteins. Anthony and co-workers have argued that oxidation of the pyrroyl groups leads to crosslinking between adducts (43–45). Biochemical evidence supports the possibility that intermolecular covalent crosslinking may occur (45), giving rise to the hypothesis that such crosslinked neurofilaments might have altered transport characteristics. 3,4-Dimethyl-2,5-hexanedione, an agent designed to form more ready pyrroyl adducts, proved to be a more potent neurotoxin and to produce axonal swellings in the proximal rather than the distal axon (43,46). It slows rather than accelerates neurofilament transport (33). Such data suggest that similar molecular mechanisms may produce quite different physiologic effects and consequent structural changes.

Aluminum Neurotoxicity

Intrathecal administration of aluminum salts produces accumulations of neurofilaments in the perikarya and proximal axon of many neuronal populations, including motor neurons. Bizzi and co-workers (47) and subsequently Troncoso et al. (48) found that aluminum retarded neurofilament transport. The axon distal to the very proximal axon underwent atrophy, as might be expected if transported neurofilaments failed to be carried beyond the proximal axon (48).

Useful as these models of altered economy of the neurofilaments have been, major questions remain. An outstanding question is the conflict between these models and the possibility that only a small proportion of the cytoskeleton is moving, with much of it stationary (49). These models of mechanical and neurotoxic injury, with their satisfying correlations between changes in the synthesis or transport of neurofilaments and changes in structure, are most compatible with a model in which most of the neurofilaments are undergoing continuous coherent proximal-to-distal transport (50,51).

The resolution of this paradox is uncertain, but it must be kept in mind that in disease there may be changes in the transport of the cytoskeleton that do not fit well with any current concept of cytoskeletal organization. For example, in the distal stump of transected C57BL/6/Ola nerves, giant neurofilamentous swellings develop distal to the transection (16). The explanation for this change is unclear. It appears to suggest retrograde transport of neurofilaments, but whatever the mechanism, it is not explained by either the "structural hypothesis" or by concepts of a predominantly stationary cytoskeleton. Proposed mechanisms of cytoskeletal organization and transport will need to take into account these toxic and disease models.

Acrylamide Intoxication and Other Models of Distal Axonal Degeneration

Distal axon degeneration (DAD) is the most commonly encountered type of axonal pathology produced by neurotoxicants. It entails the subacutely or slowly evolving sequence of changes termed "dying back" by Cavanagh (52) and "central-peripheral distal axonopathy" by Spencer and Schaumburg (53). Dying back implies initial involvement of long and large axons, with degeneration beginning in distal regions and progressing more proximally with continued exposure. In all "dying back" disorders, the final common pathway of axonal breakdown and loss resembles Wallerian degeneration of the nerve fiber that follows nerve transection. Such Wallerian-like degeneration may involve initially only the distal-most portions of the axon, but advances more proximally in a stepwise fashion. End-stage Wallerian-like axonal breakdown and nerve fiber loss are nonspecific, and give few clues as to pathogenic mechanisms. In contrast, the predegenerative structural changes in the axons can suggest alterations in specific transport systems; examples include focal collections of rapidly transported organelles and alterations in the distribution of neurofilaments along the axon.

It is likely that a variety of direct effects of toxins on the neuron can result in distal axon degeneration. Some agents interfere with fast axonal transport within the distal regions of long axons. In some of these agents the abnormalities may reflect interference with oxidative metabolism or bioenergetics within the distal axon (54–56). Other agents, including 2,5-hexanedione (see above), appear to directly interact and bind to transported constituents such as neurofilaments; their neurotoxic effect may be cumulative during passage of such constituents down the axon, contributing to a distally predominant pattern of degeneration (43). A few agents may have nerve terminal-specific effects, resulting in degeneration restricted to the axon terminals (57).

At this time acrylamide intoxication provides a prototypic model of DAD. Acrylamide, a vinyl monomer used in the polymer industry, was recognized as a neurotoxin in the 1950s and 1960s. Affected patients have classic features of distal axon degeneration, usually beginning with paresthesias in the toes and feet, and evolving to include the fingers. Sensory loss, particularly for large fiber modalities, is present in affected regions, and the tendon reflex at the ankle is lost early. Weakness initially involves intrinsic muscles of the feet and hands and the dorsiflexors of the ankle. Following cessation of exposure, there may be a period of "coasting" before the progression of the disease halts. Recovery is by axonal regeneration and consequently may be protracted.

In their classic experimental study of acrylamide toxicity, Fullerton and Barnes

(58) reproduced faithfully the basic features of the human disorder. Physiologic and pathologic studies both documented the predominance of degeneration of large fibers in their most distal regions. Both motor and sensory fibers were affected, with sensory fibers being preferentially involved. The detailed structural studies of Schaumburg, Wisniewski, and Spencer (59) documented in a definitive fashion the early changes in axonal degeneration. They found evidence of multifocal accumulation of particulate organelles, regions of accumulation of neurofilaments, and other early abnormalities which culminated in Wallerian-like degeneration in more distal regions of the nerve fiber. They proposed that repetition of this sequence at progressively more proximal regions of the nerve was the basis for the "dying back" pattern characteristic of acrylamide intoxication.

During this sequence, prominent changes occur in the nerve cell body as well (60). In many systems these changes resemble typical chromatolysis, the structural consequence of axonal transection. In cerebellar Purkinje cells, a striking feature is the formation of outpouchings containing structures which resemble microtubule organizing centers (61).

An outstanding feature of acrylamide intoxication is the impairment of the ability of transected axons to regenerate. Following axonal transection, large growth cones distended by accumulations of particulate organelles develop (62,63). They resemble in many ways the "gigantic arrested balls" described in frustrated regeneration by Ramon y Cagal (64). Cavanagh and Gysbers (65) pointed out that distal-to-proximal axonal degeneration may commence from the point of nerve transection in acrylamide-intoxicated animals, while comparable regions of the nontransected side remain intact. Axonal transection in acrylamide-treated animals is associated with marked chromatolytic changes (66).

From the standpoint of axonal transport, the acrylamide model has been the most extensively studied toxic disorder. Studies of the rate of fast anterograde transport produced conflicting results, but on balance it appears that the rate of the fastest moving materials is little altered in the proximal and mid regions of long nerves, at least at times when they are structurally intact (67,68). On the other hand, in distal nerve regions there are marked abnormalities in the fast bidirectional transport systems (69). In particular, there are changes in fast anterograde transport in regions undergoing early predegenerative structural changes (62,68) and in the turnaround of proteins and their subsequent retrograde transport (69,70). Miller et al. (71) and subsequently Moretto and Sabri (72) demonstrated that retrograde transport of exogenous materials is markedly abnormal in acrylamide-treated animals. A particularly important aspect of these studies is the demonstration that administration of single doses of acrylamide reduces the retrograde transport by a substantial degree. These effects have been achieved with doses and durations which would produce no structural change in the animals. Taken together, these studies provide compelling evidence for an early defect in turnaround and retrograde transport from the distal axon in acrylamide-intoxicated animals.

Electron microscopic autoradiography has shown abnormalities in the delivery of rapidly transported materials to distal nerve regions, and their accumulation in sites of multifocal organelle collection along the course of the axon (62,73). These changes occur in animals with substantially greater degrees of neuropathy and structural change than the animals studied by Miller and colleagues (71), and it is likely that these abnormalities reflect later consequences of acrylamide exposure than the early change in retrograde transport.

The nature of the defect in retrograde or turnaround transport is not yet clear. It is of particular interest that the turnaround and retrograde transport of anterogradely transported vesicles at nerve terminals may require a proteolytic step. Protease inhibitors block turnaround (74,75). There is no evidence that acrylamide blocks proteolysis. Several lines of evidence have raised the possibility that acrylamide might interfere with glycolysis, but the evidence remains conflicting (35,76,77).

To bring this chapter full circle, it is useful to review the changes in the axonal cytoskeleton produced by acrylamide intoxication. A variable and relatively minor feature of the predegenerative changes produced by acrylamide is the formation of distal neurofilamentous swellings (59,62,73). In addition, Gold and co-workers identified prominent axonal atrophy that began proximally and spread in a centrifugal fashion (78). The velocity of slow transport is unchanged in acrylamide-treated animals (67,78,79). Thus the explanation for the distal neurofilamentous changes remains uncertain; possibly defective neurofilament transport occurs distally. One report (80) suggests that acrylamide increases the phosphorylation of the 145 and 200 kd neurofilament proteins, even at early stages of intoxication.

The composition of the slow component in acrylamide neuropathy is altered in a fashion very similar to that produced by axotomy (78); the abundance of neurofilament proteins is reduced and that of tubulin and actin is amplified. This pattern, as after axotomy, appears to reflect changes in the cell body synthesis of cytoskeletal elements (14,25,26). Hoffman, Watson, and Griffin (unpublished observation) examined mRNA levels for specific cytoskeletal proteins following acrylamide administration. Single doses of 100 mg/kg reduced mRNA levels for neurofilament proteins and tubulin, supporting the presumption of an acute dose-related depression of neuronal synthesis. This depression may be nonselective and was seen only with very high doses. In summary, there is no evidence that direct toxin-induced defects in protein synthesis are prominent features in the neurotoxicity of acrylamide and they seem unlikely to contribute to distal axonal degeneration in this model.

In contrast, Hoffman, Gold, and Griffin (*unpublished data*) have found that in experimental acrylamide neuropathy, a pattern of changes in perikaryal synthesis similar to those of axotomy is seen: mRNAs for neurofilament proteins are reduced, and those for tubulin, actin, and GAP-43 proteins are increased. These studies were done in animals exposed to lower doses over a week; the animals had early acrylamide neuropathy at the time of study. Thus in this model, an axotomy-like response of the perikaryon appears to contribute to the changes in axonal caliber.

REFERENCES

1. Zelena J. Ribosomes in myelinated axons of dorsal root ganglia. *Z Zellforsch* 1989;124:217–229.
2. Hirokawa N. Cross-linker system between neurofilaments, microtubules, and membranous organelles in frog axons revealed by the quick-freeze, deep-etching method. *J Cell Biol* 1982;94:129–142.
3. Hirokawa N, Glicksman MA, Willard MB. Organization of mammalian neurofilament polypeptides within the neuronal cytoskeleton. *J Cell Biol* 1984;98:1523–1536.
4. Julien JP, Mushynski WE. Multiple phosphorylated sites in mammalian neurofilament polypeptides. *J Biol Chem* 1982;257:10467–10470.
5. Julien JP, Mushynski WE. The distribution of phosphorylation sites among identified proteolytic fragments of mammalian neurofilaments. *J Biol Chem* 1983;258:4019–4025.
6. Sternberger LA, Sternberger NH. Monoclonal antibodies distinguish phosphorylated and non-phosphorylated forms of neurofilaments in situ. *Proc Natl Acad Sci USA* 1983;80:6126–6130.
7. Julien J-P, Grosveld F, Yazdanbaksh K, Flavell D, Meijre D, Mushynski W. The structure of a human neurofilament gene (NF-L): a unique exon-intron organization in the intermediate filament gene family. *Biochim Biophys Acta* 1987;909:10–20.
8. Nixon RA, Brown A, Marotta CA. Posttranslational modification of a neurofilament protein during axoplasmic transport: implications for regional specialization of CNS axons. *J Cell Biol* 1982;94:150–158.
9. Wong J, Hutchison SB, Liem RKH. An isoelectric variant of the 150,000-dalton neurofilament polypeptide. Evidence that phosphorylation state affects its association with the filament. *J Biol Chem* 1984;259:10867.
10. Watson DF, Griffin JW, Fittro KP, Hoffman PN. Phosphorylation-dependent immunoreactivity of neurofilaments increases during axonal maturation and IDPN intoxication. *J Neurochem* 1989;53:1818–1829.
11. Watson DF, Griffin JW, Fittro KP, Hoffman PN. Neurofilament phosphorylation increases during axonal maturation and IDPN intoxication. *J Neurochem* 1989;53:1818–1829.
12. Watson DF, Fittro KP, Hoffman PN, Griffin JW. Phosphorylation-related immunoreactivity and the rate of transport of neurofilaments in chronic 2,5-hexanedione intoxication. *Brain Res* (*in press*).
13. Vallee RB, Bloom GS, Theurkauf WE. Microtubule-associated proteins: subunits of the cytomatrix. *J Cell Biol* 1984;99:38s–44s.
14. Hoffman PN, Lasek RJ. The slow component of axonal transport: identification of major structural polypeptides of the axon and their generality among mammalian neurons. *J Cell Biol* 1975;66:351–366.
15. Willard M, Cowan WM, Vagelos PR. The polypeptide composition of intraaxonally transported proteins: evidence for four transport velocities. *Proc Natl Acad Sci USA* 1974;71:2183–2187.
16. Glass JD, Griffin JW. Neurofilament redistribution in transected nerves: evidence for bidirectional transport of neurofilaments. *J Neurosci* (*in press*).
17. Schlaepfer WW, Zimmerman UP. Calcium-activated protease and the regulation of the axonal cytoskeleton. In: Elam JS, Cancalon P, eds. *Axonal transport in neuronal growth and regeneration.* New York: Plenum Press, 1984;261–273.
18. Schlaepfer WW, Bunge RP. The effects of calcium ion concentration on the degradation of amputated axons in tissue culture. *J Cell Biol* 1973;59:456–470.
19. Stoll G, Griffin JW, Li CY, Trapp BD. Wallerian degeneration in the peripheral nervous system: participation of both Schwann cells and macrophages in myelin degradation. *J Neurocytol* 1989;18:671–683.
20. Scheidt P, Friede RL. Myelin phagocytosis in Wallerian degeneration. Properties of millipore diffusion chambers and immunohistochemical identification of cell populations. *Acta Neuropathol* 1987;75:77–84.
21. Lunn ER, Perry VH, Brown MC, Rosen H, Gordon S. Absence of Wallerian degeneration does not hinder regeneration in peripheral nerve. *Eur J Neurosci* 1989;1:27–33.
22. Perry VH, Lunn ER, Brown MC, Cahusac S, Gordon S. Evidence that the rate of Wallerian degeneration is controlled by a single autosomal dominant gene. *Eur J Neurosci* 1990;2:408–413.
23. Perry VH, Brown MC, Lunn ER, Tree P, Gordon S. Evidence that very slow Wallerian degen-

eration in C57BL/Ola mice is an intrinsic property of the peripheral nerve. *Eur J Neurosci* 1990;2:802-808.
24. Hoffman PN, Cleveland DW, Griffin JW, Landes PW, Cowan NJ, Price DL. Neurofilament gene expression: a major determinant of axonal caliber. *Proc Natl Acad Sci USA* 1987;84:3472-3476.
25. Hoffman PN, Griffin JW, Price DL. Control of axonal caliber by neurofilament transport. *J Cell Biol* 1984;99:705-714.
26. Hoffman PN, Cleveland DW. Neurofilament and tubulin expression recapitulates the developmental program during axonal regeneration: induction of a specific beta tubulin isotype. *Proc Natl Acad Sci USA* 1988;85:4530-4533.
27. Greenberg SG, Lasek RJ. Neurofilament protein synthesis in DRG neurons decreases more after peripheral axotomy than after central axotomy. *J Neurosci* 1988;8:1739-1746.
28. Chou S-M, Hartman HA. Axonal lesions and waltzing syndrome after IDPN administration in rats. With a concept—"axostasis.". *Acta Neuropathol* 1964;3:428-450.
29. Chou S-M, Hartman HA. Electron microscopy of focal neuroaxonal lesions produced by β,β'-iminodipropionitrile (IDPN) in rats. *Acta Neuropathol* 1965;39:590-603.
30. Clark AW, Griffin JW, Price DL. The axonal pathology in chronic IDPN intoxication. *J Neuropathol Exp Neurol* 1980;39:42-55.
31. Griffin JW, Drucker N, Gold BG, et al. Schwann cell proliferation and migration during paranodal demyelination. *J Neurosci* 1987;7:682-699.
32. Griffin JW, Hoffman PN, Clark AW, Carroll PT, Price DL. Slow axonal transport of neurofilament proteins: impairment by β,β'-iminodipropionitrile administration. *Science* 1978;202:633-635.
33. Griffin JW, Anthony DC, Fahnestock KE, Hoffman PN, Graham DG. 3,4-Dimethyl-2,5-hexanedione impairs the axonal transport of neurofilament proteins. *J Neurosci* 1984;4:1516-1526.
34. Parhad IM, Swedberg EA, Hoar DI, Krekoski CA, Clark AW. Neurofilament gene expression following β,β'-iminodipropionitrile (IDPN) intoxication. *Mol Brain Res* 1988;4:293-301.
35. Sayre LM, Autilio-Gambetti L, Gambetti P-L. Pathogenesis of experimental giant neurofilamentous axonopathies: a unified hypothesis based on chemical modification of neurofilaments. *Brain Res* 1985;340:69.
36. Eyer J, McLean WG, Leterrier J-F. Effect of a single dose of β,β'-iminodipropionitrile in vivo on the properties of neurofilaments in vitro: comparison with the effect of iminodipropionitrile added directly to neurofilaments in vitro. *J Neurochem* 1989;52:1759.
37. Griffin JW, Fahnestock KE, Price DL, Cork LC. Cytoskeletal disorganization induced by local application of β,β'-iminodipropionitrile and 2,5-hexanedione. *Ann Neurol* 1983;14:55-61.
38. Griffin JW, Fahnestock KE, Price DL, Hoffman PN. Microtubule-neurofilament segregation produced by β,β'-iminodipropionitrile: evidence for association of fast axonal transport with microtubules. *J Neurosci* 1983;3:557-566.
39. Papozomenous SC, Binder LI, Bender PK, Payne MR. Microtubule-associated protein 2 within axons of spinal motor neurons: associations with microtubules and neurofilaments in normal and IDPN axons. *J Cell Biol* 1985;100:74-85.
40. Spencer PS, Schaumburg HH. Ultrastructural studies of the dying-back process. IV. Differential vulnerability of PNS and CNS fibers in experimental central-peripheral distal axonopathies. *J Neuropathol Exp Neurol* 1977;36:300-320.
41. Monaco S, Autilio-Gambetti L, Zabel D, Gambetti P. Giant axonal neuropathy: acceleration of neurofilament transport in optic axons. *Proc Natl Acad Sci USA* 1985;82:920-924.
42. Zagoren JC, Politis MG, Spencer PS. Rapid reorganization of cytoskeleton induced by a gamma diketone. *Brain Res* 1983;270:162-164.
43. Anthony DC, Boekelheide K, Graham DG. The effect of 3,4-dimethyl substitution on the neurotoxicity of 2,5-hexanedione. I. Accelerated clinical neuropathy is accompanied by more proximal axonal swellings. *Toxicol Appl Pharmacol* 1983;71:362-371.
44. Graham DG, Anthony DC, Boekelheide K, et al. Studies of the molecular pathogenesis of hexane neuropathy. II. Evidence that pyrrole derivatization of lysyl residues leads to protein crosslinking. *Toxicol Appl Pharmacol* 1982;64:629-634.
45. Graham DG, Szakal-Quin G, Priest JW, Anthony DC. In vitro evidence that covalent crosslinking of neurofilaments occurs in q-diketone neuropathy. *Proc Natl Acad Sci USA* 1984;81:4979.
46. Anthony DC, Giangaspero F, Graham DG. The spatio-temporal pattern of the axonopathy associated with the neurotoxicity of 3,4-dimethl 2,5-hexanedione in the rat. *J Neuropathol Exp Neurol* 1983;42:548-560.

47. Bizzi A, Crane RC, Autilio-Gambetti L, Gambetti P. Aluminum effect on slow axonal transport: a novel impairment of neurofilament transport. *J Neurosci* 1984;4:722–731.
48. Troncoso JC, Hoffman PN, Griffin JW, Hess-Kozlow KM, Price DL. Aluminum intoxication: a disorder of neurofilament transport in motor neurons. *Brain Res* 1985;342:172–175.
49. Nixon RA, Logvinenko KB. Multiple fates of newly synthesized neurofilament proteins: evidence for a stationary neurofilament network distributed nonuniformly along axons of retinal ganglion cell neurons. *J Cell Biol* 1986;102:647–659.
50. Lasek RJ. Studying the intrinsic determinants of neuronal form and function. In: Lasek RJ, ed. *Intrinsic determinants of neuronal form and function.* New York: Alan R. Liss, 1988.
51. Lasek RJ, Hoffman PN. The neuronal cytoskeleton, axonal transport and axonal growth. In: Goldman R, Pollard T, Rosenbaum J, eds. *Cell motility, book C, microtubules and related proteins.* Cold Spring Harbor, NY: Cold Spring Harbor Laboratory, 1976;1021–1051.
52. Cavanagh JB, Jacobs JM. Some quantitative aspects of diphtheria neuropathy. *Br J Exp Pathol* 1964;45:309–322.
53. Spencer PS, Schaumburg HH. Central-peripheral distal axonopathy—the pathogenesis of dying-back polyneuropathies. In: Zimmerman H, ed. *Progress in neuropathology, vol. 3.* New York: Grune and Stratton, 1976;253–295.
54. Sabri MI, Spencer PS. Toxic distal axonopathy: biochemical studies and hypothetical mechanisms. In: Spencer PS, Schaumberg HH, eds. *Experimental and clinical neurotoxicology.* Baltimore: Williams & Wilkins, 1980;206.
55. Sabri MI. In vitro effect of n-hexane and its metabolites on selected enzymes in glycolysis, pentose phosphate pathway and citric acid cycle. *Brain Res* 1984;297:145.
56. Sabri MI. Further observations on in vitro and in vivo effects of 2,5-hexanedione on glyceraldehyde-3-phosphate dehydrogenase. *Arch Toxicol* 1984;55:191.
57. Watson DF, Griffin JW. Vacor neuropathy: ultrastructural and axonal transport studies. *J Neuropathol Exp Neurol* 1987;46:96–108.
58. Fullerton PM, Barnes JM. Peripheral neuropathy in rats produced by acrylamide. *Br J Ind Med* 1966;23:210–221.
59. Schaumburg HH, Wisniewski HM, Spencer PS. Ultrastructural studies of the dying-back process. I. Peripheral nerve terminal and axon degeneration in systemic acrylamide intoxication. *J Neuropathol Exp Neurol* 1974;33:260–284.
60. Sterman AB. Altered sensory ganglia in acrylamide neuropathy. Quantitative evidence of neuronal reorganization. *J Neuropathol Exp Neurol* 1983;42:166–176.
61. Cavanagh JB, Gysbers MF. Ultrastructural features of the Purkinjie cell damage caused by acrylamide in the rat: a new phenomenon in cellular neuropathology. *J Neurocytol* 1983;12:413–437.
62. Griffin JW, Price JW, Drachman DB. Impaired axonal regeneration in acrylamide intoxication. *J Neurobiol* 1977;8:355–370.
63. Morgan-Hughes JS, Sinclair S, Durston JHJ. The pattern of peripheral nerve regeneration induced by nerve crush in rats with severe acrylamide neuropathy. *Brain* 1974;97:235–250.
64. Ramon y Cajal S. *Degeneration and regeneration of the nervous system (RM May, trans.).* London: Oxford University Press, 1928.
65. Cavanagh JB, Gysbers MF. Ultrastructural changes in axons caused by acrylamide above a nerve ligature. *Neuropathol Appl Neurobiol* 1981;7:315–326.
66. Prineas JW. The pathogenesis of dying-back polyneuropathies. II. An ultrastructural study of experimental acrylamide intoxication in the cat. *J Neuropathol Exp Neurol* 1969;28:598–621.
67. Sidenius P, Jakobsen J. Anterograde axonal transport in rats during intoxication with acrylamide. *J Neurochem (in press).*
68. Souyri F, Chretien M, Droz B. "Acrylamide-induced" neuropathy and impairment of axonal transport of proteins. I. Multifocal retention of fast transported proteins at the periphery of axons as revealed by light microscope radioautography. *Brain Res* 1981;205:13.
69. Sahenk Z, Mendell JR. Acrylamide and 2,5-hexanedione neuropathies: abnormal bidirectional transport rate in distal axons. *Brain Res* 1981;219:397–405.
70. Jakobsen J, Sidenius P. Early and dose-dependent decrease in retrograde axonal transport in acrylamide-intoxicated rats. *J Neurochem* 1983;40:447–454.
71. Miller MS, Miller MJ, Burks TF, Sipes IG. Altered retrograde axonal transport of nerve growth factor after single and repeated doses of acrylamide in the rat. *Toxicol Appl Pharmacol* 1983;69:96–101.
72. Moretto A, Sabri MI. Progressive deficits in retrograde axon transport precede degeneration of motor axons in acrylamide neuropathy. *Brain Res* 1988;440:18.

73. Chretien M, Patey G, Souyri F, Droz B. "Acrylamide-induced" neuropathy and impairment of axonal transport of proteins. II. Abnormal accumulations of smooth endoplasmic reticulum at sites of focal retention of fast transported proteins. Electron microscope radioautographic study. *Brain Res* 1981;205:15–28.

74. Martz D, Garner J, Lasek RJ. Protein changes during anterograde-to-retrograde conversion of axonally transported vesicles. *Brain Res* 1989;476:199.

75. Sahenk Z, Lasek RJ. Inhibition of proteolysis blocks anterograde-to-retrograde conversion of axonally transported vesicles. *Brain Res* 1988;460:199.

76. Vyas I, Lowndes HE, Howland RD. Inhibition of glyceraldehyde-3-phosphate dehydrogenase in tissues of the rat by acrylamide and related compounds. *Neurotoxicology* 1985;6:123.

77. Brimijoin S, Hammond P. Acrylamide neuropathy in the rat: effects on energy metabolism in sciatic nerve. *Mayo Clin Proc* 1985;60:3.

78. Gold BG, Griffin JW, Price DL. Slow axonal transport in acrylamide neuropathy: different abnormalities produced by single dose and continuous administration. *J Neurosci* 1985;5:1755–1768.

79. Bradley WG, Williams MH. Axoplasmic flow in axonal neuropathies. I. Axoplasmic flow in cats with toxic neuropathies. *Brain* 1973;96:235.

80. Howland RD, Alli P. Altered phosphorylation of rat neuronal cytoskeletal proteins in acrylamide-induced neuropathy. *Brain Res* 1986;363:333.

Neurotoxicology, edited by Hugh Tilson
and Clifford Mitchell. Raven Press, Ltd.,
New York © 1992.

4

Factors Influencing Morphological Expression of Neurotoxicity

Kenneth R. Reuhl and Herbert E. Lowndes

Neurotoxicology Laboratories, College of Pharmacy, Rutgers University, Piscataway, New Jersey 08855-0789

Experimental neuropathology has made major advances since the French pathologist Gombault described, more than a century ago, segmental demyelination in peripheral nerves of guinea pigs experimentally treated with inorganic lead (18,19). In the ensuing 100 years, during which neurotoxicology has matured from a minor branch of experimental pathology and neurology into a recognized discipline in its own right, descriptive morphology has remained a cornerstone in the identification of toxic injury and in the proper assessment of toxic risk. The fundamental principles of morphological evaluation of tissue and cell structure have remained largely unaltered since the time of Ramon y Cajal; excellent general discussions of neural responses to toxicants and caveats to this approach are available elsewhere (38,52,57). While descriptive morphology remains as important to understanding consequences of neurotoxicity as it was in the days of Gombault, findings stemming from new technologies have necessitated a reappraisal and, in many cases, complete revision of our concepts of nervous system responses to neurotoxicants. In particular, it is becoming necessary to reevaluate the manner in which we have traditionally interpreted the interactions between component cells of the nervous system. The burgeoning neuroscience literature on structural and functional "cross-talk" between neuron and glia, and between cells and their extracellular matrices, has revealed these interactions to be crucial to normal nervous system development and function to an extent never imagined by early neuropathologists. These crucial relationships have also emerged as primary targets for neurotoxicants. It is therefore essential in evaluating neurotoxicology to consider not only the "target" cell, but also the contributions of penultimate events in adjacent cells and structures to the ultimate morphological expressions of neurotoxicity. This chapter will examine some of the emerging concepts regarding the interactive roles of neural cells during neurotoxic insult, and will explore how these interactions may influence the idiosyncratic and often unexpected nature of neural responses to neurotoxicants.

FACTORS GOVERNING CELLULAR VULNERABILITY

The nervous system shows remarkable regional variability in response to toxicants. A specific toxic insult frequently results in involvement of highly selective regions, or specific cell types, of the brain, with relative or complete sparing of the remainder. Necrosis of the globus pallidus in carbon monoxide poisoning (28), laminar necrosis of layers III, V, and VI of the cortical gray matter in anoxia (52), and damage to the hippocampus in trimethyltin intoxication (3,10) are but three of many recognized examples of this phenomenon in the central nervous system. Occasionally, damage is so specific as to provide a virtual signature of the inducing toxicant (Fig. 1).

Selective involvement of the peripheral nervous system without CNS impairment is an equally longstanding clinical observation. While some agents damage both sensory and motor nerves, these two components of the peripheral nervous system are by no means equally vulnerable to neurotoxic insult. Motor nerve function is damaged by dithiobiuret (49) without corresponding sensory impairment. Still other toxicants predominantly target only peripheral sensory function, without known effects on motor function. Emerging evidence suggests that certain neurotoxicants are sufficiently selective as to target sensory cells which subserve restricted sensory modalities (59). It is probable that future studies of other neural structures, such as the autonomic nervous system, will similarly reveal highly cell-specific responses to neurotoxic insult.

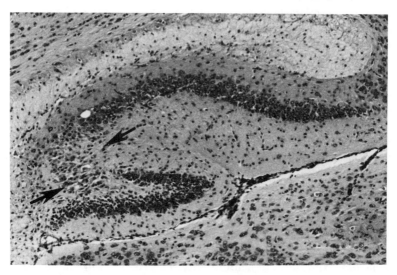

FIG. 1. Section of mouse hippocampus following treatment with trimethyltin. Loss of pyramidal neurons is marked in the CA-3 subfield of Ammon's horn (*arrows*), but other areas appear unremarkable. While no lesion is absolutely pathognomonic of a specific toxicant, this distribution of injury is characteristic of trimethyltin in mouse. Original magnification ×40.

Understanding the processes underlying selective vulnerability of different brain regions, and different cell types within a region, is perhaps the fundamental question in neurotoxicology. Nissl was among the first to recognize that certain toxicants attack distinct neuronal groups and leave others undamaged, although he later recanted this belief. In 1922, Vogt and Vogt (60) proposed that *intrinsic cellular differences in physiological activity or biochemistry underlie the distinctive cellular responses to toxicants,* a process which they termed *pathoclisis.* In a more general approach, Spielmeyer (58) postulated that regional differences in vascular supply could explain selective sensitivity of brain regions to insults like hypoxia or intoxications.

Both theories are valid in part and are by no means mutually exclusive. As noted by Shaw and Alvord (52), it is necessary to appreciate that both intrinsic metabolic capacities of different cells (true pathoclisis) and their topographical vascular anatomy govern the localization of damage within a given tissue. Additionally, it is becoming apparent that the cell's response also depends on whether toxic insult disrupts the essential intercellular communication necessary for trophic maintenance of biochemical and morphological differentiation.

CELL CONTACT AS A MODULATOR OF TOXIC EXPRESSION

Cell-cell and cell-matrix interactions have emerged as critical factors for the normal development and functioning of the nervous system. These interactions do far more than provide the static support necessary to maintain three-dimensional structure of the nervous system. In a very real way, they govern all morphological and functional processes in the nervous system, as well as the manner in which this system will respond when injured.

The extent and importance of cell contact in the biochemical "cross-talk" or communication between cells has only recently been fully appreciated. It is now recognized that virtually all aspects of neural processes are governed by the moment-to-moment interplay between cells in physical contact. These contacts provide the impetus and means for morphogenesis and cytoarchitectural sorting of different cell types, determine overall cellular morphology, dictate the distribution of organelles within the cell, and determine biochemical and physiological specializations which characterize differentiated neural cells. If these contacts are perturbed, cells respond by undergoing morphological, physiological, and biochemical changes which are either regenerative, reflecting the cell's attempt to reestablish contact, or degenerative, reflecting the deleterious effects of loss of contact on differentiated structure and function.

The importance of cell-cell contact is underscored by examining the consequences of its disturbance in the developing cerebellum. During cerebellar morphogenesis, granule cells migrate from their positions in the germinal external granule cell layer to their adult positions in the internal cell layer. This translocation is accomplished by guidance of granule cells along a thin cytoplasmic process provided by the Bergmann glia cells. The intimate relationship between the granule cell and the Bergmann glial process has led to the concept of glial contact guidance (53). This pattern is now recognized as a generalized developmental feature, in that most neuronal migrations follow physical tracts, usually provided by glia but sometimes by so-called pilot neurons, to their final destinations.

Cell migrations and eventual postmigratory differentiation have been shown to be mediated by complex and coordinated interactions between special molecules on cell membranes and on the extracellular matrix; these molecules have been referred to as cell adhesion molecules (CAMs) and substrate adhesion molecules (SAMs), respectively (17,40). Elucidation of the functions of these molecules has been one of the major achievements of recent neurobiology as it provides an elegant heuristic model of development (17,47). The relevance of cell adhesion to neurotoxicology becomes immediately apparent when one considers the general principles which govern their function. Binding between adhesion molecules occurs between either identical molecules (homophilic binding) or nonidentical binding molecules (heterophilic binding) expressed on another cell or a matrix. Interaction of CAMs and SAMs depends on close approximation of cells; beyond a critical distance, adhesion will not occur and adhesion-dependent processes will not be operative. The developing nervous system appears to employ this principle to prevent premature synergic contact between cells and resultant contact-mediated differentiation. The embryonic form of the homophilic binding neuron cell adhesion molecule (N-CAM E) contains large amounts of negatively charged sialic acid residues on its extracellular domain which repulse N-CAM E on adjacent cells, thus inhibiting close membrane associations (47). Once migration is complete, the adult form of N-CAM (N-CAM A) is expressed; this form, lacking the large number of sialic acid residues, favors stable homophilic binding and thus positional stability of the cell within the neuropil. This stability is then associated with permanent differentiation and functional maturity.

Many toxicants can change the physical relationships between cells, increasing intercellular distance and thus perturbing contact-mediated processes. For example, toxic heavy metals like methylmercury and lead have been reported to alter the expression and/or function of N-CAM in neurons (11). When cultured neurons are treated with methylmercury, there is loss of N-CAM staining and cessation of neuronal migration. Following removal of methylmercury from the medium, neuronal migration is rapidly restored coincident with the reappearance of immunofluorescently detectable N-CAM (43). Antibodies to N-CAM also inhibit cell migration. These data suggest that the well-documented disturbance of neuronal migration following developmental methylmercury exposure may be due in part to the toxicant's action on N-CAMs. The extent to which other toxicants influence development by this mechanism remains to be determined.

While the role of cell-cell contact is best documented in developing neural systems, it is also essential to the normal and pathological responses of adult

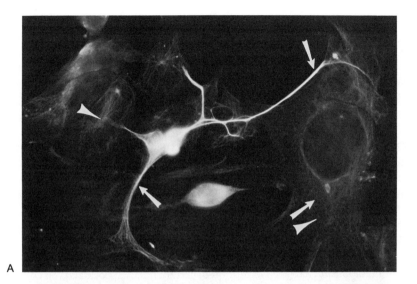

FIG. 2. Microtubules of neurons and pluripotent stem cells in tissue culture before (**A**) and after (**B**) exposure to methylmercury. **A:** In untreated controls, microtubules are well displayed in both neurons (*arrows*) and pluripotent stem cells (*arrowheads*). **B:** Following treatment with methylmercury, microtubules have disappeared from stem cells but remain relatively intact in neurons (*arrows*). Posttranslational modifications of microtubules in the differentiated, collaterally stabilized neurons account for microtubular resistance in these cells. Immunofluorescence staining. Original magnification ×1,000.

tissues. When neurons complete migration, they undergo fundamental reorganization of internal morphology and biochemistry which is facilitated, in ways yet unclear, by physical contact with adjacent cells. These changes involve alterations in metabolic and second messenger activity (51), regulation of neurotransmitter levels, compartmentation of cell organelles, and complex changes in the expression and posttranslational modification of cellular proteins (54). These modifications markedly alter the susceptibility of the cell to toxic agents. For example, the sensitivity of microtubules in cultured cells to disassembly by agents such as methylmercury is well documented (42,48,62). However, when cultured neurons are allowed to adhere to the culture dish and form neuritic contacts with adjacent neurons, a posttranslationally modified population of microtubules is formed which shows marked resistance to methylmercury-induced depolymerization (6) (Fig. 2).

The importance of contact (collateral) stabilization can also be shown by studies in which adult neural cells are disaggregated and maintained in suspension culture. In this condition mature morphology is lost, organelles change distribution, and cell metabolism changes. Disruption of certain highly specialized cell contacts will similarly result in major changes to the cell *in situ*. For example, the mature neuromuscular junction is a relatively stable specialized cell-cell contact point. When the close physical proximity of the terminal axon and the muscle endplate is lost, as occurs with chemical axotomy (14), pharmacological blockage of nerve conduction (12), or injection of anti-N-CAM antibodies (41), the axon undergoes degenerative/regenerative responses which essentially repeat the sequence of events operative during development. N-CAM A, present in low amounts to help anchor the mature neuromuscular junction, disappears and is replaced by abundant N-CAM E (44). At the same time, the distal axon develops terminal sprouts in an attempt to reestablish contact with the original muscle endplate. When contact is reestablished, the embryonic markers like N-CAM E disappear, supernumerary axon sprouts degenerate, and normal neuromuscular function is restored.

The reversion of neurons, or the targets of their innervation, to a more embryonic state appears to be a common neural response to many, if not most, forms of neural injury and repair in which physical/functional contact between cells is lost. A well-known example of targets deprived of innervation reverting to less developed states (i.e., de-differentiation) is provided by mammalian skeletal muscle. Muscle contains four major muscle fiber types, but following denervation, in which physical contact (and trophic influences) are lost, characteristics of all muscle fibers converge to those observed in undifferentiated muscle (5).

ROLE OF ENERGY DEPRIVATION IN
IDIOSYNCRATIC CELL RESPONSE

The relationship between oxygen consumption, energy state, and pathological outcome within the brain has been a persistent theme in experimental neuro-

but experimental manipulation of these factors would provide a profitable avenue of investigation.

ROLE OF NEURAL METABOLISM IN CYTOTOXIC RESPONSE

Surprisingly little is known regarding the metabolic fate of xenobiotics within the nervous system, or even about the chemical forms in which these compounds are presented to neural tissues. The ability of the blood-brain barrier to exclude compounds has focused attention primarily on the fate of small polar compounds, which can freely diffuse across the barrier, or compounds for which specific transport systems have been identified (34). The relatively low activities of metabolizing enzymes found by traditional biochemical methodologies led to the belief that the brain possessed little capacity for metabolic processing of xenobiotics. Recent immunocytochemical studies have revealed a distinctly different picture.

In accordance with the theory of pathoclisis, studies have demonstrated a heterogenous distribution of phase I and phase II enzyme systems in different brain regions and among various cell types. Hansson et al. (20) found cytochrome P-450 IIEI to be localized in astrocytes and oligodendrocytes, the striatal complex, as well as in selected populations of pyramidal neurons in frontal cortex and all pyramidal cell subfields of the hippocampus, while Philbert et al. (37) reported the isozyme in astrocytes and spinal anterior horn cells (Fig. 4). In the cerebellum,

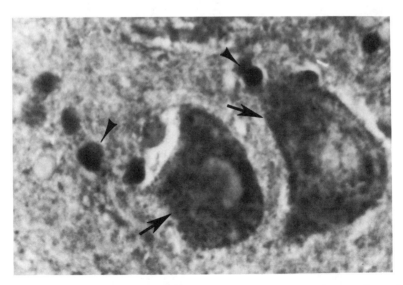

FIG. 4. Anterior horn of rat spinal cord stained for P450 IIEI. Both large neurons (*arrows*) and glia (*arrowhead*) demonstrate P450 IIEI immunoreactivity, while adjacent neuropil is essentially negative for P450 staining. Original magnification ×1,000.

P-450 IIEI was found in all layers to be restricted to glial cells. This P-450 isoform is important in the metabolism of ethanol and for the metabolic activation of chemicals, the list of which is continually expanding. The presence of high degrees of immunoreactivity of cytochrome P-450 IIEI in selected cell types in basal ganglia, frontal cortex, and hippocampus (brain regions associated with behavioral alterations by these compounds) cannot be coincidental. Although the toxicological implications of selective localization of phase I enzymes in specific cell types remain to be explored, differences in regional and/or cellular capacity to handle xenobiotics clearly form a basis for pathoclisis. Volk and co-workers (61) demonstrated marked induction of cytochrome P-450 activity in cerebellar tissues (presumably glial cells) of mice exposed to phenytoin, and suggested a role for this enzyme in producing toxic metabolites responsible for the Purkinje cell injury induced by this drug.

Similar heterogeneity has been found for "protective" systems within the nervous system. Reduced glutathione (GSH), once thought to be ubiquitous within cells, has been shown to be abundant in astrocytes and ependymal cells, but relatively sparse in central neurons (2,36,55). The distribution of enzymes in the GSH conjugating system generally correlates with the localization of GSH, with some interesting exceptions. Gamma-glutamyl transpeptidase (GGT) appears restricted to glial cells in all areas examined to date (37) (Fig. 5). Among the glutathione S-transferases (GSTs), the pi isoform Yb is localized in ependymal cells, tanycytes, subventricular zone cells, and in astrocytes, particularly in as-

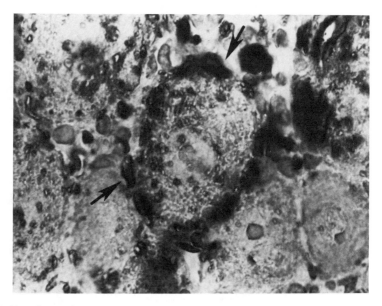

FIG. 5. Dorsal root ganglion stained for gamma glutamyl transpeptidase. Intense immunoreactivity is present in satellite cells (*arrows*) surrounding neurons. Neurons themselves demonstrate little staining. Original magnification ×1,000.

trocytic foot processes around blood vessels (1,7,8), but is not observed in oligodendroglia. Another isoform, Yp, is primarily present in oligodendroglia and is weak or absent in astrocytes and neurons. The presence of the Yb form in elements of brain barrier systems (i.e., blood-brain and cerebrospinal fluid–brain interfaces) likely reflects the role of this GST in transporting hormones, neurotransmitters, and drugs across these barriers to neurons (1) and in possible conjugation reactions of xenobiotics such as acrylamide (15) or styrene (13) with GSH prior to elimination. The role of GST-Yp in oligodendroglia is still unclear, but it may function in the detoxification of agents which target myelin. The myelin toxicant triethyltin interacts with GST-Yp (30).

The potential significance of regional distribution of metabolizing and conjugation systems becomes clear only when viewed in concert with detailed neuroanatomical and neurophysiological information regarding specific brain regions. To date, few such correlative examinations have been undertaken. Recently, Philbert, Lowndes, and co-workers (37) examined the relationship between the distribution of phase I and phase II enzymes and cell sensitivity in the dorsal root ganglion (DRG). The DRG is a particularly suitable system to initiate such work, as its relative lack of a blood-brain barrier compared to CNS structures (22) obviates problems of toxicant access to neurons, and it is one of the few areas of the central nervous system wherein neurons contain abundant GSH (36).

When DRGs were stained for α, μ, and π (7.7 form) GST, no staining could be identified in any neurons of the DRG. In contrast, strong μ and π staining was present in the satellite cells, GSH-rich glial cells which surround the DRG neurons and serve protective and nutritive roles (33). Only the large, pale-type neurons stained, albeit faintly, with antibodies to the π (7.7) form of GST (37). The lack of correlation between GSH and GSTs in the dorsal root ganglia neurons, but their co-localization in the satellite cells, raises several possibilities relating to the fate of xenobiotics to which the DRGs are exposed. This observation may also shed additional light on the long-standing observation that many environmental, occupational, and therapeutic agents initially manifest their clinical neurotoxicity as sensory dysfunction (56). In particular, the DRG contains at least four subtypes of sensory neurons, each of which shows distinct sensitivity to certain neurotoxic agents. For example, small diameter DRG neurons are sensitive to capsaicin (4,23) and 3-acetyl pyridine (59), while large, pale neurons are sensitive to acrylamide. The relationships between localization of phase II systems in glia (or neuronal subtypes) of dorsal root ganglia and expressions of neurotoxicity are of considerable research interest.

While contact mediated cell-cell interactions are essential for normal cell structure and function, the proximity of cells can also entail toxicological risk. The relationship between astrocytes and neurons provides a case in point. Astrocytes are generally thought to protect neurons by controlling fluid, amino acid, and ionic balance in the interstitium. These cells are also known to be metabolically active, and may detoxify xenobiotics. Yet recent evidence indicates that astrocytes may, under some circumstances, metabolically alter a chemical so as to render it neuronotoxic. Perhaps the best characterized example of this phenomenon is 1-methyl-4-phenyl-1,2,3,6-tetrahydropyridine (MPTP), a neurotoxic drug which causes rapid onset of a syndrome virtually undistinguishable from advanced Parkinson's disease (26). The neurotoxicity of MPTP involves bioactivation by monamine oxidase B (MAO-B) to MPP+, which is the toxic molecular species (25). This process is thought to occur in the astrocytes, which are rich in MAO-B. MPP+ released by astrocytes has high affinity for dopamine uptake sites and is thus concentrated to high levels in catecholaminergic neurons. Drugs such as deprenyl, a selective inhibitor of astrocytic MAO-B, block the formation of MPP+ and prevent the development of MPTP-induced parkinsonism in monkeys (27).

SUMMARY

Neural responses to toxic chemicals have traditionally been classified according to neuronal, astrocytic, oligodendrocytic, endothelial, and other target cells. While these classifications have been of great value in elucidation of mechanisms of action and in development of preventive/protective strategies, they tend to deflect attention away from some of the basic processes which modulate cell response to xenobiotics. This chapter has described several such processes, and provided examples of how they influence selective vulnerability within the nervous system. Proper histopathological evaluation of neurotoxic lesions can no longer be confined to mere identification and description of abnormalities, but must be firmly based on an understanding of the neurobiology of cell development, homeostatic maintenance, and interactions of cells with their immediate cellular environment. It is hoped that this chapter will underscore the growing appreciation of the importance of these events in toxicologic neuropathology.

ACKNOWLEDGMENTS

Preparation of this chapter and portions of the research reported herein were supported by USPHS NIH grants ES-04976 and NS-23325. The authors wish to thank Drs. Christine Beiswanger and Martin Philbert for materials used and constructive comments and discussion during the preparation of this manuscript.

REFERENCES

1. Abramovitz M, Homma H, Ishigaki S, Tansey F, Cammer W, Listowsky I. Characterization and localization of glutathione-S-transferases in rat brain and binding of hormones, neurotransmitters, and drugs. *J Neurochem* 1988;50:50–57.
2. Beiswanger CM, Roscoe T, Reuhl KR, Lowndes HE. Cellular localization of glutathione in the developing mouse nervous system. *Toxicologist* 1991;11:54.

3. Brown AW, Aldridge WN, Street BW, Verschoyle RD. The behavioral and neuropathologic sequelae of intoxication by trimethyltin compounds in the rat. *Am J Pathol* 1979;97:59–81.
4. Buck SH, Burks TF. The neuropharmacology of capsaicin: review of some recent observations. *Pharmacol Rev* 1986;38:179–226.
5. Burke RE. Motor units: anatomy, physiology and functional organization. In: Brooks JB, ed. *Handbook of physiology. section I: the nervous system, motor systems.* Washington DC: American Physiological Society, 1982;345–422.
6. Cadrin M, Wasteneys GO, Jones-Villeneuve EM, Brown DL, Reuhl KR. Effects of methylmercury on retinoic acid-induced neuroectodermal derivatives of embryonal carcinoma cells. *Toxicol Cell Biol* 1988;4:61–80.
7. Cammer W, Tansey F, Abramovitz M, Ishigaki S, Listowsky I. Differential localization of glutathione-S-transferase Y_p and Y_b subunits in oligodendrocytes and astrocytes of rat brain. *J Neurochem* 1989;52:876–883.
8. Carder PJ, Hume R, Fryer AA, Strange RC, Lauder J, Bell JE. Glutathione S-transferase in human brain. *Neuropathol Appl Neurobiol* 1990;16:293–303.
9. Cavanagh, JB. Lesion localisation: implications for the study of functional effects and mechanisms of action. *Toxicology* 1988;49:131–136.
10. Chang LW, Tiemeyer TM, Wenger GR, McMillan DE, Reuhl KR. Neuropathology of trimethyltin intoxication. I. Light microscopy study. *Environ Res* 1982;29:435–444.
11. Cookman GR, King W, Regan CM. Chronic low level lead exposure impairs embryonic to adult conversion of the neural cell adhesion molecule. *J Neurochem* 1987;49:399–403.
12. Covault J, Sanes JR. Neural cell adhesion molecule (N-CAM) accumulates in denervated and paralyzed skeletal muscle. *Proc Natl Acad Sci USA* 1985;82:4544–4548.
13. Das M, Seth PK, Mukhtar H. Effect of certain neurotoxins and mixed function, oxidase modifiers on glutathione S-transferase activity in rat brain. *Res Commun Chem Pathol Pharmacol* 1985;33: 377–380.
14. DeGrandchamp RL, Reuhl KR, Lowndes HE. Synaptic terminal degeneration and remodeling at the rat neuromuscular junction produced by a single high dose of acrylamide. *Toxicol Appl Pharmacol* 1990;105:422–433.
15. Dixit R, Mukhtar H, Seth PK, Murti RK. Conjugation of acrylamide with glutathione catalyzed by glutathione-S-transferases of rat liver and brain. *Biochem Pharmacol* 1981;30:1739–1744.
16. Dreyfus PM, Victor M. Effects of thiamine deficiency on the nervous system. *Am J Clin Nutr* 1961;9:414.
17. Edelman GM. Cell adhesion and the molecular processes of morphogenesis. *Ann Rev Biochem* 1984;54:135–169.
18. Gombault A. Contribution a l'histoire anatomique de l'atrophie musculaire saturnine. *Arch Physiol Norm Patholog* 1873;5:592–597.
19. Gombault A. Contribution a l'etude anatomique de la nevrite parenchymateuse subaigue et chronique-nevrite segmentaire periaxile. *Arch Neurolog* 1880;1:11–38.
20. Hansson T, Tindberg N, Ingelman-Sundberg M, Kohler C. Regional distribution of ethanol-inducible cytochrome P450 IIEI in the rat central nervous system. *Neuroscience* 1990;34:451–463.
21. Herken H, Lange K, Kolbe H, Keller K. Anti-metabolic action on the pentose phosphate pathway in the central nervous system induced by 6-aminonicotinamide. In: Herken H, Lange K, eds. *Central nervous system studies in metabolic regulation.* New York: Springer-Verlag, 1974;41.
22. Jacobs JM. Vascular permeability and neurotoxicity. In: Mitchell CL, ed. *Nervous system toxicology.* New York: Raven Press, 1982;285–298.
23. Jancso G, Kiraly E, Jancso-Gabor A. Pharmacologically-induced degeneration of chemoceptive primary sensory neurones. *Nature* 1977;270:741–743.
24. Kety SS. Discussion remark. In: *Metabolic and toxic diseases of the nervous system,* p. 58.
25. Kopin IJ, Markey SP. MPTP toxicity: implications for research in Parkinson's disease. *Ann Rev Neurosci* 1988;11:81–96.
26. Langston JW, Ballard P, Tetrud JW, Irwin I. Chronic parkinsonism in humans due to a product of meperidine-analog synthesis. *Science* 1983;219:979–980.
27. Langston JW, Irwin I, Langston EB, Forno LS. Pargyline prevents MPTP-induced parkinsonism in primates. *Science* 1984;225:1480–1482.
28. Lapresle J, Fardeau M. The central nervous system and carbon monoxide poisoning. II. Anatomical study of brain lesions following intoxication with carbon monoxide (22 cases). In: Bour H, McA

I, eds. *carbon monoxide poisoning,* vol. 24, *Progress in brain research.* Amsterdam: Elsevier-North Holland, 1967;31.

29. Leenders KL. PET studies in Parkinson's disease with new agents. In: Caine DB, et al., eds. *Parkinsonism and aging.* New York: Raven Press, 1989;187–197.
30. Mannervik B. The isozymes of glutathione transferase. *Adv Enzymol* 1985;57:357–417.
31. Melamed E, Mildworf B. Regional cerebral blood flow reductions in patients with Parkinson's disease: correlations with normal, aging, cognitive functions, and dopaminergic mechanisms. In: Caine DB, et al., eds. *Parkinsonism and aging.* New York: Raven Press, 1989;221–228.
32. Morgan KT, Gross EA, Lyght O, Bond JA. Morphological and biochemical studies of nitrobenzene-induced encephalopathy in rats. *Neurotoxicology* 1985;6:105–116.
33. Pannase E. The satellite cells of the sensory ganglia. In: Brodal A, Hild W, van Limborgh, et al., eds. *Advances in anatomy, embryology and cell biology,* vol. 65. Berlin: Springer-Verlag, 1981;1–110.
34. Pardridge WM, Connor JD, Crawford IL. Permeability changes in the blood-brain barrier: causes and consequences. *Crit Rev Toxicol* 1975;3:159–199.
35. Philbert MA, Nolan CC, Cremer JE, Tucker D, Brown AW. 1,3-Dinitrobenzene-induced encephalopathy in rats. *Neuropathol Appl Neurobiol* 1987;13:371–389.
36. Philbert MA, Beiswanger CM, Waters DK, Reuhl KR, Lowndes HE. Cellular and regional distribution of reduced glutathione in the nervous system of the rat: histochemical localization by mercury orange and o-phthaldialdehyde-induced histofluorescence. *Toxicol Appl Pharmacol* 1991;107:215–227.
37. Philbert MA, Reuhl KR, Novak RF, Primiano T, Kim SG, Manson MM, Green JA, Lowndes HE. Immunolocalization of phase I and II enzymes in rat dorsal root ganglia and spinal cord. *Toxicologist* 1991;11:216.
38. Price DL, Griffin JW. Neurons and ensheathing cells as targets of disease processes. In: Spencer PS, Schaumburg HH, eds. *Experimental and clinical neurotoxicology.* Baltimore: Williams & Wilkins, 1980;2–23.
39. Ray DE, Brown AW, Cavanagh JB, Nolan CC, Richards HK, Wylie SP. Functional/metabolic modulation of the brain stem lesions caused by 1,3-dinitrobenzene. *Neurotoxicol* 1991 (in press).
40. Reichardt LF, Tomaselli KJ. Extracellular matrix molecules and their receptors: functions in neural development. *Ann Rev Neurosci* 1991;14:531–570.
41. Remsen LG, Strain GM, Newman MJ, Satterlee N, Daniloff JK. Antibodies to the neural cell adhesion molecule disrupt functional recovery in injured nerves. *Exp Neurol* 1990;110:268–273.
42. Reuhl KR. Post-translational modification of neuronal microtubules alters susceptibility to methylmercury. *Toxicologist* 1989;9:96.
43. Reuhl KR, Borgeson B. Methylmercury, N-CAM expression and dysmorphogenesis. *Toxicologist* 1990;10:136.
44. Rieger F, Nicolet M, Pincon-Raymond M, Murawsky M, Levi G, Edelman GM. Distribution and role in regeneration of N-CAM in the basal laminae of muscle and schwann cells. *J Cell Biol* 1988;107:707–719.
45. Riggs HE, Boles RS. Wernicke's disease: a clinical and pathological study of 42 cases. *Q J Stud Alcohol* 1944;5:361.
46. Robertson DM, Wasan SM, Skinner DB. Ultrastructural features of early brain stem lesions of thiamine-deficient rats. *Am J Pathol* 1968;52:1081.
47. Rutishauser U, Acheson A, Hall AK, Mann DM, Sunshine J. The neural cell adhesion molecule (NCAM) as a regulator of cell-cell interactions. *Science* 1988;240:53–57.
48. Sager PR, Syversen TL. Disruption of microtubules by methylmercury. In: Clarkson TW, Sager PR, Syversen TL, eds. *The cytoskeleton. A target for toxic agents.* New York: Plenum Press, 1986;97–116.
49. Sahenk Z. Distal terminal axonopathy produced by 2,4-dithiobiuret: effects of long-term intoxication in rats. *Acta Neuropathol* 1990;81:141–147.
50. Schneider H, Navarro JC. Acute gliopathy in spinal cord and brain stem induced by 6-aminonicotinamide. *Acta Neuropathol (Berl)* 1978;27:11.
51. Schuch U, Lohse MJ, Schachner M. Neural cell adhesion molecules influence second messenger systems. *Neuron* 1989;3:13–20.
52. Shaw C-M, Alvord EC, Jr. Nervous system injury. In: Mottet NK, ed. *Environmental pathology.* New York: Oxford University Press, 1985;368–453.

53. Sidman RL, Rakic P. Neuronal migration with special reference to developing human brain: a review. *Brain Res* 1973;62:1–35.
54. Sinclair GI, Baas PW, Heidemann SR. Role of microtubules in the cytoplasmic compartmentation of neurons. II. Endocytosis in the growth cone and neurite shaft. *Brain Res* 1988;450:60–68.
55. Slivka A, Mytilneau C, Cohen G. Histochemical evaluation of glutathione in brain. *Brain Res* 1987;409:275–284.
56. Spencer PS, Schaumburg HH. Classification of neurotoxic disease: a morphological approach. In: Spencer PS, Schaumburg HH, eds. *Experimental and clinical neurotoxicology*. Baltimore: Williams & Wilkins, 1980;92–99.
57. Spencer PS, Bischoff MC. Contemporary neuropathological methods in toxicology. In: Mitchell CL, ed. *Nervous system toxicology*. New York: Raven Press, 1982;259–275.
58. Spielmeyer W. Zur pathogenese ortlich electiven Gehirnveranderungen. *Z Gesamte Neurol Psychiatr* 1925;99:756.
59. Valle AL, Beiswanger CM, Lowndes HE, Reuhl KR. Neuron-specific degeneration in dorsal root ganglia following 3-acetylpyridine. *Toxicologist* 1991;11:313.
60. Vogt C, Vogt O. Erkrankungen der Grosshirnrinde in Lichte der Topistik, Pathoklise und Pathoarchitektonik. *J Psychiatr Neurol* 1922;28:1.
61. Volk B, Hettmannsperger U, Amelizad Z, Oesch F, Knoth R. Distribution of cytochrome P-450 in the brain and its induction by antiepileptic drugs (phenytoin). Immunomorphological and neurochemical investigations in the mouse central nervous system. Second Meeting of the Internatl Neurotoxicol Assoc, Sitges, Spain, May 22–26, 1989;94.
62. Wasteneys GO, Cadrin M, Jones-Villeneuve EM, Reuhl KR, Brown DL. The effects of methylmercury on the cytoskeleton of murine embryonal carcinoma cells. *Toxicol Cell Biol* 1988;4: 41–60.

Neurotoxicology, edited by Hugh Tilson
and Clifford Mitchell. Raven Press, Ltd.,
New York 1992.

5

Assessment of Neurotoxicity Using Assays of Neuron- and Glia-Localized Proteins: Chronology and Critique

James P. O'Callaghan

*Neurotoxicology Division, US Environmental Protection Agency,
Research Triangle Park, North Carolina 27711*

The achievements in neuroscience research over recent years have greatly advanced our understanding of nervous system structure and function. Yet, with each increment in knowledge, we are faced with the increasing realization of the overwhelming complexity of this organ system. Gone are the days when the brain can be viewed as a complexly wired collection of a few neuronal cell types surrounded by an apparently inert contingent of supporting glia. We now know that neurons exist in a seemingly endless array of subtypes, each of which has been revealed by the presence of an identifying constituent. Classically, these cell type "markers" took the form of specific neurotransmitter/neuromodulator candidates. More recently, these neuron-specific tags have grown to encompass an enormous number of well-characterized peptides and macromolecules. That glia also exist as a large collection of specific subtypes, each with a potentially unique function, has only recently been appreciated (1).

COMPLEXITY OF THE CENTRAL NERVOUS SYSTEM: IMPEDIMENT TO NEUROTOXICITY ASSESSMENT

Although the cellular and molecular heterogeneity of the mammalian central nervous system is now a widely accepted neurobiological fact, this is of little consolation to the neurotoxicologist. Diverse cell types provide diverse and unpredictable targets of neurotoxic insult. In the absence of *a priori* knowledge of the nature and location of the affected cell type, as would be the case in any screening context, it is difficult to know what to measure or even where to look. It seems obvious that this situation has led neurotoxicologists to rely heavily on morphological evaluations for assessing damage to the central nervous system. Indeed, the classic textbook on experimental neurotoxicology by Spencer and

The research described in this article has been reviewed by the Health Effects Research Laboratory, U.S. Environmental Protection Agency, and approved for publication. Approval does not signify that the contents necessarily reflect the views and policies of the Agency nor does mention of trade names or commercial products constitute endorsement or recommendation for use.

Schaumburg (2) is replete with examples of morphological validation of the "selectively vulnerable" targets of neurotoxicity. Neuroanatomy remains the "gold standard" for verification of toxicant-induced damage to the nervous system. Yet, recent advances in neuroanatomical techniques for estimating cell volume and number (3), techniques for cell type identification and characterization such as immunocytochemistry and *in situ* hybridization, and improved methods for identifying degenerating neurons (4) have only recently being applied to neurotoxicology (5–7). What remains the accepted practice (e.g., see USEPA Neurotoxicity Assessment Guidelines (8)) is the application of classic Nissl stains to detect, qualitatively, cell loss or damage. While this approach is adequate for assessing toxicant-induced damage to structures with large cell types organized in easily recognized layers, such as the hippocampus and cerebellum, it seems likely that damage to other equally vulnerable but more heterogenous structures may go undetected.

The sheer volume of potentially neurotoxic chemicals (9) and chemical mixtures combined with the time and expense required to perform adequate morphological assessments has led to the development of alternative strategies for neurotoxicity assessment. Chief among them is the use of functional assessments of the whole animal to detect toxicant-induced alterations, no matter what the cellular targets (10–15). This approach deals with the problem of cell type vulnerability by assuming that a battery of functional assessments will reveal neurotoxicity as alterations in the output of the whole system, regardless of the cellular target. Unfortunately, even in the face of extensive damage to the CNS, functional techniques will not necessarily be sensitive enough to reveal neurotoxicity. The lack of behavioral deficits following exposure to the dopaminergic neurotoxicant MPTP serves as one example (16,17). This does not imply that the approach is inherently insensitive. It may merely reflect the reserve capacity of specific regions of the nervous system to absorb toxic insults without an alteration in behavior (18).

Indeed, functional observation batteries by design (12) were intended to, and have proven to be, sensitive enough to detect pharmacological actions on the nervous system. These batteries, however, will not necessarily be able to differentiate pharmacological effects from those of chemicals that cause subtle to extensive damage of the nervous system (19). Moreover, compounds which are primarily toxic to organs other than the nervous system may falsely be labeled as neurotoxicants (20).

The expense associated with *in vivo* screens of neurotoxicity has prompted recommendations for developing *in vitro* testing methods (21–24). Proponents of this approach, rather than viewing the complexity of the nervous system as a formidable obstacle, argue that the complexity itself justifies the use of "simplified model systems" (22). Such suggestions beg the question: which model system or systems? How many cell lines will be required to model the nervous system, *in vitro?* What about compounds or mixtures that are bioactivated to neurotoxicants by the liver or brain cells that were omitted from a particular test system?

These and similar issues will need to be addressed before *in vitro* approaches will serve as first-line screens for neurotoxicity (25).

ASSESSMENT OF NEUROTOXICITY IN THE FACE OF CNS CELLULAR HETEROGENEITY: AN ALTERNATE STRATEGY

We recognized that the cellular heterogeneity of the nervous system was likely the basis for the cell type-specific damage (selective vulnerability) associated with toxicant exposures. Thus, rather than ignore the complexity of the nervous system, we reasoned that a strategy to deal with it must be adopted. Because proteins that distinguish the diverse cell types comprising the mammalian nervous system have been described, we proposed that assays of these nervous system-specific proteins could be used to detect and characterize the cellular responses to toxicant-induced alterations of the developing and adult nervous system (26). This neurotoxicity assessment strategy has been reviewed in brief (27) and more extensively in a paper written several years ago that only recently has been scheduled for publication (28). Its use in the context of assessing developmental neurotoxicity also has been reviewed (29) and will not be covered in length here. One protein in the battery of proteins examined, glial fibrillary acidic protein (GFAP), has proven particularly useful for initial neurotoxicity evaluations. Reviews on the use of GFAP in neurotoxicity assessments also have been published (30,31). The purpose of the present chapter will be to provide a brief chronology of the development and refinement of the use of nervous system protein assays for assessing neurotoxicity. In so doing, the central (positive) findings will emerge, as will the drawbacks of the approach. The latter will serve as a critique to underscore future research needs.

ORIGINS OF OUR APPROACH: THE SEARCH FOR A BIOCHEMICAL DEFINITION OF ADVERSE (NEUROTOXIC) EFFECT

If common biochemical pathways of cell damage and death in the CNS were known, there would be no point in developing a neurotoxicity assessment strategy that took the cellular heterogeneity of the nervous system into consideration. Unfortunately, biochemical changes underlying all types of chemical injury of the CNS are as yet unknown. In recognition of this fact, we felt it was critical to link any broadly applicable biochemical approach for neurotoxicity assessment to the most widely accepted criteria for defining neurotoxicity, that being morphology. By so doing we could deal with the problem of cell type vulnerability to neurotoxic insult on a biochemical basis. Moreover, we could define a neurotoxic response with biochemical endpoints that could be linked, quantitatively, to changes in the cellular and subcellular elements of the nervous system. A related purpose for adopting this strategy was to avoid the pitfalls associated with differentiating neuropharmacological effects from those associated with neurotoxic responses. In this context, short-acting, reversible, and presumably nonharmful effects of drugs, chemicals, or chemical mixtures would not be considered neurotoxic and would not be expected to alter the content of brain proteins. By documenting neurotoxicant effects in this manner, we hoped to establish a molecular and cellular, rather than a pharmacological, basis for interpreting changes in whole animal function (behavior or physiology).

What has not been stated in previous reviews is that the ultimate goal of this research was not to establish a molecular basis for neuropathology. Rather, the intent was to establish a quantitative basis for defining neuronal and glial responses to toxicant-induced injury which could be used as a starting point for discovering common biochemical pathways underlying these cell type-specific changes. Our finding that toxicant-induced changes in one protein, GFAP, appear to be a manifestation of the existence of such pathways (30,31), serves as the best example of the success of our approach to date (see below).

EXPERIMENTAL APPROACH: ADMINISTER KNOWN NEUROTOXICANTS AND ASSAY KNOWN NEURON- AND GLIA-LOCALIZED PROTEINS

To validate a broadly applicable biochemical approach for assessing neurotoxicity, we reasoned that a large variety of known neurotoxicants and injurious conditions should be used as positive controls. Examples of the agents and conditions we have used over the years are shown in Table 1. Toxicants were chosen that were known (based on morphological studies) to affect different regions of the CNS and different cell types within an affected region. Initially our intent was to focus on developmental neurotoxicants. We did this for two reasons. First, short of exposures that resulted in cell death, we felt it would be difficult to observe changes in brain proteins in the adult nervous system. Second, the developing CNS had been shown to be particularly vulnerable to a diverse variety of toxic insults and a number of proteins associated with the affected processes

TABLE 1. *Prototype neurotoxicants or injurious conditions used to validate neuron- and glia-localized proteins as biochemical indicators of neurotoxicity*

trimethyltin	MPTP
triethyltin	6-OH dopamine
cadmium	5,7-dihydroxytryptamine
methylmercury	kainate
colchicine	ibotenate
iminodipropionitrile	advanced age
3-acetylpyridine	stab wounds
hyperbilirubinemia	brain heating
Nocardia asteroides	neurological mutations
aminoadipic acid	ethyl alcohol (neonatal)

MPTP, 1-methyl-4-phenyl-1,2,3,6-tetrahydropyridine.

had already been characterized (29,32–34). We anticipated that it would be more feasible to use brain protein assays to model toxicant-induced changes in brain development than it would be to model cell damage and death in the adult. The chronology of our efforts to validate brain proteins as biochemical indicators of neurotoxicity reflects our adherence to this strategy (27,29). Surprisingly, several unexpected but positive findings emerged from our work on developmental neurotoxicants which permitted the application of our approach to studies of adult neurotoxicity (see below).

A cartoon depiction of neuron- and glia-localized proteins associated with various cellular and subcellular elements of the central nervous system is shown in Fig. 1. We have used assays for most of these proteins to successfully model neuronal and glial responses to toxicant injury of the developing and adult CNS. Some of the proteins chosen, like the cell types with which they are associated, are found in all areas of the nervous system, while others are restricted to a single cell type found only in a discrete brain region. For example, synapsin I is present

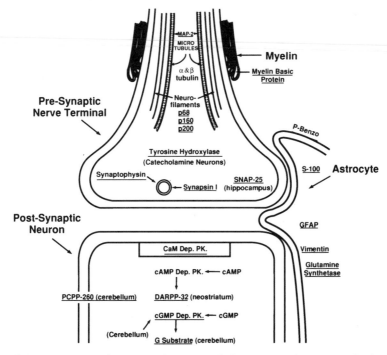

FIG. 1. Schematic diagram of a presynaptic nerve terminal, a postsynaptic neuron, and an impinging astrocytic process. Neuron- and glia-localized proteins associated with their respective cellular and subcellular elements are underlined. The abbreviations used to designate each protein, the CNS cell type specificity of each protein, their cellular distribution, and their associated developmental process or cell type response have been previously presented (26–29,35). The exceptions are as follows: SNAP-25: synaptosomal protein of 25 kD (62), P-Benzo: peripheral type benzodiazepine receptor (63); glutamine synthetase (64) and S-100 (65).

in neurons throughout the CNS in association with synaptic vesicles (35,36), whereas myelin basic protein is found in the myelin sheath elaborated by oligodendroglia (37). In the developing nervous system, we made extensive use of these two proteins to model toxicant effects on the processes with which they are associated: synaptogenesis and myelinogenesis, respectively (26,29,33). In the adult CNS we assayed these same two proteins to assess neuronal damage and demyelination, respectively. Proteins enriched in only a single CNS cell type such as the soluble protein of Purkinje cells, G-substrate (38), or the soluble protein of medium spiny neurons of the neostriatum, DARPP-32 (35,39), have been successfully used to assess toxicant-induced damage to the cerebellum (34) and neostriatum (40).

In laying out our experimental strategy for validating specific brain proteins as biochemical indicators of neurotoxicity, we used both known toxicants and known neuron and glial proteins. Using this scheme we have found our expected and observed findings to be in good agreement. This has given us the confidence to implement assays for several proteins in a screening context for determining the potential neurotoxic effects of compounds for which little neurotoxicity data exist (e.g., tributyltin; see ref. 41). Our initial approach has been to assay specific proteins in a tiered fashion. We started by assaying generally distributed neuronal and glial proteins in discrete (dissected) regions. In samples where toxicant-induced changes were observed, proteins known to be localized to cell types within this region were then assayed. Ideally this reductionist approach would be pursued until the affected cell(s) were identified. By identifying targets of neurotoxicity in this manner, we ultimately intend to evaluate mechanisms of toxicity using *in vitro* modeling of the affected cell type (17).

Although the battery of proteins we can assay is large enough to permit an assessment of toxicant effects anywhere in the nervous system, it is certainly not large enough to model damage to all cell types. Current estimates indicate that there are 30,000 distinct mRNAs in brain, most are in low relative abundance, and roughly 20,000 may be brain-specific (42). These observations indicate that there are likely to be thousands of unique brain proteins, most of which have not been discovered and many of which may be associated with cells targeted by unknown neurotoxicants. Clearly then, one limit to using assays of neuron- and glia-localized proteins to assess neurotoxicity is the limit of known neural cell type proteins and the probability that thousands of such proteins exist. Krady et al. (7) recently provided one elegant demonstration that this is likely to be the case. Using a novel molecular biological approach, they showed that neurons vulnerable to damage from exposure to the organometallic compound, trimethyltin, possess unique and previously uncharacterized species of mRNA. The neurons destroyed by trimethyltin are not linked to the loss of any known neuronal pathway or constituent (7). Thus, the mRNAs isolated by Krady et al. (7) likely code for previously unknown gene products related to the neuronal subtypes susceptible to trimethyltin-induced neurotoxicity. While we were able to use generally distributed neuronal and glial proteins to identify the regions affected by trimethyltin exposure of the neonate (29,43) and adult (5,6,44), based on the data obtained by Krady et al. (7), it is not surprising that we failed to identify transmitters or proteins specific to the affected neurons. Thus, the approach delineated by Krady et al. (7) should prove useful in the future for identification of proteins unique to specific neuronal populations affected by diverse neurotoxic exposures.

METHODOLOGY FOR MEASURING BRAIN PROTEINS: NECESSITY FOR RADIOIMMUNOASSAYS

To validate the use of specific neural proteins as biochemical indicators of neurotoxicity requires methodology for measuring the proteins. Gel electrophoresis represents a powerful technique for resolving specific proteins from complex mixtures such as a brain homogenate. We (32) agreed with other investigators (45) that this technique, combined with subfractionation of nervous system tissue (32,45), would be sufficient to reveal the effects of known neurotoxicants on a number of known brain proteins. The data shown in Fig. 2 provide a dramatic illustration that this is not the case. A single administration of trimethyltin to the rat destroys a large portion of the neurons in the dorsal hippocampus (Fig. 2). When membrane fractions were prepared from homogenates of hippocampus and resolved on SDS-polyacrylamide gels, no differences were observed between samples obtained from saline and trimethyltin (TMT)-treated rats. Radiometric assays (44) and radioimmunoassays (5,44) subsequently revealed that neuron-specific proteins were indeed decreased in the samples from TMT-treated rats. With the benefit of hindsight, we realized our failure to see protein changes using SDS-PAGE reflected the low relative abundance of neuron-specific proteins in subtractions of nervous tissue composed largely of proteins common to all cell types (42). This sobering lesson combined with the successful use of specific radioimmunoassays led us to adopt the latter as the method of choice for measuring neuron- and glia-localized proteins. Radioimmunoassays also allowed more samples to be processed in far less time, they were less labor-intensive, and they were far more cost-effective. Unfortunately, assays for most of the proteins in Fig. 1 were not available, therefore they had to be developed.

The procedure we chose was based on the dot-immunobinding technique of Jahn et al. (46). Several modifications to this procedure have been developed in my laboratory (5,34,40) which now have been described in detail (47). A cartoon representation of the modified procedure, referred to as the slot-immunobinding assay (47), is shown in Fig. 3. The assay is based on the propensity of proteins to bind to nitrocellulose paper. A detergent homogenate of a brain region is applied to the nitrocellulose sheets with the aid of a slotted template. The sheets then are incubated with antibodies raised against the specific protein of interest. The amount of bound antibody and, therefore, the amount of the specific neuron or glia protein is determined by binding ^{125}I protein A to the anti-neuron or

Protein Staining

Histology

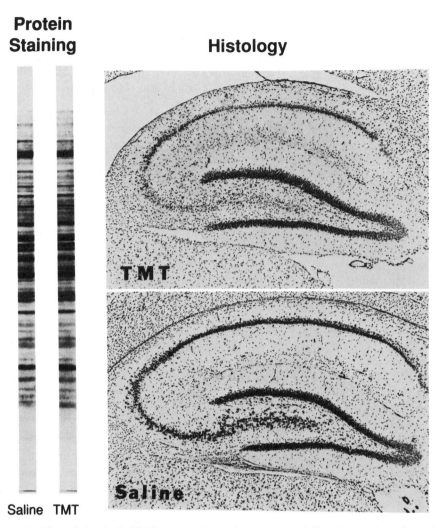

Saline TMT

FIG. 2. Effect of trimethyltin (TMT) on synaptic membrane proteins (*left*) and morphology of dorsal hippocampus (*right*). Crude synaptic membrane fractions were prepared from rats treated with saline or TMT (9.0 mg/kg 3 weeks earlier). The proteins in these fractions then were resolved by SDS-polyacrylamide gel electrophoresis and stained with Coomassie blue R-250. Photomicrographs are of dorsal hippocampus from rats treated with saline or TMT (8.0 mg/kg 3 weeks earlier). (Adapted from O'Callaghan and Miller, ref. 44, Brock and O'Callaghan, ref. 5.)

anti-glia specific antibody (protein A binds to the Fc portion of most IgG subclasses). A main advantage of this assay system is that detergent homogenates of nervous tissue can be assayed directly without further fractionation or purification. Thus, the data obtained are indicative of effects that occurred in the intact tissue, not an extract or an operationally defined membrane or cytosol

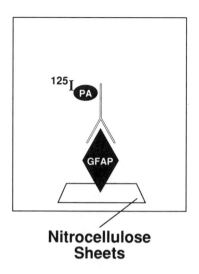

**Nitrocellulose
Sheets**

^{125}I **PA** = ^{125}I **Protein A**

λ = **Rabbit** α **GFAP**

FIG. 3. Schematic diagram of the slot-immunobinding assay for glial fibrillary acidic protein. (Adapted from O'Callaghan, refs. 31,47.)

fraction. Because of the inherent sensitivity of radioimmunoassays, only small amounts of tissue are needed. This permits analysis of discrete regions and cell layers. Furthermore, because the sample amount generally is not limiting, aliquots of a given homogenate (which can be stored frozen) can be assayed for as many proteins as there are antibodies available (31). A final advantage to these assays is that the proteins to be assayed do not need to be purified. Data can simply be presented as a percent of control. Recently I developed a novel sandwich ELISA for the glial-localized protein, GFAP (47). The simplicity, cost-effectiveness, safety, and speed of this procedure may make it preferable to the slot-immunobinding assay. With the exception of tyrosine hydroxylase (48), however, the applicability of the sandwich ELISA format for measuring other brain proteins remains to be demonstrated.

MAJOR FINDINGS: DEVELOPMENTAL EXPOSURES

In the developing rat, a cellular growth spurt occurs shortly after birth that peaks on postnatal day 5 (29,33). The various cell types that proliferate during this event, combined with their subsequent migration and maturation, constitute targets thought to be particularly vulnerable to toxic intervention (29,33). To exploit this "window of vulnerability," we focused much of our work on an examination of the effects of toxicants administered acutely on postnatal day 5

(29). Several studies involved an examination of the effects of three organotin compounds: trimethyltin (TMT), triethyltin (TET) and tributyltin (TBT). These chemicals were of interest because we could group them into three categories: (a) a prototype "positive control" that causes overt cytopathology (TMT); (b) a myelin toxicant in the adult rat that lacks prominent cytopathic effects in the neonate (TET), and (c) a compound with no known effects on the nervous system of the adult or neonate (TBT). Because TMT was a prototype cytotoxicant, we reasoned that it could be used as a denervation tool to validate specific neuronal and glial proteins as indicators of altered brain development. TET and TBT, on the other hand, would be used as test compounds to assess whether effects of these agents could be demonstrated in the absence of cytopathology. The major findings resulting from these studies are listed below:

1. *Assays of brain proteins can be used to model processes associated with brain development.* We assayed proteins associated with the ontogeny of specific neuronal and glial processes (29,33). For example, we assayed synapsin I to evaluate synaptogenesis and we assayed neurofilament 200 to examine the ontogeny of the axonal cytoskeleton. Myelin basic protein was used to monitor myelinogenesis, and glial fibrillary acidic protein was used to follow the ontogeny of astrocytes. Our control data indicated that these proteins served as faithful indicators of the ontogeny of the processes with which they were associated. Thus, brain regions that myelinate earlier than others were shown to accumulate myelin basic protein at an earlier point in development. Likewise, regions showing an earlier onset of synaptogenesis showed an earlier accumulation of synapsin I.

2. *Toxicant-induced changes in brain proteins were observed in the absence of cytopathology.* As expected, TMT caused a permanent loss of neuronal proteins in the hippocampus (29). This effect was shared by TET (29,33), which was not expected because this compound is viewed predominantly as a myelin toxicant and, unlike TMT, it does not cause hippocampal cytopathology (29,33). Both TMT and TET permanently reduced the concentration of myelin basic protein in hippocampus and cerebellum (29). In this case, the former observation was surprising because developmental exposures to TMT are not usually associated with myelin damage in regions that do not show cytopathology, such as the cerebellum. Although neonatal exposures to TBT did not cause permanent changes in specific proteins (29,41), transient but large changes in neuronal and glial proteins were seen in forebrain and cerebellum (29,41). Together, these data indicate that assays of neuronal and glial proteins not only reveal the expected targets of a known cytopathic compound, TMT, but they reveal unexpected targets not revealed by classic histology. Moreover, the permanent and transient effects of the noncytopathic compounds, TET and TBT, also are revealed by assays of neuron and glial proteins associated with specific aspects of brain development.

3. *Astrogliosis occurs in neonates.* A dominant response of the adult nervous system to trauma and chronic neurological disease is proliferation and hypertrophy of astrocytes, a condition that results in the so-called "glial scar" (49).

The ability of the developing nervous system to respond to injury in a similar fashion is thought to be limited (50). Therefore, we did not expect to see evidence of an astrocyte response to neurotoxicant administration on postnatal day 5. What we observed, however, was just the opposite.

Using assays of the major intermediate filament protein of astrocytes, glial fibrillary acidic protein (GFAP), we found that exposure to neurotoxicants early in postnatal development results in large dose-, time-, and region-dependent increases in this protein (29). For example, TMT-induced damage to the hippocampus initiated by its administration on postnatal day 5 resulted in an increase in GFAP that reached 400% of control by postnatal day 13, the earliest time point sampled (29). This increase did not simply reflect an enhanced ontogeny of astroglia because the subsequent increases observed by postnatal day 22 exceeded values seen in adult control animals (29). Moreover, induction of GFAP due to exposure to developmental neurotoxicants did not require concomitant induction of cytopathology (29). Developmental exposures resulting in cytopathology, like TMT, did indeed cause large increases in GFAP. Noncytopathic dosages of these same compounds, however, also caused an increase in GFAP as did exposures, such as TET, that did not result in overt cytopathology at any dose tested (29,33).

In aggregate, our findings indicated that assays of GFAP represented a potentially useful approach for assessing brain damage that results from exposure to developmental neurotoxicants. The unexpected successes achieved with assays of GFAP in a developmental context led us to apply the same approach for assessing neurotoxic responses in the adult (see below). Our studies of developmental neurotoxicants to date have been limited to postnatal exposure regimens. Recently, a positive GFAP response to prenatal stab injuries of the brain has been reported (51). This raises the possibility that assays of GFAP during the pre- or early postnatal period may be used to detect neurotoxicity caused by prenatal toxic exposures.

MAJOR FINDINGS: ADULT EXPOSURES

As mentioned earlier, we did not expect to observe decrements in neuron-localized proteins following exposure of adult animals to neurotoxicants unless there was substantial cell damage or loss. In general, the results obtained with several neurotoxicants were consistent with this expectation. For example, with the prior knowledge that TMT destroys a major portion of pyramidal cells in the dorsal hippocampus (e.g., Fig. 2), we were able to dissect this brain region and demonstrate that TMT caused a dose-dependent decrease in the synaptic protein, synapsin I (5,44). From the standpoint of using TMT as a chemical tool for denervating the hippocampus, our findings were important because they indicated that the degree of damage to specific cell types in this structure could be quantified using assays of synapsin I. From the standpoint of screening "real-world" chemicals for neurotoxicity, however, these findings were not a significant advance because hippocampal damage would have been picked up in routine neuropathological examinations.

Detection of neuronal damage in the adult nervous system by finding a decrease in neuron-localized proteins seemed an unlikely prospect unless you already know where to look, as was the case for TMT. Benefiting from the hindsight we had achieved from our studies of developmental neurotoxicants, however, we realized that assays of GFAP, a protein that increased with injury, might represent a more successful first-tier (31) approach for detecting neurotoxicity. Thus, most of our work on the adult nervous system has been directed toward the use of this protein in the assessment of neurotoxicity. The major findings resulting from these studies are listed below.

1. GFAP serves as a sensitive indicator of neurotoxicity. We evaluated a wide range of compounds and conditions known to have an adverse effect on the CNS (Table 1). In selecting these insults, we purposely included toxicants or conditions that ranged from those causing overt cell loss (e.g., TMT) to those that did not cause Nissl-based evidence of cell damage or loss (e.g., MPTP, aging). By so doing we could start with compounds or conditions that were obviously neurotoxic and then proceed to an examination of toxicants representative of those likely to be found in "real world" scenarios. If the GFAP-based approach was to be successful, we felt that a positive GFAP response must be observed under conditions where damage was not manifested by overt cytopathology, i.e., conditions which would be detected by current neuropathology screens.

As expected, compounds that caused overt neuronal damage, such as TMT, caused large [as great as 600% of control for TMT (5,6)] increases in GFAP. However, dosages of these compounds below those necessary to cause Nissl-based evidence for neuronal damage still resulted in an increase in GFAP (5,6). Thus, assays of GFAP were sensitive enough to reveal underlying damage in the absence of cytopathology. Moreover, prototype neurotoxicants or injurious conditions that do not cause Nissl-based cytopathology, such as MPTP, still cause large [350% of control for MPTP (40)] increases in this astrocyte protein. The possibility has been raised (1) that astrocytes in particular brain regions may not respond to injury. We had this possibility in mind when we chose prototype toxicants to evaluate region-specific GFAP responses. All brain regions examined to date, from brain stem to olfactory bulbs, have been shown to contain astrocytes that respond to neurotoxic injury by increasing GFAP (30,31).

2. Toxicant-induced increases in GFAP are confined to sites of damage. Trauma-induced glial scarring, as revealed by GFAP immunocytochemistry, has been reported to occur at sites far removed from the wound (52). Therefore, we expected to see the increases in GFAP spread to brain regions outside those targeted by the toxicant in question. Instead, we found that toxicant exposures cause increases in GFAP that are limited to the sites of damage (6,53). In the few cases where regions not known to be affected by a prototype neurotoxicant (based on Nissl stains) showed an increase in GFAP (e.g., frontal cortex after TMT), they were later found to contain damaged cells as evidenced by sensitive degeneration stains (6). These data indicated that assays of GFAP were sensitive enough to reveal neurotoxicity in the absence of cytopathology and selective enough to be limited to the regions containing the damaged cells. Unfortunately, the latter characteristic is not necessarily a desirable one if GFAP assays are to used as a primary screen, because whole brain levels of this protein might not be increased. This would occur if region-specific increases in GFAP were diluted out by tissue not affected by the toxicant in question. On the other hand, because regional increases in GFAP reveal the affected brain areas, this "homing" response can serve as the basis for subsequent studies designed for determining the damaged cell types. One example of how this tiered approach would work, again with the benefit of hindsight, is the case of MPTP. If one had been fortunate enough to examine this compound in the mouse (the rat being insensitive), this would have led to the discovery of an increase in striatal GFAP (40,53,54). This observation might in turn have led to an examination of the nigral striatal pathway and the subsequent demonstration of a long-lasting dopamine depletion, as well as a loss of the rate limiting enzyme in its biosynthesis, tyrosine hydroxylase (40).

   3. Toxicant-induced increases in GFAP are not permanent. The astrocyte re-action to physical injury of the CNS or to neurological disease states is thought to result in a persistent glial scar that is GFAP positive (49). On the basis of this long-held view, we expected toxicant-induced astrogliosis to result in similar scarring and, therefore, a long-lasting or permanent increase in GFAP. Based on our experience with several of the compounds and conditions listed in Table 1, however, we found just the opposite to be the case. The increase in GFAP often was transient, with the time-course of the decline in its levels varying markedly from toxicant to toxicant [for example, TMT and MPTP; (5,40)]. The reason for the disparity between the longevity of disease-induced gliosis versus toxicant-induced gliosis is that the stimulus responsible for the continued increase in GFAP is apparently maintained in disease states. It subsides, however, after termination of neurotoxic exposures, or as one moves away from the sites of traumatic injury to the CNS (1,55).

   4. Toxicant-induced increases in GFAP reflect astrocytic hypertrophy, not hy-perplasia. Astrocytic proliferation, not hypertrophy, often is viewed as the dom-inant response to all types of brain injuries (1), including those induced by toxic exposures (30,31). This notion has gained acceptance due to the use of ^{3}H-thymidine autoradiography in combination with GFAP immunocytochemistry to reveal dividing astrocytes. For example, in a widely cited study, Latov and co-workers (56) demonstrated that stab injuries of the brain caused astrocytes surrounding the wound site to divide. However, what is often overlooked is that Latov et al. (56) reported that only a very small percentage of total astrocytes actually divided. Indeed, using the same methodology reported by Latov et al. (56), we found that only a very small percentage (<1%) of astrocytes in hippo-campus divided in response to TMT-induced neurotoxicity (5). Likewise, the

TABLE 2. *Pharmacological agents that do not increase GFAP*

phenobarbital	pargyline
pentobarbital	amfonelic acid
ketamine	clorgyline
diethylether	apomorphine
ethyl alcohol (1% vehicle)	scopolamine
reserpine	atropine
indomethacin	parachlorophenylalanine
difluoromethylornithine	alphamethyltyrosine
prazosin	fluoxetine
MK-801	diethyldithiocarbamate
butyrolactone	muscimol
nomifensine maleate	SKF 38, 393
SCH 23390	L-DOPA

enhanced expression of GFAP in response to MPTP-induced damage to neo-striatum (40) does not appear to involve significant astrocyte division (J. P. O'Callaghan, K. F. Jensen, D. B. Miller, *unpublished observation*). Thus, at least for the neurotoxicants we have examined, hypertrophy, not hyperplasia, appears to be the dominant response to chemically induced brain injuries (30,31). The same conclusion was reached by Hatten et al. (1) in reviewing data obtained from studies of noninvasive and genetically programmed brain injuries, such as those resulting from anoxia or a genetic lesion in copper uptake. If these disparate observations can be applied to the general case, we expect that astrocytic hyperplasia will not be a major factor underlying most increases in GFAP. Given this possibility, it is doubtful that GFAP immunocytochemistry-based cell counts will be of much value in assessment of general neurotoxic responses because more GFAP per astrocyte, not more astrocytes, will be the predominant neurotoxic response.

5. *Pharmacological agents at therapeutic dosages do not increase GFAP.* If astrogliosis and the attendant increase in GFAP result from cellular damage caused by chemical exposures, drugs administered at therapeutic and presumably nondamaging dosages should not cause the same response. We have not conducted a comprehensive search for drug-induced (i.e., false positive) increases in GFAP. Several drugs, however, have been used in the course of investigations aimed at determining mechanisms of specific neurotoxic responses. These compounds, several of which are listed in Table 2, serve as negative controls for assessing increases in GFAP. Although complete dose- and time-effect studies have not been attempted with these compounds, it is clear that agents from several pharmacological classes do not increase GFAP.

FUTURE DIRECTIONS: FIND THE SIGNALS FOR ASTROGLIOSIS

If astrogliosis represents a common response to all types of neurotoxic exposures, then the possibility exists that a common set of "damage" signals initiates this response. No such signals have as yet been identified. Traditionally, the induction of gliosis has been regarded as a relatively late step in the cascade of events that follows neural cell damage. Thus, the search for common pathways underlying toxicant-induced damage to the nervous system has focused on events presumably far upstream from the factors responsible for astrogliosis and the concomitant induction of GFAP. Not surprisingly, therefore, putative mechanisms of neurotoxicity such as excess oxygen radical generation (57), altered calcium homeostasis (58,59), glutamate-induced excitotoxicity (59), or induction of immediate early genes (60) are not often considered as features of the neurotoxic state that overlap temporally with those that underlie gliosis.

What is clear from our studies of GFAP, however, is that induction of gliosis can coincide with the earliest evidence of cell damage, often within hours of toxicant exposure (e.g., MPTP). Thus, it is possible that factors generated in the very earliest stages of neural cell damage may be those responsible for initiating the glial reaction. Pharmacological manipulations can modify the expression of putative CNS damage mechanisms [e.g., see (57,61)]. Given the overlap in events associated with neural cell damage and the induction of astrogliosis, it is possible that these same manipulations may affect not only target cell damage but the initiation of astrogliosis as well. Few studies have examined the potential for pharmacological agents, *in vivo,* to modify astroglial responses to toxicant-induced injuries of the CNS. This may represent a fruitful avenue of investigation for determining common mediators of neurotoxicity.

ACKNOWLEDGMENTS

The author thanks Dr. D. B. Miller for valuable suggestions. This work was supported in part by the National Institutes of Drug Abuse (IAG RA-ND-89-4).

REFERENCES

1. Hatten ME, Liem RKH, Shelanski ML, Mason CA. Astroglia in CNS injury. *Glia* 1991;4:233–243.
2. Spencer PS, Schaumburg HH. *Experimental and clinical neurotoxicology.* Baltimore: Williams and Wilkins, 1980.
3. Møller A, Strange P, Gundersen HJG. Efficient estimation of cell volume and number using the nucleator and the disector. *J Micros* 1989;159:61–71.
4. Carlsen J, de Olmos J. A modified cupric silver technique for the impregnation of degenerating neurons and their processes. *Brain Res* 1981;208:426–431.
5. Brock TO, O'Callaghan JP. Quantitative changes in the synaptic vesicle proteins synapsin I and p38 and the astrocyte-specific protein glial fibrillary acidic protein are associated with chemical-induced injury to the rat central nervous system. *J Neurosci* 1987;7:931–942.
6. Balaban CD, O'Callaghan JP, Billingsley ML. Trimethyltin-induced neuronal damage in the rat: comparative studies using silver degeneration stains, immunocytochemistry and immunoassay for neuronotypic and gliotypic proteins. *Neuroscience* 1988;26:337–361.
7. Krady JK, Oyler GA, Balaban CD, Billingsley ML. Use of avidin-biotin subtractive hybridization to characterize mRNA common to neurons destroyed by the selective neurotoxicant trimethyltin. *Mol Brain Res* 1990;7:287–297.

8. US Environmental Protection Agency. Pesticide assessment guidelines: subdivision F, hazard evaluation: human and domestic animals, addendum 10, neurotoxicity, series 81, 82, and 83. National Technical Information Services, Springfield, VA, 1991.
9. US Congress, Office of Technology Assessment. *Neurotoxicity: identifying and controlling poisons of the nervous system.* Washington: US Government Printing Office, 1990.
10. Gad SC. A neuromuscular screen for use in industrial toxicology. *J Toxicol Environ Health* 1982;9:691–704.
11. Moser VC. Screening approaches to neurotoxicity: a functional observational battery. *J Am Coll Toxicol* 1989;8:85–93.
12. Irwin S. Comprehensive observational assessment: Ia. A systematic quantitative procedure for assessing the behavioral and physiologic state of the mouse. *Psychopharmacology* 1968;13:222–257.
13. Tilson HA, Cabe PA. Strategy for the assessment of neurobehavioral consequences of environmental factors. *Environ Health Perspec* 1978;26:287–299.
14. Tilson HA, Mitchell CL, Cabe PA. Screening for neurobehavioral toxicity: the need for and examples of validation of testing procedures. *Neurobehav Toxicol* 1979;1:137–148.
15. Tilson HA, Mitchell CL. Neurobehavioral techniques to assess the effects of chemicals on the nervous system. *Ann Rev Pharmacol Toxicol* 1984;24:425–450.
16. Heikkila RE, Sonsalla PK. The use of the MPTP-treated mouse as an animal model of parkinsonism. *Can J Neurol Sci* 1987;14:436–440.
17. Miller DB, Reinhard JF, Jr, Daniels AJ, O'Callaghan JP. Diethyldithiocarbamate potentiates the neurotoxicity of in vivo 1-methyl-4-phenyl-1,2,3,6-tetrahydropyridine and of in vitro 1-methyl-4-phenylpyridinium. *J Neurochem* 1991;57:541–549.
18. Zigmond MJ, Abercrombie ED, Berger TW, Grace AA, Stricker EM. Compensations after lesions of central dopaminergic neurons: some clinical and basic implications. *Trends Neurosci* 1990;13:290–296.
19. Broxup B, Robinson K, Losos G, Beyrouty P. Correlation between behavioral and pathological changes in the evaluation of neurotoxicity. *Toxicol Appl Pharmacol* 1989;101:510–520.
20. Gerber GJ, O'Shaughnessy D. Comparison of the behavioral effects of neurotoxic and systemically toxic agents: how discriminatory are behavioral tests of neurotoxicity? *Neurobehav Toxicol Teratol* 1986;8:703–710.
21. Atterwill CK, Walum E. Neurotoxicology in vitro: model systems and practical applications. *Toxicol in vitro* 1989;3:159–161.
22. Walum E, Hansson E, Harvey AL. In vitro testing of neurotoxicity. *ATLA* 1990;18:153–179.
23. Hooisma J. Tissue culture and neurotoxicology. *Neurobehav Toxicol Teratol* 1982;4:617–622.
24. Goldberg AM, Frazier JM. Alternatives to animals in toxicity testing. *Sci Am* 1989;261:24–30.
25. Tilson HA. Neurotoxicology in the 1990s. *Neurotoxicol Teratol* 1990;12:293–300.
26. O'Callaghan JP, Miller DB. Nervous-system specific proteins as biochemical indicators of neurotoxicity. *Trends Pharmacol Sci* 1983;4:388–390.
27. O'Callaghan JP. Neurotypic and gliotypic proteins as biochemical markers of neurotoxicity. *Neurotoxicol Teratol* 1988;10:445–452.
28. O'Callaghan JP. Neurotypic and gliotypic proteins as biochemical indicators of neurotoxicity. In: Abou-Donia MB, ed. *Textbook on neurotoxicology. (in preparation).*
29. O'Callaghan JP, Miller DB. Assessment of chemically-induced alterations in brain development using assays of neuron- and glia-localized proteins. *Neurotoxicol* 1989;10:393–406.
30. O'Callaghan JP. Assessment of neurotoxicity: use of glial fibrillary acidic protein as a biomarker. *Biomed Environ Sci* 1991;4:197–206.
31. O'Callaghan JP. The use of glial fibrillary acidic protein in first-tier assessments of neurotoxicity. *J Am Coll Toxicol (in press).*
32. O'Callaghan JP, Miller DB, Reiter LW. Acute postnatal exposure to triethyltin in the rat: effects on specific protein composition of subcellular fractions from developing and adult brain. *J Pharmacol Exp Ther* 1983;242:466–472.
33. O'Callaghan JP, Miller DB. Acute exposure of the neonatal rat to triethyltin results in persistent changes in neurotypic and gliotypic proteins. *J Pharmacol Exp Ther* 1988;244:368–378.
34. O'Callaghan JP, Miller DB. Cerebellar hypoplasia in the Gunn rat is associated with quantitative changes in neurotypic and gliotypic proteins. *J Pharmacol Exp Ther* 1985;234:522–533.

Neurotoxicology, edited by Hugh Tilson
and Clifford Mitchell. Raven Press, Ltd.,
New York © 1992.

6

Effect of Neurotoxicants on Brain Neurochemistry

Lucio G. Costa

*Department of Environmental Health, University of Washington
Seattle, Washington 98195*

Neurochemistry plays a central role in the neurosciences by acting as an integrating science representing molecular and mechanistic aspects. By utilizing biochemical methods, the goal of neurochemistry is to understand the mechanisms and pathways involved in normal and altered nervous system functioning. In pharmacology, the development of new therapeutic drugs is greatly enhanced by our increasing knowledge of new possible molecular targets, such as new receptor subtypes or specific steps in second messenger pathways. Within neurotoxicology, neurochemistry can also play a pivotal role in the understanding of the molecular and biochemical mechanisms which underlie the responses of the nervous system to toxic insults. By providing information on the biochemical substrates involved in behavioral changes and/or correlates of neuropathological lesions, neurochemical approaches to neurotoxicity are relevant to the process of assessing the impact of human exposure to neurotoxicants and, in particular, to the design and development of proper antidotes. Knowledge of the molecular target of neurotoxicants can also be extremely useful in predicting possible interactions with other exogenous chemicals (e.g., drugs, food constituents), as well as interaction with preexisting alterations (due, for example, to disease states or genetic predisposition).

A number of reviews have been published on the application of neurochemical techniques to neurotoxicology (3,27,52,60,67,68,70,109–111). This chapter is not intended to replicate their content, but to introduce new and different examples, and novel perspectives and points of view. A review of all aspects of neurochemistry that have been the subject of investigation within neurotoxicology is beyond the scope of this chapter. Metabolism of amino acids, lipids, carbohydrates, proteins, and nucleic acids, and aspects of energy metabolism, etc., have been investigated over the years from a neurotoxicological perspective, but will not be reviewed here. Instead, this chapter will focus on neurochemical approaches related in particular to the study of neurotransmission and intracellular signaling mechanisms. The problems and limitations of neurochemical approaches to neurotoxicology are presented. A discussion of new possible biochemical approaches to the study of neurotoxicity in humans is also presented. Other relevant information related to neurochemical approaches can be found elsewhere in this volume (e.g., see Verity, Griffin, O'Callaghan, and Mattsson).

CHEMICAL NEUROTRANSMISSION

Transmission of messages between cells in the nervous system involves the release of a chemical neurotransmitter from one cell and its subsequent recognition by a second cell. Specific enzymes synthesize the neurotransmitter from one or more precursors. The neurotransmitter is usually stored in vesicles in the presynaptic ending and released in the synaptic cleft upon arrival of a stimulus. After their interaction with the second cell, neurotransmitters are rapidly inactivated by an uptake mechanism into the neuron that released them or into glial cells, or by specific enzymes. The interaction with a specific receptor protein will generate a transmission signal which will ultimately lead to a physiological response.

Several toxic agents are known to affect one or more parameters of chemical neurotransmission. While a few neurotoxic compounds have a specific action on certain biochemical processes, similarly to most pharmaceuticals or many natural toxins, most neurotoxic chemicals are likely to exert their effect by interacting with more than one biological site. Additionally, several effects on neurotransmission can derive from an indirect action of the neurotoxicant, whose primary target might be elsewhere in the nerve cell. Certain compounds could affect neurotransmitter systems *in vitro* but not *in vivo,* for example, because they do not cross the blood-brain barrier, or they are rapidly metabolized to inactive molecules. Conversely, compounds which are inactive *in vitro* might be biotransformed *in vivo* to neurotoxic or more neurotoxic metabolites. Furthermore, acute exposure to a neurotoxicant might cause certain alterations in neurotransmitter systems, which are not present following chronic exposure because of adaptive or compensatory reactions. On the other hand, only chronic exposure to certain chemicals might elicit alterations in neurotransmission.

The complexity of the potential interactions of neurotoxicants with neurotransmitter systems is obvious, due to the array of factors which could play a role in these processes. Nevertheless, these biochemical studies are of invaluable importance for the understanding of the mechanism of action of neurotoxic substances, and, when properly conducted and interpreted, have provided very relevant information. The number of examples of neurotoxicants that have been shown to interact with neurotransmission is rather large. Here there will be discussed only selected examples of chemicals whose target(s) are one or more parameters of neurotransmission, with special attention to those neurochemical investigations which have yielded useful mechanistic information and where correlations between biochemical targets and behavioral findings could be found.

Synthesis, Uptake, and Degradation of Neurotransmitters

Enzymes devoted to the synthesis of neurotransmitters do not appear to be a common target for neurotoxicants. One of the few examples is carbon disulfide which increases the level of brain dopamine while decreasing norepinephrine content, possibly as a consequence of its inhibition of dopamine beta-hydroxylase. Indeed, a metabolite of carbon disulfide, dithiocarbamate, which is formed after its reaction with amino acids, chelates copper, a metal necessary for the functioning of this enzyme (119). Another compound, phenylacetate, one of the main metabolic products of phenylalanine (a constituent of aspartame) in phenylketonuria, has neurotoxic properties which have been ascribed to inhibition of choline acetyltransferase by its metabolite phenylacetyl-coenzyme A (CoA) (139). In general, measurements of neurotransmitter-synthesizing enzymes, by either biochemical or immunohistochemical methods, are of most value in neurotoxicology for establishing the loss of a particular neuronal population, which may have been caused by other mechanisms.

A number of chemicals can affect neurotransmitter movements, including both the calcium-dependent release and the energy-dependent uptake. In addition to botulinum toxin (155), several heavy metals including lead, cadmium, and mercury inhibit acetylcholine release, possibly because of a reduction of voltage-gated calcium entry into presynaptic nerve terminals (11,46). Uptake of neurotransmitters has been shown to be inhibited by a food color, erythrosin B (101), by other metals such as lead, copper, mercury, and organotins (48,93,166), or by pesticides such as chlordecone (71). Many of these effects have been observed *in vitro* and the exact role that they may have in these compounds' neurotoxicity is unclear at most. For example, inhibition of dopamine uptake by erythrosin B was initially seen as the way this food color, which had been suggested to be involved in the etiology of childhood hyperkinesis, might affect behavior. However, further studies challenged this hypothesis and ascribed the effect of erythrosin B to its nonspecific interaction with the cell membrane; these studies also showed that only very low concentrations of erythrosin B were present in the brain following *in vivo* administration, thus making inhibition of dopamine uptake an unlikely event to occur *in vivo* (112). Inhibition of GABA uptake by trimethyltin (TMT) has been observed *in vitro* and following *in vivo* administration, and might be involved in certain aspects of its neurotoxicity (58,72). This effect appears to be due in part to inhibition of Na^+/K^+-ATPase, but also to other mechanisms (48). However, the finding that TMT inhibits the uptake of other neurotransmitters, such as norepinephrine and serotonin (72), and that other organotins have a similar or greater effect than TMT on GABA uptake, at least *in vitro* (48), leaves unanswered the question of the role, if any, of such biochemical effects in TMT neurotoxicity.

Degradation of neurotransmitters is affected by various classes of pesticides. Organophosphates and carbamates inhibit acetylcholinesterase, and signs of their acute toxicity are consistent with a hyperstimulation of the cholinergic system by endogenous acetylcholine (49,122). The mechanism of action of organophosphates was discovered in the late 1930s, almost at the same time as their synthesis (51). This unusual rapid finding was certainly facilitated by the knowledge of the effects and of the mechanism of action of physostigmine, whose mode of action as a cholinesterase inhibitor was found by Loewi and Nevratil in 1926 (35). Inhibition of acetylcholinesterase is certainly responsible for the acute toxic effects of these insecticides, since blockage of cholinergic muscarinic receptors with atropine and reactivation of the enzyme with oximes represent the antidotes for their intoxication. Another class of pesticides, the insecticide/acaricide formamidines, inhibits monoamine oxidase, one of the enzymes involved in the degradation of catecholamines, (19); in this case, however, this action does not appear to be related to their acute toxicity or neurotoxicity (29).

Neurotransmitter Receptors

Several neurotoxic compounds have been shown to affect neurotransmitter receptors, either directly or indirectly. In this section, the term receptor is used in a narrow sense to indicate the specific protein that recognizes and binds the neurotransmitter, though it can be argued that a recognition site, a transmission signal, and a physiological response must be present to apply the term receptor to this system (95). The availability of radiolabeled ligands and the development of radioligand binding assays have been the key to the past 15 years' success in measuring small amounts of receptors in the nervous system. Receptor binding assays have been, and still are, useful in neurotoxicology in two ways: they can be used to determine whether a neurotoxicant interacts with a receptor directly, or whether it alters other components of the nervous system which, in turn, will lead to changes in receptor binding (60,67,111). *In vitro* experiments allow us to establish whether a toxicant interacts with a receptor site directly (either as an agonist or as an antagonist; 60), and this approach has been utilized successfully to investigate new mechanisms of action of insecticides. Type II pyrethroid insecticides, those containing a cyano substituent such as deltamethrin or fenvalerate, produce signs of poisoning that resemble picrotoxin, and include salivation, hyperactivity, chloroathetosis, and clonic/tonic convulsions. Though the target for toxic action for both Type I and Type II pyrethroids has been identified as the axonal sodium channel (124), binding studies with ^{35}S-TBPS (t-butylbicyclophosphorothionate), a ligand for the chloride channel on the GABA receptor complex, have shown that Type II pyrethroids potently interact with this site (2,98). Further radioligand studies have shown that Type II pyrethroids have an even higher affinity for the site labeled by ^{3}H-Ro5-4864, a convulsant benzodiazepine ligand (82). It is now generally accepted that the interaction of Type II pyrethroids with the GABA receptor-ionophore complex can play a relevant role in the acute toxicity of this group of compounds (79). A similar series of studies revealed that several organochlorine insecticides (e.g., gamma lindane, dieldrin, chlordane, but not DDT or chlordecone) interact with the TBPS site on the chloride channel (1,97). For both classes of insecticides, a good correlation was also shown between *in vitro* potency in radioligand binding assays and *in vivo* acute toxicity (97,98). Another example of application of radioligand binding techniques for the identification of targets of neurotoxicity is shown by studies on the formamidine insecticides. In insects, formamidines such as chlordimeform or amitraz interact with octopamine receptors and activate an adenylate cyclase (74). In mammals, the sign and symptoms of formamidine intoxication are sympathomimetic in nature (18); however, the postulated mechanism of action [inhibition of monoamine oxidase; (19)] does not appear to be an important factor in their neurotoxicity (29). *In vivo* studies have indicated that several effects of formamidines (e.g., mydriasis, bradycardia, changes in visual evoked potential) are antagonized by the alpha$_2$-adrenergic antagonist yohimbine (30,87) and *in vitro* binding studies have shown that formamidines interact competitively, and with high potency and specificity, with alpha$_2$-adrenoceptors in mouse and rat brain (62,65). This was further confirmed by *in vivo* studies which showed a dose- and time-dependent correlation of changes in alpha$_2$-adrenoceptor binding and behavioral effects (65). Altogether, these experiments provide examples of how neurochemical studies of neurotransmitter receptors have allowed identification of novel and important sites of action for these classes of relevant pesticides.

Certain neurotoxicants may not interact directly with specific neurotransmitter receptors but might nevertheless cause alterations in receptor systems as a result of a primary interaction at another target. A good example of this is represented by the effect of organophosphorus insecticides (123). Indeed, repeated exposures to these compounds cause a decrease in the density of cholinergic muscarinic and nicotinic receptors in brain and peripheral tissues (41,56,60,64). Although a direct interaction of organophosphates with cholinergic receptors (14,91), and the recently reported direct neurotoxic effect (162) could contribute to this receptor decrease, this is probably due, for the most part, to a phenomenon of down regulation in response to excessive stimulation by acetylcholine, which accumulates because of inhibition of acetylcholinesterase by the organophosphates. These receptor changes are of interest because they may offer an at least partial explanation of the phenomenon of tolerance to organophosphate toxicity (63) and of cognitive deficits found in chronically organophosphate-treated animals (117). Another example is represented by trimethyltin, which has been reported to cause a 21% decrease of muscarinic receptors in the hippocampus of rats 2 weeks following administration (107). However, trimethyltin does not interact with muscarinic receptors *in vitro,* or affect muscarinic binding measured in forebrain 14 hr after administration (58,72). It is therefore most probable that the observed decrease is due to loss of pyramidal cells observed in the hippocampus. Loss of cells and receptors may be involved in the marked deficits in retention of passive avoidance observed in trimethyltin-treated rats (107).

Future studies in the area of receptor alterations will most probably utilize some of the most recent techniques of neurobiology and molecular biology. For example, genes for five different subtypes of muscarinic receptors have been cloned (120). Through the use of subtype-specific compounds in receptor binding and/or autoradiography studies, and the use of cDNA probes, it will be possible to increase our knowledge of more specific changes linked to exposure to certain neurotoxicants.

Second Messenger Systems

The interaction of a neurotransmitter with a specific receptor activates a series of mechanisms which allow transmission of the message from the cell surface to the inside of cell and its possible propagation to other cells. These mechanisms might consist of the opening of an ion channel (e.g., chloride, sodium, potassium, or calcium channel), the activation of specific membrane associated enzymes (e.g., phosphoinositidase C, adenylate cyclase), or a combination of the two. Ionotropic receptors are usually better studied utilizing electrophysiological approaches. On the other hand, the study of the events associated with activation of metabolotropic receptors is most often carried out with biochemical techniques. Several receptors, e.g., beta adrenergic or dopaminergic, are coupled through a G protein (Gs) to the enzyme adenylate cyclase which increases the production of cyclic adenosine $3',5'$ monophosphate (cyclic AMP) from adenosine triphosphate (ATP). Cyclic AMP, in turn, activates a protein kinase (PKA), which can phosphorylate a number of intracellular substrates (146). Cyclic AMP is then metabolized to $5'$-AMP by the action of phosphodiesterases. Some receptors (e.g., muscarinic M_2, alpha$_2$-adrenergic) are "negatively coupled" to adenylate cyclase and their activation inhibits its activity. Receptor-mediated inhibition of this enzyme involves another G protein known as Gi, and leads to a decrease in the intracellular levels of cyclic AMP. Activation of some receptors (e.g., cholinergic muscarinic, histaminergic) also causes the elevation in intracellular cyclic GMP (guanosine $3'5'$-monophosphate) levels. This action is generally considered to be an indirect effect involving an additional second messenger, possibly calcium ions or a metabolite of arachidonic acid. Despite the great importance of cyclic nucleotides in a large number of cell functions, there has been only a limited number of studies investigating their interaction with neurotoxicants. For example, organophosphorus insecticides, certain pyrethroids, and DDT have been shown to increase levels of cyclic GMP (6,25,104), while other pesticides (e.g., organochlorines) might enhance the brain cyclic AMP content (25). Lead, on the other hand, has been shown to inhibit adenylate cyclase (125). In all cases, however, the significance of these effects with regard to neurotoxicity is unclear at best.

Several receptors for neurotransmitters, hormones, and growth factors are coupled to a distinct second messenger system, the phosphoinositide/protein kinase C pathway. Through one or more still unidentified G proteins, activation of these receptors (e.g., muscarinic M_1, excitatory amino acids) stimulates hydrolysis of membrane phosphoinositides by phosphoinositidase C to generate primarily inositol 1,4,5-triphosphate, responsible for the release of calcium ions from intracellular stores into the cytosol, and 1,2-diacylglycerol, which activates protein kinase C (PKC) (22,40,76,128,141). Because of the presence of several receptor systems coupled to this response in both nerve and glial cells and the high concentration of PKC in brain tissue, phosphoinositide metabolism in the nervous system has been receiving much attention recently (40,76). Some studies have investigated the interaction of neurotoxic chemicals with the phosphoinositide/PKC system and these have been reviewed recently (53). A number of metals (mercury, lead, aluminum, cadmium) have been found to interact with PKC, mostly in an inhibitory fashion. Interestingly, however, lead has been shown to activate PKC partially purified from rat brain at less than nanomolar concentrations (113). Several chlorinated hydrocarbons (chlordane, DDT, lindane) also stimulate PKC from rat brain, an effect which may be related to the hepatic tumor promoter action of some of the nongenotoxic organochlorines (121). Alterations in inositol phosphate metabolism may also play a role in the acute neurotoxicity of certain organophosphates (54). Pyrethroid insecticides stimulate phosphoinositide metabolism, at least in part in a manner analogous to other sodium channel agents (85), while aluminum ions cause a similar effect through a possible interaction with G proteins (33,118). In all cases, the possible relevance of such interactions with phosphoinositide metabolism or protein kinase C is inferred by our knowledge of the multiple functions that this system has in nerve tissue. However, an exact understanding of how this interaction may be involved in the mechanisms of toxicity of these compounds is still incomplete.

Recent studies have suggested the hypothesis that the muscarinic receptor-stimulated phosphoinositide metabolism in the brain of the developing rat might represent a relevant target for the developmental neurotoxicity of ethanol (15,16). The age-, brain region-, and receptor-specific effects of ethanol on this system are indeed of interest, and might relate to some of the effects found in the fetal alcohol syndrome (e.g., microencephaly), though the complex mechanisms of these ethanol-phosphoinositide interactions need to be elucidated before any firm conclusion about a possible causal relationship can be drawn.

A link between excitatory amino acid receptors, phosphoinositide metabolism, and neurotoxicity has also been proposed (53), which might involve intracellular calcium. Calcium ions play a most relevant role in synaptic transmission, particularly for their action at neurotransmitter release (12). The role of calcium as an intracellular second messenger system is also well established (141,163). There is increasing evidence, however, that calcium may play a key role as an intracellular toxin (132,140,143), and more recent studies have suggested that it may be a significant determinant in neurotoxicity (28,92). Significant and persistent changes in intracellular calcium levels are induced by neurotoxic organometals and by organochlorine insecticides (28). However, the best evidence for a calcium involvement has been found in neurotoxicity of excitatory amino acids (39,81,154), since removal of extracellular calcium or chelation of intracellular calcium (149) prevents their neurotoxicity.

Finally, what might be considered a "third messenger" system, the phosphorylation of specific proteins, could represent a relevant biochemical target for neurotoxicants. Protein phosphorylation systems consist of three primary components: a protein kinase, a protein phosphatase, and a substrate protein (31). Many protein kinases have been identified in nerve tissue, which are dependent on cyclic nucleotides, calcium, or calmodulin. A protein kinase phosphorylates a substrate protein by transferring a terminal phosphate from ATP; the so-formed phosphoprotein changes its function, presumably by changing its structure, and is then reversed through removal of the phosphate group by phosphatases (31). A large number of proteins are regulated by phosphorylation. These include, for example, enzymes involved in neurotransmitter biosynthesis, neurotransmitter receptors, ion channels, cytoskeletal proteins, and proteins involved in regulation of transcription and translation. Several unknown proteins have been detected by phosphorylation, such as the neuron-specific proteins synapsin I or B-50, a substrate for PKC (31). Surprisingly, while phosphorylation of proteins by protein kinases has been long recognized as a primary dynamic regulatory process for posttranslational modifications in the nervous system (31), its possible role in neurotoxicity has only recently received some attention (3,4,127). Some of the recent studies on the effects of neurotoxicants on protein phosphorylation and phosphoproteins are described elsewhere (see O'Callaghan).

   In summary, several steps in chemical neurotransmission can be affected by neurotoxicants. In some cases there is clear evidence that a certain component of neurotransmission represents the primary target for neurotoxicity of a class of compounds. For many other compounds, further studies are needed to draw firm conclusions on the relevance of biochemical effects in their overall neurotoxicity. Furthermore, some aspects of neurotransmission (e.g., signal-transduction mechanisms from the cell surface recognition site to the gene) have not been investigated to their full potential for their possible implications in neurotoxicity.

OTHER NEUROCHEMICAL APPROACHES

Oxidative Stress

   The nervous system has not yet received broad and intensive study of the actions of hydrogen peroxide or derived oxy radicals. Defense mechanisms of nerve cells (e.g., glutathione) have also not received the same degree of attention that they have in other organs, such as the liver. Most of our knowledge of superoxide, peroxide, and hydroxyl radical production comes from studies of catecholaminergic neurotoxins such as 6-hydroxydopamine and 6-aminodopamine (43). A working hypothesis is that an increased generation of hydrogen peroxide formed by oxidative deamination of dopamine by monoamineoxidase might be involved in the neurochemical changes found in Parkinson's disease (43,157). An hypothesis on the mechanism of action of 2-methyl-4-phenyl-1,2,3,6-tetrahydropyridine (MPTP) also involves the generation of oxidative stress (148). Production of reactive oxygen radicals may arise as a result of a redox reaction occurring between MPP$^+$ and MPDP$^+$ (two metabolites of MPTP) (116). This view has received support from observations that administration of antioxidants partially prevented the loss of striatal dopamine in MPTP-treated mice (134), whereas the superoxide dismutase inhibitor diethyldithiocarbamate (DDC) potentiated the effects (47). This potentiating effect of DDC could also be due to an altered biodisposition of MPP$^+$ (88). Furthermore, a strong alternative hypothesis for the mechanisms of MPTP neurotoxicity also exists, involving an inhibition of mitochondrial respiration by MPP$^+$ (148).

Several metals, including lead, thallium, and triethyllead, increase lipid peroxidation in brain, measured by the thiobarbituric acid method (7). Recently, methyl mercury has been shown to increase the rate of formation of oxygen reactive species *in vitro,* particularly in the cerebellum, a brain region which represents a target for methyl mercury morphological damage (100). A novel neurochemical approach used in this last study is the use of 2',7'-dichlorofluorescein diacetate, which, after being hydrolyzed, can be oxidized in the presence of oxygen reactive species to highly fluorescent 2',7'-dichlorofluorescein (99). Using this technique, trimethyltin also has been shown to increase the formation of oxygen reactive species *in vitro* (100).

A major neuroprotective function in nerve tissue is exerted by reduced glutathione (GSH), which detoxifies electrophylic substances via glutathione transferase conjugation reactions, reduces peroxides via GSH peroxidase reactions, and reduces, or conjugates with, aminoquinones formed by oxidation of biogenic amines (157). GSH also acts as a scavenger of free radicals and provides reducing equivalents for reduction of peroxides in the ischemic rat brain (129). Although brain GSH levels are generally quite stable and GSH is not easily depleted (115), recent studies have suggested that modification of GSH homeostasis in the nervous system might play some role in selective neurotoxicity. Manganese, whose neurotoxic effects partially resemble Parkinson's disease, reduces GSH and GSH peroxidase, particularly in the striatum, a dopamine-rich area (103). This is in agreement with the postulated mechanism of action of manganese, which involves an increased production of catecholamine oxidation products (80,84). In brains from parkinsonians and in MPTP-treated mice, a decrease in GSH levels have also been reported (145,167). Other neurotoxicants such as cadmium (153), styrene, and styrene oxide (90,161) have been shown to alter brain levels of glutathione. Oxidative stress might represent a mechanism of neurotoxicity of a number of environmental neurotoxicants (137), which might be relevant in the etiology of certain neurodegenerative diseases. Furthermore, in addition to the aforementioned functions, brain glutathione, which is present in high concentrations in glial cells (135,142,161), appears to be involved in other important intracellular events (e.g., protein synthesis; 131) and extracellular actions (e.g., modulation of glutamate neurotransmission and neurotoxicity; 130,142). The finding that most neuronal cell bodies do not contain appreciable amounts of GSH indicates that other mechanisms must exist in neurons for preventing oxidative damage to their cytoplasmic constituents (135). Ascorbate may provide an alternative to GSH in neuronal somata, and preliminary evidence of this role in the neurotoxicity of cadmium and methamphetamine has been presented (69,152). Further studies in the area of mechanisms of cytotoxicity and cellular protection in an organ like the brain, which consists of nonhomogenous cell types, might offer further insights into the mechanisms of selective neurotoxicity.

Protein Synthesis

The rate of protein synthesis is usually estimated by measuring the rate of incorporation of a radioactive amino acid (most often valine) into proteins that are precipitable by trichloroacetic acid. The large number of variables that can influence measurements of protein synthesis (cerebral blood flow, amino acid transport, body temperature, etc.) indicates that great care should be exerted in utilizing the proper methodology (96). A number of chemicals have been shown to inhibit protein synthesis *in vivo* and/or *in vitro;* these include acrylamide (86), carbon disulfide (147), trimethyltin (57), and methyl mercury (38). Inhibition of protein synthesis does not appear to be involved in the mechanism of neurotoxicity of acrylamide; in fact, methylene bis acrylamide, a nonneurotoxic analog, also exerts an inhibitory effect (150). On the other hand, a more substantial role of protein synthesis inhibition in methyl mercury neurotoxicity might be present, although whether this action is involved to some degree in the neurotoxicity of other compounds is still unknown. The multiplicity of the targets for many neurotoxicants (see also below) complicates the definition of the exact mechanism by which protein synthesis is inhibited. For example, in the case of trimethyltin, its effect on protein synthesis could be direct, or a consequence of its inhibition of ATPase, oxidative phosphorylation, or changes in zinc or ammonia content (57).

PROBLEMS AND LIMITATIONS

It is generally recognized that a number of factors have to be considered in the acquisition and interpretation of neurochemical data of neurotoxicity. It is probably always useful to reiterate that the best application of neurochemical studies is in the investigation of the mechanism(s) of neurotoxicity as an indispensable support to behavioral, electrophysiological, and pathological studies. It should also be remembered that only the full integration of these multidisciplinary approaches can provide satisfactory answers to neurotoxicological questions. A large number of variables can confound neurochemical approaches and these should be considered. For example, body temperature has a profound effect on protein synthesis and if not controlled for, will give false positive results (96). Nutritional status, diet, sex, endocrine status, circadian rhythms, and environmental factors such as housing conditions or handling of animals can all add variables which need to be taken into account very carefully (68).

In a previous section, problems related to the interpretation of neurochemical data obtained with erythrosin B and TMT were reviewed. Here, a series of examples are illustrated for a number of additional important factors: lack of specificity of neurotoxicants, primary versus secondary effects, *in vitro* versus *in vivo* effects, acute versus chronic exposure, and age.

One of the great difficulties in studying and interpreting the effects of neurotoxicants on neurotransmission is the multiplicity of their targets. Metals, for example, are a class of compounds which have been shown to affect several of the parameters of neurotransmission on which they have been tested. Mercury is chosen as an example of multiplicity of effects, but the same considerations apply to other neurotoxic metals, lead in particular (166). There is no doubt that mercury, particularly in its organic form, is a potent neurotoxicant (36). One of the main effects believed to be involved in its neurotoxicity appears to be inhibition of protein and RNA synthesis (38). However, its high affinity for sulfhydro groups renders it very reactive with a variety of enzymes and other proteins, some of which are involved in neurotransmission. In a number of experiments, mercury, both inorganic and as methyl mercury, both *in vitro* and *in vivo,* has been shown to affect such diverse processes as GABA, dopamine, and choline uptake; acetylcholinesterase, cholineacetyltransferase, monoamine-oxidase, muscarinic, nicotinic, beta-adrenergic, and benzodiazepine receptors; and protein kinase C, to cite only a few (see ref. 52 for a full reference list). It is apparent that the multiple interactions of mercury with these systems complicate an assessment of the relative relevance of each effect. Possibly, a detailed re-evaluation of the pathological and behavioral effects of mercury and/or methyl mercury would be useful for discerning the relevant biochemical mechanisms from less important neurochemical effects.

In several situations, observed effects on neurotransmission might be the result of nonspecific actions of neurotoxicants on the cell membrane or on cellular metabolism (70). For example, several solvents are known to alter neurotransmitter receptors or the levels of neurotransmitters and/or their metabolites. Yet it is unclear whether there are specific targets for the solvents, or whether modifications of these and other parameters could be due, for example, to nonspecific alterations in membrane fluidity, which would alter the lipid microenvironment of several macromolecules, including uptake systems and receptors (42). Similarly, compounds which affect the cell's energy production could perturb any of the energy-dependent processes in the neurons; for example, inhibition of oxidative phosphorylation by organotins is possibly involved in several of their observed neurochemicals effects (5).

A loss of various parameters of neurotransmission in certain brain areas would most probably be found following administration of neurotoxicants which cause neuronal death through yet unspecified mechanism(s), as in the case of trimethyltin or MPTP, for example (94,148). While neurochemical measurements would in this case be very useful "biomarkers" of neurotoxicity and would offer information for correlative behavioral studies, the observed neurochemical changes should be seen as secondary events and the primary target(s), those responsible for the neuronal necrosis, might remain elusive.

Many neurochemical studies are carried out *in vitro* and often the neurotoxicant is used at rather high concentrations, in the micromolar to millimolar range. Any positive finding should always be confirmed by the determination of the same effects following *in vivo* administration of a given neurotoxicant and/or by determination of its level in brain tissue. An example of this potential problem is given by a recent study in which the interaction of a number of known neurotoxicants with the Na^+/K^+-ATPase was investigated (108). Inhibition of the sodium pump by triethyltin (TET) has been suggested to be involved in TET-induced brain edema in the white matter (165), and indeed TET was shown to inhibit rat brain Na^+/K^+-ATPase *in vitro* confirming previous reports (48); however, following administration of a high neurotoxic dose of TET, no inhibition of brain Na^+/K^+-ATPase activity could be detected, also in accordance with previous results (160). Dialysis experiments demonstrated that dilution from tissue preparation was not responsible for the lack of detectable inhibition (108). However, when the average total tin level in brain following administration of 10 mg/kg was measured, it was found to be only 10 μM, again consistent with the levels of tin observed in previous studies (44,45), but not sufficient to inhibit Na^+/K^+-ATPase, as inferred from concentration-response studies *in vitro*. Thus, whether inhibition of the pump represents a primary mechanism for TET neurotoxicity remains to be determined.

In several cases acute and chronic exposures to neurotoxicants may generate different biochemical effects. For example, acute exposure to an organophosphate will cause significant inhibition of acetylcholinesterase, which is readily associable with cholinergic signs of intoxication (122). In chronically exposed animals, however, the same degree of cholinesterase inhibition may not be associated with signs of toxicity, since tolerance has developed (63). Furthermore, changes in muscarinic and nicotinic receptors are present following chronic exposure to organophosphates, but not after acute administration (64). These receptor changes represent for the most part a homeostatic adaptive response which may contribute to tolerance. While contributing to the adaptation of the organism to organophosphate intoxication, these receptor changes may underlie other problems, such as altered responses to cholinergic drugs (63,64) or possible cognitive disorders (117).

A very important variable in the study of any neurochemical effects is age. Many toxic substances appear to be most damaging during early development and again during senescence. The reasons for this could be multiple. The neurochemical profile of the brain changes with development: receptors, uptake systems, enzymes, and second messenger systems vary with age. For the most part, there is an increase in the content and/or functions of these parameters which often take place at different rates in different brain regions. However, there are also systems which show a decline associated with the development of the brain, such as, for example, the metabolism of phosphoinositides activated by glutamate or muscarinic receptors (53). The particular susceptibility of the newborn to neurotoxicants *in vivo* can often be ascribed to the lack of a well-developed blood-brain barrier or to a diminished ability to detoxify neurotoxicants. Sometimes *in vitro* experiments also reveal intrinsic higher sensitivity of the developing brain to neurotoxicants. Such appears to be the case, for example, for the muscarinic receptor-stimulated phosphoinositide metabolism which is significantly affected by ethanol *in vitro,* as well as *in vivo,* in the neonatal rat but not in adult animals (15). Disruption of the delicate balance of brain mechanisms during development often leads to long-term effects which persist well after termination of exposure. For example, rats exposed to lead only through the dam's milk from parturition to weaning (day 21), and tested on day 90, when blood and brain lead levels were down to control values, displayed a supersensitivity to the disrupting effects of scopolamine on the visual system and a parallel selective decrease of muscarinic receptors in the visual cortex (55,77).

The aging process is also accompanied by changes which might affect neurotoxicity and neurochemical measurements. Altered patterns of xenobiotic and drug metabolizing enzymes might change the bioavailability of neurotoxicants (105). Reactions causing oxidative stress increase with age, while protective mechanism (e.g., glutathione) decrease in the aging brain (20,66,144). Synthesis of protein and RNA, cerebral energy metabolism, and many parameters of neurotransmission also show an age-related decline (10,75,156), and this could lead to an altered response to neurotoxicants. For example, aged rats do not recover as well as young ones from the changes in muscarinic receptor which result from prolonged exposure to organophosphorus insecticides (136,138), possibly because of a reduced rate of receptor resynthesis.

BIOCHEMICAL MARKERS FOR NEUROTOXICITY ASSESSMENT IN HUMANS

Progress in biochemical toxicology has led to the discovery of molecular targets and mechanisms of action of many environmental chemicals. Nevertheless, the knowledge of the biochemical changes occurring in humans exposed to environmental and occupational toxicants has been limited. During the last few years great attention has been given to the concept of "biomarkers," indicators of exposure and/or health effects which could be used (preferably by minimally invasive procedures) in humans. Although the term biomarker is often overused, its real collocation should be in the feasibility of its application to biochemical and molecular epidemiology studies in exposed populations. Despite its obvious importance within toxicology, the area of neurotoxicity seems to be progressing more slowly than other fields with regard to biological monitoring and markers of exposure and of health effects (26,50). The complexity of the nervous system and its distinctive peculiarities together with problems associated with the determination of the precise target for neurotoxic action are certainly responsible for this limited advancement. Studies in neuroepidemiology are most often relying on neurobehavioral testing (8,126) or on electrophysiological measurements (102,151). The use of novel noninvasive techniques, such as computerized imaging, may be of a certain usefulness in particular neurotoxicity cases, however, their widespread application to neurotoxicological investigations is doubtful (17). Neurochemical measurements for detecting neurotoxicity in humans are limited by the inaccessibility of target tissue. Thus, a necessary approach for identifying and characterizing neurotoxicity is the search for neurochemical parameters in peripheral tissues, which can be easily and ethically obtained in humans and which could represent a marker for the same parameters in nerve tissue.

A neurochemical approach for studying neurotoxicity in humans could be justified by the progression of neurotoxic damage. The events leading to neurotoxic effects by exogenous chemicals could be divided into various phases: (a) penetration of the neurotoxicant into the organism and distribution to reach the target cells; (b) interaction of the neurotoxicants with the critical targets; (c) induction of biochemical and functional modification of the nerve cells; and (d) appearance of structural and/or functional changes in the nervous system (114). Thus, neurotoxic injury is thought to proceed from one or more primary biochemical alterations through reversible cellular injury, to eventual irreversible cytotoxicity and permanent organic lesions (24). Markers for neurotoxic effects should, therefore, provide reliable indications of early-stage effects related to biochemical alterations that might precede irreversible cell damage.

The most conventional tests, exploring phase (a), consist of measuring the toxicant in body fluids, typically blood or urine, to predict the possibility of occurrence of adverse effects in the nervous system. In order to get a better understanding of the target dose, *in vivo* dosimetry techniques are being developed. Binding to macromolecules, particularly covalent adducts to hemoglobin, have proven useful in monitoring exposure to genotoxic agents in order to determine the dose of toxic compound which escapes detoxication and reaches its target protein or DNA (73). An application of this approach to neurotoxicology is represented by acrylamide, whose adduct to hemoglobin has been recently characterized (13,21,32). Measurement of hemoglobin adducts has also been proposed as a way to monitor *in vivo* doses of methyl bromide (89). Similarly, it would be possible to monitor pyrrole adducts of gamma-diketones in some serum or erythrocyte protein as a measure of exposure to *n*-hexane or methyl *n*-butyl ketone (9).

Since only a limited number of compounds might exert neurotoxicity by forming covalent bonds with specific proteins, another line of research aims at identifying and validating peripheral markers which could be used as "surrogate indicators" of the neurochemical targets in the central and/or peripheral nervous system (158). The best example of such strategy can be found in the field of organophosphorus pesticides. Innumerable animal and human studies have confirmed the validity and usefulness of red cell and plasma cholinesterases as peripheral markers for detecting organophosphate exposure and effects (122). Additionally, for those organophosphates causing delayed polyneuropathy, measurement of the activity of neuropathy target esterase (NTE) in lymphocytes appears to offer a useful marker for biological monitoring and for predicting purposes (106).

Several additional biochemical parameters of neurotransmission are present in blood cells; these include receptors (e.g., cholinergic muscarinic and beta-adrenergic in lymphocytes, alpha$_2$-adrenergic receptors in platelets), enzymes (e.g., monoamineoxidase B in platelets, Na^+/K^+-ATPase in erythrocytes), or uptake systems (e.g., serotonin uptake in platelets). Changes in these parameters might reflect similar changes occurring in the nervous system (50). For example, as discussed earlier, chronic exposure of rats to organophosphates is known to cause a decrease in the density of brain muscarinic receptors which has been associated with cognitive disorders (52,117). A similar muscarinic receptor change has been found recently in circulating lymphocytes (59). Alterations of beta-adrenoceptors have been observed in brain capillaries of lead-exposed rats (83), as well as in lymphocytes from lead-exposed workers (133). Activity of the enzyme dopamine-beta-hydroxylase, a putative target for carbon disulfide neurotoxicity (34,119), was reported to be lower than control in serum of a group of occupationally exposed workers (164). Exposure-related changes in monoamineoxidase B activity and serotonin uptake in platelets have also been found in workers exposed to neurotoxic solvents (23,37).

An additional area in which peripheral measurements might be useful in biochemical and molecular neuroepidemiology is that addressing the issue of "individual susceptibility." Genetically determined and acquired variations in metabolism or biodisposition might explain interindividual variability in neurotoxic responses. For example, the activity of plasma paraoxonase, an enzyme which detoxifies paraoxon and other organophosphorus insecticides, is polimorphically distributed in most human populations, with almost half of the people displaying low activity (78). This has generated the hypothesis that people with low serum paraoxonase activity might be more susceptible to the neurotoxic effects of organophosphates. Recent animal studies showing that levels of plasma paraoxonase can modulate the toxicity of paraoxon and chlorpyrifos-oxon (61) have provided preliminary confirmation of this hypothesis. Various defects in metabolism have been found in parkinsonian patients and it has been suggested that these might lead to changes in detoxication pathways, making Parkinson's disease patients more susceptible to exogenous or endogenous toxins (159).

Each of these neurochemical approaches to investigating neurotoxicity in humans obviously has some shortcomings. For example, adduct measurements are mostly compound-specific, and can be considered markers of exposure since they do not provide information on the biochemical consequences of such binding. Also, the methods used for measuring adducts might be cumbersome (particularly if gas chromatography–mass spectrometry is utilized) and not easily applicable to large epidemiological investigations. Use of markers of effects, such as parameters of neurotransmission in blood as "surrogate markers," is also limited to investigations of compounds which affect neurotransmission. They could also generate "false positives" (e.g., compounds that do not cross the blood-brain barrier). Most importantly, this area requires much characterization of these peripheral markers and validation of their suitability as "mirrors" of the nervous system.

On the more positive side, it seems clear that as our knowledge of the mechanism of action of neurotoxicants and their molecular targets increases, new biomarkers of effect will most certainly be developed. However, because of the multiplicity of the potential targets for neurotoxicants, some aspects of neurotoxicity cannot be identified by the proposed methods. For example, presently no specific biochemical measurements have been identified that may represent a surrogate marker for detecting effects of neurotoxicants on axoplasmic transport. In summary, the use of biochemical markers for detection of neurotoxicity in human populations is still at a very early stage and few opportunities are available for biochemical monitoring at this time. However, the increasing concern about environmental and occupational exposure to neurotoxicants and their possible role in the etiology of some neurodegenerative diseases will, it is hoped, foster the development of new biochemical measurements to be used in epidemiological studies.

CONCLUSION

Neurochemical studies have provided neurotoxicologists with much information on the mechanism of action of neurotoxicants, despite the complex and difficult task of determining the exact and relevant molecular target underlying responses of the nervous system to toxic insults. Further progress in this area will likely be linked to progress in the molecular neurosciences. Growing research on novel areas such as growth factors, gene regulation, immune system–nervous system interactions, and genetic susceptibility will offer a new input of ideas for neurochemical studies in neurotoxicology. An increased utilization of molecular biology techniques will complement and integrate current neurochemical approaches and greatly contribute to advances in neurotoxicology.

ACKNOWLEDGMENTS

The author's work was supported by grants from NIOSH, NIEHS, EPA, the Alcohol and Drug Abuse Institute of the University of Washington, and the Fondazione Clinica del Lavoro, Pavia. Ms. Chris Sievanen provided secretarial assistance.

REFERENCES

1. Abalis IM, Eldefrawi ME, Eldefawi AT. High affinity stereospecific binding of cyclodiene insecticides and gammahexachlorocyclohexane to gamma-aminobutric acid receptors in rat brain. *Pestic Biochem Physiol* 1985;24:95–102.
2. Abalis IM, Eldefrawi ME, Eldefrawi AT. Effects of insecticides on GABA-induced chloride influx into rat brain microsacs. *J Toxicol Environ Health* 1986;18:13–23.
3. Abou-Donia MB, Lapadula DM, Carrington CD. Biochemical methods for assessment of neurotoxicity. In: Ballantyne B, ed. *Perspectives in basic and applied toxicology.* London: Wright, 1988;1–30.
4. Abou-Donia MB, Patton SE, Lapadula DM. Possible role of endogenous protein phosphorylation in organophosphorus compound-induced delayed neurotoxicity. In: Narashashi T, ed. *Cellular and molecular neurotoxicology.* New York: Raven Press, 1984;265–283.

5. Aldridge WN, Street BW. Oxidative phosphorylation: the relation between the specific binding of trimethyltin and triethyltin to mitochondria and their effects on various mitochondrial functions. *Biochem J* 1971;124:221–234.
6. Aldridge WN, et al. The effect of DDT and pyrethroids cismethrin and decamethrin on the acetylcholine and cyclic nucleotide content of rat brain. *Biochem Pharmacol* 1978;27:1703–1706.
7. Ali SF, Bondy SC. Triethyllead-induced peroxidative damage in various regions of the rat brain. *J Toxicol Environ Health* 1989;26:235–242.
8. Anger WK. Worksite behavioral research: results, sensitive methods, test batteries and the transition from laboratory data to human health. *Neurotoxicology* 1990;11:629–720.
9. Anthony DC, Boekelheide K, Anderson CW, Graham DG. The effect of 3,4-dimethyl substitution on the neurotoxicity of 2.5-hexanedione. *Toxicol Appl Pharmacol* 1983;71:371–382.
10. Arauyo DM, Lapchack PA, Meaney MJ, Collier B, Quirion A. Effects of aging on nicotinic and muscarinic autoreceptor function in the rat brain: relationship to presynaptic cholinergic markers and binding sites. *J Neurosci* 1990;10:3069–3078.
11. Atchison WA, Clark AW, Narahashi T. Presynaptic effects of methylmercury at the mammalian neuromuscular junction. In: Narahashi T, ed. *Cellular and molecular neurotoxicology.* New York: Raven Press, 1984;23–42.
12. Augustine GJ, Charlton MP, Smith SJ. Calcium action in synaptic transmitter release. *Ann Rev Neurosci* 1987;10:633–693.
13. Bailey E, Farmer PB, Bird J, Lamb JH, Peal JA. Monitoring exposure to acrylamide by determination of S-(2-carboxyethyl) cysteine in hydrolyzed hemoglobin by gas chromatography-mass spectrometry. *Anal Biochem* 1986;157:241–248.
14. Bakry NMS, El-Rashiday AM, Eldefrawi AT, Eldefrawi ME. Direct actions of organophosphate anticholinesterases on nicotinic and muscarinic acetylcholine receptors. *J Biochem Toxicol* 1988;3:235–259.
15. Balduini W, Costa LG. Developmental neurotoxicity of ethanol: *in vitro* inhibition of muscarinic receptor-stimulated phosphoinositide metabolism in brain from neonatal but not adult rats. *Brain Res* 1990;512:248–252.
16. Balduini W, Costa LG. Effects of ethanol on muscarinic receptor-stimulated phosphoinositide metabolism during brain development. *J Pharmacol Exp Ther* 1989;250:541–547.
17. Barrio JR, Huang SC, Phelps ME. In vivo assessment of neurotransmitter biochemistry in humans. *Ann Rev Pharmacol Toxicol* 1988;28:213–230.
18. Beeman RW, Matsumura F. Chlordimeform: a pesticide acting upon amine regulatory mechanisms. *Nature* 1973;242:273–274.
19. Benezet JJ, Chang KM, Knowles CO. Formamidine pesticides: metabolic aspects of neurotoxicity. In: Hankland DL, Hollingworth RM, Smith T, eds. *Pesticides and venom neurotoxicity.* New York: Plenum Press, 1988;189–206.
20. Benzi G, Pastoris O, Marzatico F, Villa RF. Age-related effect induced by oxidative stress on the cerebral glutathione system. *Neurochem Res* 1989;14:473–481.
21. Bergmark E, Calleman CJ, Costa LG. Formation of hemoglobin adducts of acrylamide and its epoxide metabolite glycidamide in the rat. *Toxicol Appl Pharmacol* (in press).
22. Berridge MJ. Inositol triphosphate and diacylglycerol: two interacting second messengers. *Ann Rev Biochem* 1987;56:159–193.
23. Beving H, Kristensson J, Malgrem R, Olsson P, Unge G. Effect on the uptake kinetics of serotonin (S-hydroxytryptamine) in platelets from workers with longterm exposure to organic solvents. *Scand J Work Environ Health* 1984;10:229–234.
24. Blecker ML, Agnew J. The assessment of dose-response relationship for low-level exposure to neurotoxicants in man. In: Foa V, Emmett EA, Maroni M, Columbi A, eds. *Occupational and environmental chemical hazards.* Chichester UK: Ellis Horwood, 1987;508–515.
25. Bodnaryk RP. The effects of pesticide and related compounds on cyclic nucleotides metabolism. *Insect Biochem* 1982;12:589–597.
26. Bondy SC. Especial considerations for neurotoxicological research. *CRC Crit Rev Toxicol* 1985;14:381–402.
27. Bondy SC. The biochemical evaluation of neurotoxic damage. *Fund Appl Toxicol* 1986;6:208–216.
28. Bondy SC. Intracellular calcium and neurotoxic events. *Neurotoxicol Teratol* 1989;11:527–531.
29. Boyes W, Moser VC, MacPhail RC, Dyer RS. Monoamineoxidase inhibition cannot account

for changes in visual evoked potentials produced by chlordimeform. *Neuropharmacology* 1985;24: 853–860.
30. Boyes WK, Moser VC. An alpha$_2$-adrenergic mode of action of chlordimeform on rat visual function. *Toxicol Appl Pharmacol* 1988;92:402–418.
31. Browning MD, Huganir R, Greengard P. Protein phosphorylation and neuronal function. *J Neurochem* 1985;45:11–23.
32. Calleman CJ, Bergmark E, Costa LG. Acrylamide is metabolized to glycidamide in the rat: evidence from hemoglobin adduct formation. *Chem Res Toxicol* 1990;3:406–412.
33. Candura SM, Castoldi AF, Manzo L, Costa LG. Interaction of aluminum ions with phospho-inositide metabolism in rat cerebral cortical membranes. *Life Sci* 1991 (in press).
34. Caroldi S, Jarvis JAE, Magos L. In vivo inhibition of dopamine beta hydroxylase in rat adrenals during exposure to carbon disulphide. *Arch Toxicol* 1986;55:265–267.
35. Casida JE. Esterase inhibitors as pesticides. *Science* 1964;146:1011–1017.
36. Chang LW. Neurotoxic effect of mercury. A review. *Environ Res* 1977;14:329–373.
37. Checkoway H, Costa LG, Camp J, Coccini T, Daniell W, Dills R. Peripheral markers of neu-rochemical function among styrene-exposed workers (*submitted*).
38. Cheung MK, Verity MA. Experimental methylmercury neurotoxicity: similar in vivo and in vitro perturbation of brain cell-free protein synthesis. *Exp Mol Pathol* 1983;38:230–242.
39. Choi DW. Glutamate neurotoxicity in cortical cell cultures is calcium dependent. *Neurosci Lett* 1985;58:293–297.
40. Chuang DM. Neurotransmitter receptors and phosphoinositide turnover. *Ann Rev Pharmacol Toxicol* 1989;29:71–90.
41. Churchill L, Pazdernik TL, Samson F, Nelson SR. Topographical distribution of down-regulated muscarinic receptors in rat brains after repeated exposure to disopropyl phosphoroflouridate. *Neuroscience* 1984;11:463–472.
42. Ciofalo F. Effects of some membrane perturbers on alpha$_1$-adrenergic receptor binding. *Neurosci Lett* 1981;21:313–318.
43. Cohen G. Oxidative stress in nervous system. In: Sies H, ed. *Oxidative stress.* New York: Academic Press, 1985;383–402.
44. Cook LL, Stine K, Reiter LW. Tin distribution in adult tissues after exposure to triethyltin and trimethyltin. *Toxicol Appl Pharmacol* 1984a;76:344–348.
45. Cook LL, Stine-Jacobs K, Reiter LW. Tin distribution in adult and neonatal rat brain following exposure to triethyltin. *Toxicol Appl Pharmacol* 1984b;72:75–81.
46. Cooper GP, Suszkiw JB, Manalis RS. Heavy metals: effects on synaptic transmission. *Neuro-toxicology* 1984;5:247–266.
47. Corsini GU, Pintus S, Chiuch CC, Weiss JF, Kopin IJ. 1-Methyl-4-phenyl-1,2,3,6-tetrahydro-pyridine (MPTP) neurotoxicity in mice is enhanced by pretreatment with diethyldithiocarbamate. *Eur J Pharmacol* 1985;119:127–128.
48. Costa LG. Inhibition of ^{3}H-GABA uptake by organotin compounds in vitro. *Toxicol Appl Pharmacol* 1985;79:471–479.
49. Costa LG. Interaction of insecticides with the nervous system. In: Costa LG, Galli CL, Murphy SD, eds. *Toxicology of pesticides: experimental clinical and regulatory perspectives.* Heidelberg: Springer-Verlag, 1987;77–91.
50. Costa LG. Peripheral models for the study of neurotransmitter receptors: their potential appli-cations to occupational health. In: Foa V, Emmett EA, Maroni M, Colombi A, eds. *Occupational and environmental chemical hazard.* Chichester, UK: Ellis Horwood, 1987;524–528.
51. Costa LG. Toxicology of pesticides: a brief history. In: Costa TG, Galli CL, Murphy SD, eds. *Toxicology of pesticides: experimental, clinical and regulatory perspectives.* Heidelberg: Springer-Verlag, 1987;1–10.
52. Costa LG. Interaction of neurotoxicants with neurotransmitter systems. *Toxicology* 1988;49: 359–366.
53. Costa LG. The phosphoinositide/protein kinase C system as a potential target for neurotoxicity. *Pharmacol Res* 1990;22:393–408.
54. Costa LG. Role of second messenger systems in response to organophosphates. In: Chambers JE, Levi PE, eds. *Organophosphates: chemistry, fate and effects.* New York: Academic Press (*in press*).
55. Costa LG, Fox DA. A selective decrease of cholinergic muscarinic receptors in the visual cortex of adult rats following developmental lead exposure. *Brain Res* 1983;276:259–266.

56. Costa LG, Murphy SD. ³H-Nicotine binding in rat brain: alteration after chronic acetylcholinesterase inhibition. *J Pharmacol Exp Ther* 1983;226:392–397.
57. Costa LG, Sulaiman R. Inhibition of protein synthesis by trimethyltin. *Toxicol Appl Pharmacol* 1986;86:189–196.
58. Costa LG, Doctor SV, Murphy SD. Antinociceptive and hypothermic effect of trimethyltin in mice. *Life Sci* 1982;31:1093–1102.
59. Costa LG, Kaylor G, Murphy SD. *In vitro* and *in vivo* modulation of cholinergic muscarinic receptors in rat lymphocytes and brain by cholinergic agents. *Int J Immunopharmacol* 1990;12: 67–76.
60. Costa LG, Marinovich M, Galli CL. Receptor binding techniques in neurotoxicology. In: Galli CL, Manzo L, Spencer PS, eds. *Recent advances in nervous system toxicology.* New York: Plenum Press, 1988;307–349.
61. Costa LG, McDonald BE, Murphy SD, Omenn GS, Richter RJ, Motulsky AG, Furlong CE. Serum paraoxonase and its influence on paraoxon and chlorpyrifosoxon toxicity in rats. *Toxicol Appl Pharmacol* 1990;103:66–76.
62. Costa LG, Olibet G, Murphy SD. Alpha₂-adrenoceptors as a target for formamidine pesticides: in vitro and in vivo studies in mice. *Toxicol Appl Pharmacol* 1988;93:319–328.
63. Costa LG, Schwab BW, Murphy SD. Tolerance to anticholinesterase compounds in mammals. *Toxicology* 1982;25:79–97.
64. Costa LG, Schwab BW, Murphy SD. Differential alterations of cholinergic muscarinic receptors in acute and chronic tolerance to organophosphorus insecticides. *Biochem Pharmacol* 1982;31: 3407–3413.
65. Costa LG, Wu DS, Olibet G, Murphy SD. Formamidine pesticides and alpha₂-adrenoceptors: studies with amitraz and chlordimeform in rats and development of a radioreceptor binding assay. *Neurotoxicol Teratol* 1989;11:405–411.
66. Cutler RG. Peroxide-producing potential of tissues: inverse correlation with longevity of mammalian species. *Proc Natl Acad Sci USA* 1985;82:4798–4802.
67. Damstra T, Bondy SC. The current status and future of biochemical assays for neurotoxicity. In: Spencer PS, Schaumburg HH, eds. *Experimental and clinical neurotoxicology.* Baltimore: Williams and Wilkins, 1980;820–833.
68. Damstra T, Bondy SC. Neurochemical approaches to the detection of neurotoxicity. In: Mitchell CL, ed. *Nervous system toxicology.* New York: Raven Press, 1982;349–373.
69. De Vito MJ, Wagner GC. Functional consequences following methamphetamine-induced neuronal damage. *Psychopharmacology* 1989;97:432–435.
70. DeHaven DL, Mailman RB. The interaction of behavior and neurochemistry. In: Annau Z, ed. *Neurobehavioral toxicology.* Baltimore: Johns Hopkins University Press, 1986;214–243.
71. Desaiah D. Chlordecone interaction with catecholamine binding and uptake in rat brain synaptosomes. *Neurotoxicology* 1985;6:159–166.
72. Doctor SV, Costa LG, Kendall DA, Murphy SD. Trimethyltin inhibits uptake of neurotransmitters into mouse forebrain synaptosomes. *Toxicology* 1982;25:213–221.
73. US Environmental Protection Agency. Protein adduct forming chemicals for exposure monitoring: literature summary and recommendation. 1990;134.
74. Evans PD, Gee JD. Action of formamidine pesticides on octopamine receptors. *Nature* 1980;287: 60–62.
75. Finch CE, Morgan DG. RNA and protein metabolism in the aging brain. *Ann Rev Neurosci* 1990;13:75–87.
76. Fisher SK, Agranoff BW. Receptor activation and inositol lipid hydrolysis in neural tissues. *J Neurochem* 1987;48:999–1017.
77. Fox DA, Wright AA, Costa LG. Visual acuity defects following neonatal lead exposure: cholinergic interactions. *Neurobehav Toxicol Teratol* 1982;4:689–693.
78. Furlong CE, Richter RJ, Seidel SL, Motlusky AG. Role of genetic polymorphism of human plasma paraoxonase/arylesterase in hydrolysis of the insecticide metabolites chlorpyrifos oxon and paraoxon. *Am J Hum Genet* 1988;43:230–238.
79. Gammon DW, Sanders G. Two mechanisms of pyrethroid action: electrophysiological and pharmacological evidence. *Neurotoxicology* 1985;6:63–86.
80. Garner CD, Nachtman JP. Manganese-catalyzed auto-oxidation of dopamine to 6-hydroxydopamine in vitro. *Chem Biol Interact* 1989;69:345–351.
81. Garthwaite G, Garthwaite J. Aminoacid neurotoxicity: intracellular sites of calcium accumulation

associated with the onset of irreversible damage to rat cerebellum neurons in vitro. *Neurosci Lett* 1986;71:53–58.

82. Ghiasuddin SM, Soderlund DM. Pyrethroid insecticides: potent stereospecific enhancers of mouse brain sodium channel activation. *Pestic Biochem Physiol* 1985;24:200–206.

83. Govoni S, Lucchi L, Magnoni MS, Spano PF, Trabucchi M. Decreased density of beta-receptors in brain capillaries of lead-exposed rats. In: Hayes AW, Schnell CC, Miya TS, eds. *Developments in the science and practice of toxicology.* Amsterdam: Elsevier, 1983;521–524.

84. Graham DG. Catecholamine toxicity: a proposal for the molecular pathogenesis of manganese neurotoxicity and Parkinson's disease. *Neurotoxicology* 1984;5:83–96.

85. Gusovsky F, Secundo SI, Daly JW. Pyrethroids: involvement of sodium channels in effects on inositol phosphate formation in guinea pig synaptoneurosomes. *Brain Res* 1989;692:72–78.

86. Hashimoto K, Ando K. Alteration of aminoacid incorporation into proteins of the nervous system *in vitro* after administration of acrylamide to rats. *Biochem Pharmacol* 1973;22:1057–1066.

87. Hsu WH, Kakuk TJ. Effect of amitraz and chlordimeform on heart rate and pupil diameter in rats: mediated by alpha$_2$-adrenoceptors. *Toxicol Appl Pharmacol* 1984;73:441.

88. Irwin I, Wu EY, DeLanney LE, Trevor A, Langston JW. The effect of diethyldithiocarbamate on the biodisposition of MPTP: an explanation for enhanced neurotoxicity. *Eur J Pharmacol* 1987;141:209–217.

89. Iwasaki K. Individual differences in the formation of hemoglobin adducts following exposure to methyl bromide. *Indust Health* 1988;26:257–262.

90. Katoh T, Higashi K, Inoue N. Sub-chronic affects of styrene and styrene oxide on lipid peroxidation and the metabolism of glutathione in rat liver and brain. *J Toxicol Sci* 1989;16:1–9.

91. Katz LS, Marquis JK. Modulation of central muscarinic receptor binding in vitro by ultralow levels of the organophosphate paraoxon. *Toxicol Appl Pharmacol* 1989;101:114–123.

92. Komulainen H, Bondy SC. Increased free intracellular Ca^{2+} by toxic agents: an index of potential neurotoxicity. *Trends Pharmacol Sci* 1988;9:154–156.

93. Komulainen H, Tuomisto J. Effects of heavy metals on monoamine uptake and release in brain synaptosomes and blood platelets. *Neurobehav Toxicol Teratol* 1982;4:647–649.

94. Krigman MR, Cranmer JM. (eds.) Mechanisms and manifestations of tin neurotoxicity. *Neurotoxicology* 1984;5:127–299.

95. Laduron PM. Criteria for receptor sites in binding studies. *Biochem Pharmacol* 1984;33:833–839.

96. Lajtha A, Dunlop D. Turnover of protein in the nervous system. *Life Sci* 1981;29:755–767.

97. Lawrence LJ, Casida JE. Interaction of lindane, toxaphene and cyclodienes with brain specific t-butylbyclophosphorothionate receptors. *Life Sci* 1984;35:171–178.

98. Lawrence LJ, Casida JE. Stereospecific action of pyrethroid insecticides on the gamma-aminobutyric acid receptor-ionophore complex. *Science* 1983;221:1399–1401.

99. LeBel CP, Bondy SC. Sensitive and rapid quantitation of oxygen reactive species formation in rat synaptosomes. *Neurochem Int* 1990;17:435–440.

100. LeBel CP, Ali SF, McKee M, Bondy SC. Organometal-induced increases in oxygen reactive species: the potential of 2',7' dichloroflourescein diacetate as an index of neurotoxic damage. *Toxicol Appl Pharmacol* 1990;104:17–24.

101. Lefferman JA, Silbergeld EK. Erythrosin B inhibits dopamine transport in rat caudate synaptosomes. *Science* 1979;205:410–412.

102. LeQuesne PM. Clinically used electrophysiological end-points. In: Lowndes HE, ed. *Electrophysiology in neurotoxicology.* Boca Raton, FL CRC Press, 1987;103–116.

103. Liccione JJ, Maines MD. Selective vulnerability of glutathione metabolism and cellular defense mechanisms in rat striatum to manganese. *J Pharmacol Exp Ther* 1988;247:156–161.

104. Liu DD, Watanabe HK, Ho IK, Hoskins B. Acute effects of soman, sarin and tabun on cyclic nucleotide metabolism in rat striatum. *J Toxicol Env Health* 1986;19:23–32.

105. Loi CM, Vestal RE. Drug metabolism in the elderly. *Pharmacol Ther* 1988;36:131–149.

106. Lotti M. Organophosphate-induced delayed polyneuropathy in humans: perspectives for biomonitoring. *Trends Pharmacol Sci* 1987;8:175–178.

107. Loullis CC, Dean RL, Lippa AS, Clody DE, Coupet J. Hippocampal muscarinic receptor loss following trimethyltin administration. *Pharmacol Biochem Behav* 1985;22:147–151.

108. Maier WE, Costa LG. Na$^+$/K$^+$-ATPase in rat brain and erythrocytes as a possible target and

marker, respectively, for neurotoxicity: studies with chlordecone, organotins and mercury compounds. *Toxicol Lett* 1990;51:175–188.

109. Mailman RB. Neurotoxicants and membrane-associated functions. In: Hodgson E, Bend JR, Philpot RM, eds. *Reviews in biochemical toxicology,* vol. 4. Amsterdam: Elsevier, 1982;213–255.

110. Mailman RB. Mechanisms of CNS injury in behavioral dysfunction. *Neurotoxicol Teratol* 1987;9:417–426.

111. Mailman RB, Mileson BE, Lewis MH. Neurotoxicity expressed through alterations of cell-cell interaction. In: Milman HA, Elmore E, eds. *Biochemical mechanisms and regulation of intercellular communication.* Princeton: Princeton Scientific, 1987;87–111.

112. Mailman RB, DeHaven DL. Responses of neurotransmitter systems to toxicant exposure. In: Narahashi T, ed. *Cellular and molecular neurotoxicology.* New York: Raven Press, 1984;207–224.

113. Markovac J, Goldstein GW. Picomolar concentrations of lead stimulate brain protein kinase C. *Nature* 1988;334:71–73.

114. Maroni M. Summary report. Biochemical indices of nervous tissue toxicity and exposure to neurotoxicants. In: Foa V, Emmett EA, Maroni M, Colombi A, eds. *Occupational and environmental chemical hazards.* Chichester UK: Ellis Horwood, 1987;539–543.

115. Masukawa T, Sai M, Tochino Y. Methods for depleting brain glutathione. *Life Sci* 1989;44:417–424.

116. McCrodden JM, Tipton KF, Sullivan JP. The neurotoxicity of MPTP and the relevance to Parkinson's disease. *Pharmacol Toxicol* 1990;67:8–13.

117. McDonald BE, Costa LG, Murphy SD. Spatial memory impairment and central muscarinic receptor loss following prolonged treatment with organophosphates. *Toxicol Lett* 1988;40:47–56.

118. McDonald LJ, Mamrack MD. Aluminum affects phosphoinositide hydrolysis by phosphoinositide C. *Biochem Biophys Res Commun* 1988;155:203–208.

119. McKenna MJ, Distefano V. Carbon disulfide II A. A proposed mechanism for the action of carbon disulfide on dopamine beta hydroxylase. *J Pharmacol Exp Ther* 1977;202:253–266.

120. Mei L, Roeske WR, Yamamura HI. Molecular pharmacology of muscarinic receptor heterogeneity. *Life Sci* 1989;45:1831–1851.

121. Moser GI, Smart RC. Hepatic tumor promoting chlorinated hydrocarbons stimulate protein kinase C activity. *Toxicologist* 1989;9:125.

122. Murphy SD. Toxicology of pesticides. In: Klaassen CD, Amdur MO, Doull J, eds. *Casarett and Doull's Toxicology: the basic science of poisons.* New York: MacMillan, 1986;519–581.

123. Murphy SD, Costa LG, Wang C. Organophosphate insecticide interaction at primary and secondary receptors. In: Narashashi T, ed. *Cellular and molecular neurotoxicology.* New York: Raven Press, 1984;165–176.

124. Narahashi T. Nerve membrane sodium channels as the target of pyrethroids. In: Narahashi T, ed. *Cellular and molecular toxicology.* New York: Raven Press, 1984;85–108.

125. Nathanson JA, Bloom FE. Lead-induced inhibition of brain adenyl cyclase. *Nature* 1975;255:419–420.

126. Needleman HL. Epidemiological studies. In: Annau Z, ed. *Neurobehavioral toxicology* Baltimore: The Johns Hopkins University Press, 1986;279–287.

127. Neuman PE, Taketa F. Effects of triethyltin bromide on protein phosphorylation in subcellular fractions from rat and rabbit brain. *Mol Brain Res* 1987;2:83–87.

128. Nishizuka Y. Studies and perspectives on protein kinase C. *Science* 1986;233:305–312.

129. Noguchi K, Higuchi S, Matsui H. Effects of glutathione isopropyl ether on glutathione concentration in ischemis rat brain. *Res Comm Chem Pathol Pharmacol* 1989;64:165–168.

130. Ogita K, Kitage T, Mikamuta H, Fukada Y, Koida M, Okawa Y, Yoneda Y. Glutathione-induced inhibition of Na^+ independent and dependent bindings of L-^{3}H-glutamate in rat brain. *Life Sci* 1986;39:2411–2418.

131. Orlowski M, Kerkowsky A. Glutathione metabolism and some possible functions of glutathione in the nervous system. *Int Rev Neurobiol* 1976;19:75–121.

132. Orrenius S, McConkey DJ, Bellomo G, Nicotera P. Role of Ca^{2+} in toxic cell killing. *Trend Pharmacol Sci* 1989;10:281–285.

133. Padovani A, Govoni S, Magnoni MS, Fernicola C, Coniglio L, Trabucchi M. Lead decreases

lymphocyte beta-adrenergic receptors in humans. In: Foa V, Emmet EA, Maroni M, Colombi A, eds. *Occupational and environmental chemical hazards.* Chichester, UK: Ellis Horwood, 1987;534–538.

134. Perry TL, Yong VW, Clavier RM, Jones K, Wright JM, Foulks JG, Wall RA. Partial protection from the dopaminergic neurotoxin N-methyl-4-phenyl-1,2,3,6,-tetrahydropyridine by four different antioxidants in the mouse. *Neurosci Lett* 1985;60:109–114.

135. Philbert MA, Beiswanger CM, Waters DK, Reuhl KR, Lowndes HE. Cellular and regional distribution of reduced glutathione in the nervous system of the rat: histochemical localization by mercury orange and o-phtaldiadehyde-induced histofluorescence. *Toxicol Appl Pharmacol* 1991;107:205–227.

136. Piantanelli L, Viticchi C, Fattoretti P, Basso A. Impaired receptor regulation: an index of aging? *Arch Gerontol Geriat* 1986;5:325–332.

137. Pinsky C, Bose R. Pyridine and other coal tar constituents as free radical-generating environmental neurotoxicants. *Mol Cell Biochem* 1988;84:217–222.

138. Pintor A, Fortuna S, De Angelis S, Michalek M. Impaired recovery of brain muscarinic receptor sites following an adaptive down-regulation induced by repeated administration of disopropylfluorophosphate in aged rats. *Life Sci* 1990;46:1027–1036.

139. Potemska A, Leo YH, Wisniewski HM. On the possible mechanism of phenylacetate neurotoxicity: inhibition of choline acetyltransferase by phenylacetyl-CoA. *J Neurochem* 1984;42:1499–1501.

140. Pounds JG, Rosen JF. Cellular Ca^{2+} homeostasis and Ca^{2+}-mediated cell processes as critical targets for toxicant action: conceptual and methodological pitfalls. *Toxicol Appl Pharmacol* 1988;94:331–341.

141. Putney JW, Jr. A model for receptor-regulated calcium entry. *Cell Calcium* 1986;7:1–12.

142. Raps SP, Lei JCK, Hertz L, Cooper AJL. Glutathione is present in high concentrations in cultured astrocytes but not in cultured neurones. *Brain Res* 1989;493:398–401.

143. Rasmussen H, Barett P, Smallwood J, Ballag W, Isales C. Calcium ion as intracellular messenger and cellular toxin. *Environ Health Persp* 1990;84:17–25.

144. Ravindranath V, Shivakumar B, Anandathearthavarada HK. Low glutathione levels in brain regions of aged rats. *Neurosci Lett* 1989;101:187–190.

145. Riederer P, et al. Transition metals, ferritin, glutathione and ascorbic acid in Parkinsonian brains. *J Neurochem* 1989;52:d515–520.

146. Robison GA, Butcher RW, Sutherland EW. *Cyclic AMP.* New York: Academic Press, 1971;316.

147. Savolainen H, Jarvisalo J. Effects of acute CS_2 intoxication on protein metabolism in rat brain. *Chem Biol Interac* 1977;17:51–59.

148. Sayre LM. Biochemical mechanism of action of the dopaminergic neurotoxin 1-methyl-4-phenyl-1,2,3,6-tetrahydropyridine (MPTP). *Toxicol Lett* 1989;48:121–144.

149. Scharfman HE, Schwartzkroin PA. Protection of dentate hilar cells from prolonged stimulation by intracellular calcium chelation. *Science* 1989;246:257–260.

150. Schotman P, Gipon L, Jennekens FGI, Gispen WH. Polyneuropathies and CNS protein metabolism changes in protein synthesis induced by acrylamide intoxication. *J Neuropathol Exp Neurol* 1978;37:820–837.

151. Seppalainen AMH. Neurophysiological approaches to the detection of early neurotoxicity in humans. *CRC Crit Rev Toxicol* 1988;18:245–298.

152. Shukla GA, Chandra SB. Cadmium toxicity and bioantioxidants: status of vitamin E and ascorbic acid of selected organs in rat. *J Appl Toxicol* 1989;9:119–122.

153. Shukla GS, Srivastava RS, Chandia SV. Glutathione status and cadmium neurotoxicity: studies in discrete brain regions of growing rats. *Fund Appl Toxicol* 1988;11:229–235.

154. Siesjo BK. Calcium, excitotoxins and brain damage. *NIPS* 1990;5:120–125.

155. Simpson LL. The origin, structure and pharmacological activity of botulinum toxin. *Pharmacol Rev* 1981;33:155–188.

156. Smith CB. Aging and changes in cerebral energy metabolism. *Trends Neurosci* 1984;2:203–208.

157. Spina MB, Cohen G. Dopamine turnover and glutathione oxidation: implication or Parkinson disease. *Proc Natl Acad Sci USA* 1989;86:1398–1400.

158. Stahl SM. Peripheral models for the study of neurotransmitter receptors in man. *Psychopharmacol Bull* 1985;21:663–671.

159. Steventon GB, Heafield MTE, Waring RH, Williams AC. Xenobiotic metabolism in Parkinson's disease. *Neurology* 1989;39:883–887.
160. Stine KE, Reiter LW, Lemaster JJ. Alkyltin inhibition of ATPase activities in tissue homogenates and subcellular fractions from adult and neonatal rats. *Toxicol Appl Pharmacol* 1988;94:394–406.
161. Trenga CA, Kunkel DO, Eaton DL, Costa LG. Effect of styrene oxide on rat brain glutathione. *Neurotoxicology* 1991;12:165–178.
162. Veronesi B, Jones K, Pope C. The neurotoxicity of subchronic acetylcholinesterase inhibition in rat hippocampus. *Toxicol Appl Pharmacol* 1990;104:440–456.
163. Villereal ML, Palfrey HC. Intracellular calcium and cell function. *Ann Rev Nutr* 1989;9:347–376.
164. Wasilewska E, Stanosz S, Bangiel S. Serum dopamine-beta-hydroxlase activity in women occupationally exposed to carbon disulphide. *Industr Health* 1989;27:89–93.
165. Watanabe I. Organotins. In: Spencer PS, Schaumburg HH, eds. *Experimental and clinical neurotoxicology.* Baltimore: Williams and Wilkins, 1980;545–557.
166. Winder C, Kitchen I. Lead neurotoxicity: a review of the biochemical, neurochemical and drug-induced behavioral evidence. *Prog Neurobiol* 1984;22:59–87.
167. Wong VW, Perry TL, Krisman AA. Degeneration of glutathione in brainstem of mice caused by N-methyl-4-phenyl-1,2,3,6-tetrahydropyridine is prevented by antioxidant pretreatment. *Neurosci Lett* 1986;63:56–60.

Neurotoxicology, edited by Hugh Tilson
and Clifford Mitchell. Published by
Raven Press, Ltd., New York 1992.

7

Incorporating Evoked Potentials into Neurotoxicity Test Schemes

Joel L. Mattsson,* W. K. Boyes,† and J. F. Ross‡

*The Dow Chemical Company, Health and Environmental Sciences, Midland,
Michigan 48674; †US EPA, Health Effects Research Laboratory, Neurotoxicology
Division, Research Triangle Park, North Carolina 27711; ‡The Procter and Gamble
Company, Miami Valley Laboratories, Cincinnati, Ohio 45239-8707

Evoked potentials (EPs) are a generally accepted diagnostic tool in human and veterinary neurology, and have been applied successfully in many neurotoxicity studies in both humans and animals (1–12). Despite this success, relatively few neurotoxicology laboratories include facilities for measuring EPs, and most toxicologists are inexperienced in planning neurotoxicology programs which include EPs. Current United States Environmental Protection Agency (USEPA) guidelines (Table 1) (13) do not include assessment of animals for neurotoxicity using EPs, although draft guidelines have been written.

Contemporary electrophysiological tests have enormous potential to complement routine neurotoxicity assessments. For example, at some dose most chemicals will decrease responsiveness to stimuli, decrease locomotor activity, and cause weakness. Nonspecific depression from systemic toxicity minimally affects subcortical sensory input systems (peripheral nerves, spinal cord, brain stem, thalamus, retina, cochlea), while many neurotoxicants do affect sensory input systems. It would seem logical, therefore, to use a well-recognized diagnostic technology such as EPs to help interpret nonspecific clinical signs. Utilization of tests listed in Table 2 can provide critical clues to a neurologic diagnosis that otherwise would have remained obscure.

EPs can be utilized as a screen (i.e., less detail, but more breadth of data across sensory modalities and anatomic areas of the neuraxis) or as a tool to localize and characterize effects already identified. Part of the success of EPs in the neurology clinic comes from their use within an explicit diagnostic algorithm which indicates the type of EP assessed based on signs/history of the patient. The goal of this chapter is to provide a similar outline for using EPs in both the discovery (screening) and characterization phases of neurotoxicology testing.

The research described in this article has been reviewed by the Health Effects Research Laboratory, U.S. Environmental Protection Agency, and approved for publication. Approval does not signify that the contents necessarily reflect the views and policies of the Agency nor does mention of trade names or commercial products constitute endorsement or recommendation for use.

TABLE 1. *USEPA Neurotoxicity Test Guidelines: Federal Insecticide, Fungicide, and Rodenticide Act[a]*

Delayed neurotoxicity of organophosphorus substances following acute and 28-day exposures (hens).
Neurotoxicity screening battery (functional observational battery, motor activity, neuropathology). Appendix 1: Guideline for assaying glial fibrillary acidic protein.
Developmental neurotoxicity study.
Schedule controlled operant behavior.
Peripheral nerve function (electrophysiology).

[a] Also applies to the Toxic Substances Control Act (TSCA) (13).

TABLE 2. *Systems evaluated by evoked potential screen[a]*

System	Functional tests
Visual pathway, retina to visual cortex	Flash evoked potentials (FEPs, strobe stimulus, record from visual cortex; electroretinogram optional)
Visual function (acuity, contrast sensitivity, resolution of patterns)	Pattern reversal evoked responses (PREPs, TV screen with shifting pattern, varying rates of reversal, varying contrasts from gray to light and dark)
Auditory function (acuity, brain stem integrity)	Auditory brain stem responses (ABRs, tested at low to high frequencies); cochlear microphonics (hair-cell potentials)
Somatosensory pathway (with details on spinal cord, brain stem, and thalamus)	Somatosensory evoked potentials (SEPs, mild electrical stimulus to a peripheral nerve, transit spinal cord, and thalamus; record from sensory cortex and cerebellum)
Sensorimotor (optional) (Sherrington's reflex arc)	(H-reflex; electric stimulus at ankle, action potentials ascend sensory fibers, synapse on motor neurons in spinal cord, descend motor axons to cross the neuromuscular synapse, then record the muscle action potential)
Mixed nerve	Caudal nerve action potential (CNAP) (stimulate tip of tail, record from base of tail); [Optional Sciatic nerve action potentials (SNAP) and H-reflex]
Motor axons (optional)	(CNAP–isolate motor components) (SNAP)—isolate motor components) (Electromyograms (EMGs) for denervation potentials)
Sensory axons (optional)	(CNAP—isolate sensory components) (SNAP—isolate sensory components)
Neuromuscular (optional)	(H-reflex; M-response, stimulate peripheral nerve, record from muscle; EMGs)
Lower motor neurons (optional)	(M-response: H-reflex amplitude ratio)

[a] Optional means the test can be added as needed. Animals are fully conscious for most tests. Light anesthesia necessary for SNAPs, ERGs, and H-reflexes. Time involved for epidural implants is about 45 min/rat, and about 40 min to test for FEP, SEP, ABR (clicks and tones), and SNAP. PREPs (one setting) add about 20 min. Data management time depends on efficiency of waveform analyses, statistical analyses, and report writing.

WHAT ARE EVOKED POTENTIALS?

The nervous system generates a continuous background of electrical activity which reflects the asynchronous firing of individual neurons or unrelated groups of neurons. When stimulated by a discrete event, such as a brief noise or flash of light, large numbers of neuronal elements fire synchronously. This pattern of neural activity generates an electrical potential which represents the summed electrical activity of many neurons, and is referred to as an evoked potential.

A discrete sensory stimulus causes receptors to generate electrical activity. This electrical activity is propagated up the sensory pathways through successive nuclei leading to the cerebral cortex and other sensory areas. Signals then are distributed to cortical-cortical loops, and cortical-subcortical-cortical loops. An electrical recording of this summed activity is a complex waveform with several peaks and valleys reflecting activity generated along the serial and parallel sensory pathways. The specific neural generators of particular peaks or valleys are in some cases known, and in others not (as will be discussed below). In general, the early-latency peaks of sensory EPs reflect the initial stages of sensory information processing and are closely related to the physical parameters of the sensory stimulus such as brightness or loudness. In comparison, the later-latency peaks reflect more the associative aspects of sensory processing and may be influenced by cognitive or behavioral factors.

EPs also may be elicited in, and recorded from, areas of the nervous system other than sensory pathways. Controlled electrical pulses delivered via electrodes can be used to stimulate almost any neural structure. For example, motor systems frequently are studied through stimulation of motor neural pathways and recording the evoked muscle action potentials. Thus, EP technology can be targeted to study virtually any area of the nervous system which may be of interest.

INTERPRETATION OF EP CHANGES

EPs often enable localization of neural injury. In general, the early components of evoked waveforms are generated by electrical activity in the pathway *en route* to the cortex. Correlations between waveform components and anatomy are well-known for the early components of the auditory response (auditory brain stem responses, Fig. 1), are reasonably well-known for the somatosensory evoked potential (Fig. 2), and are less well-known for the visual system (Fig. 3). Thus, when toxic effects occur in early components of EPs, a substantial amount of information is available that identifies not only the sensory system or systems, but often the specific areas or even the specific tracts and nuclei.

Changes in amplitude, timing, and shape. EP waveforms can be altered in amplitude (microvolts), become slower or faster, and the overall waveform shape

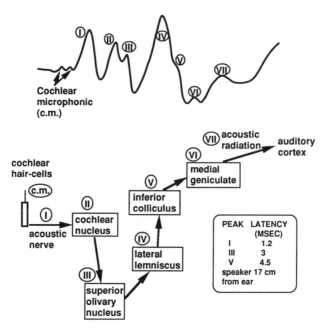

FIG. 1. The auditory pathway and a typical click-induced auditory brainstem response (ABR) which occurs in the first 10 msec of the auditory response. The neural generators of peaks I through VII are more complex than illustrated. The diagram is linear, from hair cell to cortex, when in fact much parallel processing also occurs. Most waves are generated by more than one anatomic structure, and each structure may contribute to more than one wave. For clinical purposes, wave I represents activity in the acoustic nerve. Ototoxicity causes a smaller and slower wave I. The I–V interval represents conduction from the acoustic nerve to upper midbrain. Alterations in the I–III interval suggest lesions in the lower brain stem, and the III–V interval suggests lesions in the upper brain stem (56).

may change (Fig. 4). Damage to nerve cells or nerve connections is seen as a loss of voltage in anatomically correlated peaks and valleys of the waveform. For instance, injury to the auditory nerve will decrease the voltage of peak I of the auditory evoked response. Voltage loss is the result of a reduction in the number of firing units (fewer cells or cell connections), or to a loss of firing synchrony (fewer units firing at any one time). Slowing of waveforms occurs with damage to white matter (myelin). Since pathological changes commonly are mixed and involve nerve cells, connections, and white matter to various degrees, abnormal EP waveforms generally are smaller than normal, poorly shaped, and slow.

The importance of determining adverse effects. Neurotoxicity commonly is defined as any *adverse* effect on the structure or function of the nervous system

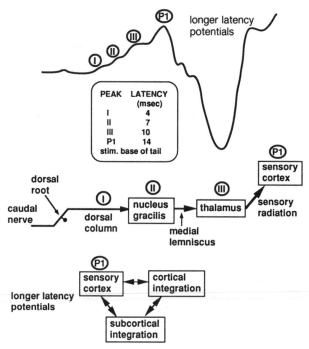

FIG. 2. The dorsal column pathway and the early components of the rat somatosensory evoked potential (SEP). The SEP early components of the rat were mapped by Wiederholt and Iragui-Madoz (57). Although the SEP is primarily conducted along ascending fibers in the dorsal column, parallel processing also occurs in spinocerebellar and spinothalamic pathways. Like the auditory system, most waves are generated by more than one anatomic structure, and each structure may contribute to more than one wave. Evaluation of the SEP allows rapid localization to areas below the brain stem, to the brain stem and midbrain, or to cortical and cortical/subcortical integration areas of the brain.

related to exposure to a chemical substance. The need to differentiate adverse effects from other treatment-related events was highlighted recently in a USEPA Science Advisory Board Report on cholinesterase inhibition (14). The board concluded that there was little support for the assumption that reduction in blood cholinesterase constituted an adverse effect. Rather, adverse effects should be defined on the basis of functional (behavioral, electrophysiological) measures, accompanied, where feasible, by morphological indices such as those provided by both newer and established histochemical techniques.

The importance of this Science Advisory Board recommendation is that toxicologists are challenged to differentiate adverse effects from other treatment-related effects. Too often a simple change in a dependent variable is taken as an

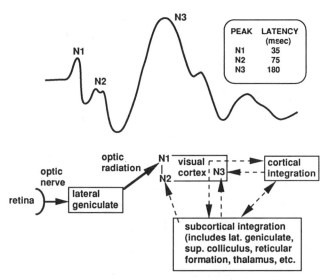

FIG. 3. The visual pathway and a typical rat flash evoked potential (FEP). The source of the rat FEP appears to be limited to the cerebral cortex, with no waves linked directly to input pathways or nuclei (unlike the auditory brainstem response and early waves of the somatosensory evoked potential that have direct links to input tracts and nuclei) (58). Therefore, alterations at N1 reflect changes either peripheral to the first cortical neurons (i.e., retina, optic nerve, lateral geniculate), or at the first cortical neuron itself. Altered N2 or later components indicate functional changes in the cortical/subcortical networks. Thus, FEPs provide a simple functional evaluation of the visual system, but limited anatomic localization. Similarly, pattern reversal evoked potentials (PREPs) provide more sophisticated functional evaluations, but little anatomic localization. PREPs and FEPs may be differentially sensitive to visual toxicity, depending on the test compound, illustrating that they measure different aspects of visual function. FEPs are preferred in albino rats, infants, and other subjects who either have poor ability to see shapes, or will not (or cannot) focus on a pattern on a TV screen. PREPs are easily recorded in subjects who will focus on a pattern, and are easily recorded from pigmented rats because of their fixed focal length (3).

indicator of toxicity, when in fact the actual consequence of the effect is unknown. The "adverse" nature of an effect can be evaluated in the context of ability to adapt to the environment (15). For EPs a reasonably strong correlation between a treatment-related EP change and a known maladaptive consequence should be present to consider the EP change as adverse. For example, loss of hearing obviously is maladaptive to the environment. Since a diminished amplitude or slow acoustic nerve potential in the auditory brain stem response (ABR) (peak I) is highly predictive of loss of hearing, this ABR change can be defined as an adverse effect. A significantly larger and faster peak I would be difficult to classify, however, and in the absence of a clear association with known maladaptive consequences, probably should be classified as an effect of unknown (or indifferent) consequence, and should not be automatically classified as adverse.

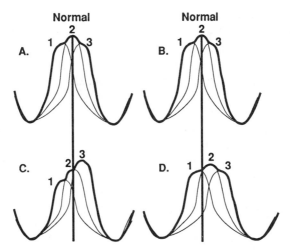

FIG. 4. Two major types of effects on evoked potentials. This representation of an evoked potential has three subpeaks. In a peripheral nerve, the subpeaks represent fast, medium and slow fiber populations. In the CNS the subpeaks represent different electrical generators (e.g., nuclei). The area under the curve of each subpeak, and the combination of subpeaks into a compound action potential, is a reflection of the number of firing units. **A** and **B** are identical, and are references for **C** and **D,** respectively. **C** shows a loss of firing units in subpeaks 1 and 2, causing a loss of amplitude and a diminished area under the curve. Latency, however, is unaffected. A careful reading of **C** is necessary to avoid misinterpretation. Peak 3 was unaffected, and is now the largest peak. Latency measurements to the highest point on **C** could lead one to erroneously believe that latency was prolonged. **D** shows actual slowing of all peaks. The slowing can be due to a slow conduction time in fibers, slower synapses, or in the CNS to changes in the synchronization of activity. The amplitude of **D** is reduced due to spreading out of the subpeaks, but the area under the curve is unchanged (all units firing). Neurotoxic effects usually are mixed, and affect the number of firing units, the speed of conduction, and the synchrony of the response. Thus, neurotoxicity usually causes evoked potentials to become smaller, slower, and of distorted shape (but other patterns are possible, including increases in amplitude and faster waveforms).

NEED FOR EPs IN NEUROTOXICITY TESTING

There are two compelling reasons for including electrophysiological testing in a neurotoxicological testing framework: (a) important sensory effects may be missed by routine testing procedures; and (b) current neurotoxicity testing procedures may yield results which cannot be adequately interpreted.

Important sensory effects may be missed. Sensory effects are frequent consequences of exposure to neurotoxic and other compounds. Although there are no reliable data on the proportion of compounds which affect sensory systems, rough estimates of the percentage of compounds which are likely to affect sensory systems range from 1.5 to 16% of all compounds (16). Grant (17) has listed more than 2,800 compounds which are toxic to the visual system. Given that compromised vision, audition, and somatosensory function are unacceptable consequences of exposure to chemical compounds, and that sensory systems of animals can be evaluated quantitatively in the laboratory, it is logical to perform sensory evaluations in the process of hazard identification.

Standard histopathological examination of the eye and visual pathway may not detect subtle changes produced by chemical exposure and do not reveal the functional status of the tissue. For example, humans who abuse toluene sometimes have serious visual problems (18), but rats exposed to a simulated toluene "abuse" paradigm appeared to be healthy and had no alterations in ocular or CNS histology in perfusion-fixed tissues stained with hematoxylin and eosin, luxol fast blue/periodic acid-Schiff-hematoxylin, and Sevier-Munger silver (19). Flash EPs in the toluene "abuse" rats were, however, very slow and had significant alterations in waveform shape. Thus, an EP procedure detected neurotoxicity which was missed using routine toxicity testing procedures.

Perhaps the greatest deficiency in current screening concerns evaluation of the auditory system. A large number of toxicity studies had been conducted on aromatic solvents before it was reported (20) that toluene induced hearing loss in rats. This hearing loss was substantial, and was more severe at higher frequencies. Subsequently, ototoxicity evaluations were conducted on styrene and xylene, which also were found to be ototoxic in rats (21).

In another study (60), the startle response to a sharp noise, as used in functional observational batteries, did not reveal the ototoxicity that occurred in rats exposed to high levels of an aromatic hydrocarbon for 3 weeks (Table 3). Although the exposed rats had normal startle responses to noise, they generally lacked ABRs to 16 kHz and 30 kHz tone-pips, had severely altered 8 kHz ABRs, but had normal responses at 4 kHz (Fig. 5). Thus, low frequency ABRs were intact, which indicated that exposed rats could respond to the low frequency startle noise in spite of severe high frequency hearing loss.

It is easy to understand why hearing loss may go undetected in animals. Most ototoxicity occurs at high frequency, while clinical evaluations test responsiveness to noise (claps, clicks, bangs, etc.) which contains low frequency energy. Therefore, animals may respond relatively normally to the lower frequencies. This problem of high frequency ototoxicity is exacerbated because the cochlea lies within the dense temporal bone, making cochlear histopathology difficult to perform. Because routine behavioral and neuropathological assessments may miss ototoxicity in animals, it has been recommended that either behavioral or electrophysio-

TABLE 3. *Startle response to sharp noise in rats exposed to exaggerated levels of aromatic hydrocarbon vapors for 3 weeks*

Startle response	Control	Treated
Clear response	8	8
Mild response	3	3
No response	1	1

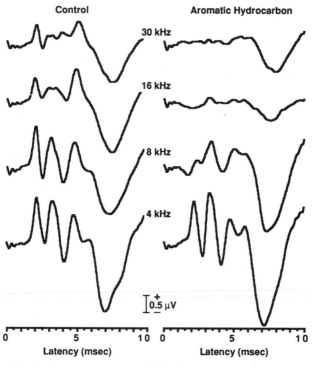

FIG. 5. Rat auditory brain stem responses (ABRs) after exaggerated exposure to an aromatic hydrocarbon for 3 weeks (*N* = 12 controls and 12 treated Fischer 344 male rats). The tone-pip stimuli were about 50 dB above ABR threshold at each frequency. The first major peak is produced by the acoustic nerve. Note the nearly flattened ABRs at 16 kHz and 30 kHz, the severely altered ABR at 8 kHz, but the virtually normal ABR at 4 kHz. The presence of a startle response to a sharp noise (Table 3) is believed to be due to sparing of low (4 kHz) frequency hearing.

logical ototoxicity *screening* methods be included at the end of rat subchronic studies (61). Furthermore, it was recommended that responses to low, middle and high frequency stimuli be evaluated.

There are several measures designed to assess sensory function in the typical Functional Observational Battery (FOB). These include measures such as pupil response to light, and behavioral responses to stimuli such as a click, tail pinch, touch, and approach of an object. Although these endpoints may be rapid and inexpensive, they lack comprehensiveness, sensitivity, and specificity. The FOB assessment of sensory function is not designed to test discriminatory ability of sensory systems and therefore these tests are likely to be less sensitive than more sophisticated procedures (22–24). In addition, FOB tests rely on the observed response of the animal. For this reason, deficits anywhere in sensory-integration-motor response pathways may be mistaken for sensory deficits. It is possible to alleviate many of these concerns with newer behavioral procedures which have not yet been incorporated into the mainstream of neurotoxicity testing, such as pre-pulse inhibition of the acoustic startle reflex (16). However, the startle reflex, like other tests based on sensory-motor reflexes, is mediated at the spinal and lower brain areas. Higher areas of the brain which are critical for sensory perception are not tested by these procedures.

Current neurotoxicity testing may yield results which cannot be adequately interpreted. A major goal in the screening phase of classic toxicity testing is identification of the target organ. Target organs are those that are preferentially affected by a xenobiotic (affected at lower doses than other organs, or affected more severely at the same dose). Within the target organ context, it is no more appropriate to regard the nervous system as a single target organ than to consider "abdominal organs" as such. The following analogy will illustrate. The abdominal cavity contains many organs including the liver, kidney, intestines, spleen, and pancreas which have diverse functions such as digestion, metabolism, and fluid/electrolyte balance. So, too, the brain contains many integrated structures controlling such disparate functions as vision, hearing, emotionality, learning, movement, and energy balance.

Many routine toxicity studies contain simple anatomic (i.e., routine histopathology) and functional (i.e., hematology, serum chemistry, urinalysis) measures to identify the specific target organ. The USEPA neurotoxicity screen is designed to respond to many kinds of neurobehavioral dysfunction, but may not allow differentiation among different neural target organs. For example, weakness, decreased locomotor activity, and decreased acoustic startle response are common observations which could represent specific dysfunction in central or peripheral motor systems, or general CNS depression. Decreased acoustic startle also could reflect hearing loss. It is possible that these neurobehavioral observations reflect simple illness or malaise. It is important, therefore, to interpret these findings in the context of other signs of systemic toxicity.

If neurobehavioral alterations are detected by FOB and the effects cannot be explained by systemic toxicity or by neuropathological examination, then a more specific examination can be implemented. EP tests of muscle, nerve, spinal cord, brain stem, thalamus, and auditory function can be implemented immediately with small needle electrodes as in human clinical neurology. Those EPs that optimize evaluation of cortical function require surgical implantation of electrodes and can be implemented in later, more definitive studies.

EPs SUPPLEMENT NEUROTOXICITY GUIDELINES

Under Federal Insecticide, Fungicide, and Rodenticide Act (FIFRA), the current USEPA neurotoxicity screening battery requires behavioral tests (FOB, motor activity) and routine light microscopic neuropathology (Table 1). The Office of Pesticides and Toxic Substances (OPTS) guidelines describe generic approaches for operant behavior, developmental neurotoxicity, and peripheral nerve function (13). A draft of a sensory EP guideline also has been written for OPTS.

Supplementation of neuropathology. In addition to identifying the presence of neurotoxicity, a goal of neurotoxicity evaluations is to identify a specific neural target organ. This goal is met if there is functional-anatomic agreement between the location of neural lesions and the type of behavioral changes observed. Often, however, behavioral changes are nonspecific (decreased responsiveness to stimuli and decreased locomotor activity), and either no neural lesions are present, or the neuropathological lesions do not reasonably account for the behavioral changes.

While routine light microscopy is a very effective tool for screening of neurons, axons, and myelin, special techniques are necessary to evaluate the small structures such as dendrites and synapses that make up the neuropil. The literature on "neuropilopathies" is growing. Ethanol (25–27), lead (28,29), drugs (30,31), disease states (32–35), and subclinical undernutrition (36,37) are known to affect dendrites and/or other small CNS structures. EPs have been shown to be sensitive indices of such neurotoxicants (38–43), and are a logical intermediate step before embarking on detailed specialized morphological tests when faced with nonspecific neurobehavioral signs and no definitive neural lesions.

EPs as prelude to operant testing. Schedule Controlled Operant Behavior (SCOB) evaluates the effects of chemical exposure on rate and pattern of responding (e.g., lever presses) under schedules of reinforcement (e.g., performance requirements to be met to get a food pellet). Auditory or visual (light) cues commonly are used for training and later to signal that a particular schedule of reinforcement is in effect. Because of the necessary reliance on sensory cues and alterations in overall behavior in rats with sensory dysfunction, a subchronic operant test would not be prudent without first knowing something about sensory function. An analogy is that of school children; before concluding that a child is mentally slow, we need to know whether the child can see the chalk board and hear the teacher. EPs can be used to screen rapidly for auditory or visual dysfunction, as well as to provide information on other essential functions of the nervous systems and muscle.

EPs, electromyography, and the peripheral nerve function guideline. The goal of the USEPA peripheral nerve function guideline is to evaluate amplitude and conduction velocity changes produced by substances *known* to be peripheral neurotoxicants and to determine dose-effect. Generally, identification of peripheral neurotoxicants is by clinical signs (weakness, diminished responsiveness to tactile/nociceptive stimuli) and histopathological lesions in the peripheral nerves. The diagnosis, however, sometimes is difficult to make with confidence. Weakness and diminished responsiveness may be due to mechanisms other than peripheral neuropathy (e.g., systemic toxicity, myopathy, neuromuscular junction dysfunction), or the neuropathy may be focal and not in the area of the tissue sample. Alternatively, peripheral neuropathy may be diagnosed, but may be only one of multiple causes of sensorimotor dysfunction.

A diagnosis of sensorimotor dysfunction, therefore, can require more data than are available from routine clinical examinations and routine sections of nerve and muscle. Peripheral nerve electrophysiological evaluation can provide useful data about the limb segment of peripheral nerves, but does not address the spinal cord segment of sensory axons. The most distal parts of peripheral nerves are in the digits and in the upper spinal cord, at the synapse of the dorsal column with the nucleus gracilis at the brain stem. Distal axonopathies sometimes are more severe in the upper spinal cord than in the limbs, and somatosensory EPs can be a very sensitive tool for making this diagnosis (44,45).

Measurements limited to the peripheral nerve do not address muscle or neuromuscular junctions. In fact, amplitude and conduction velocities of the pe-

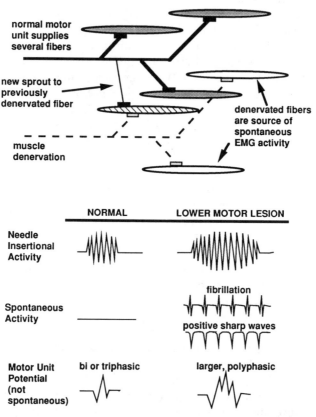

FIG. 6. Sources of denervation potentials of muscle, and typical electromyographic (EMG) waveforms from spinal cord motor neuron deficits or from peripheral neuropathy (derived from Kimura, 59). Muscles normally have a minimal burst of activity upon insertion of the electrode needle, and then are electrically quiet. Spontaneous electrical activity in relaxed muscles is abnormal. Small voluntary contractions of muscle yield simple motor unit potentials. A motor unit is the collection of muscle fibers supplied by a single motor axon. Individual muscle fibers within a normal motor unit fire nearly at the same time, creating a simple bi- or triphasic action potential. During denervation, motor units with healthy axons send axonal sprouts to nearby denervated muscle fibers, thereby reinnervating these muscle fibers. As a consequence, the donor motor unit gains muscle fibers. The axon sprouts usually are thinner than normal, and therefore are slower than normal, and the reinnervated muscle fiber itself may fire slowly. The net result is a loss of synchrony, and an enlarged, polyphasic motor unit potential.

ripheral nerve may be quite normal in spite of significant clinical weakness and frank neurotoxicity (11). Electromyograms (EMGs) provide direct evidence of muscle pathology, or evidence of muscle denervation (Fig. 6). It takes only a few minutes to perform EMGs on several muscles. One also can stimulate nerves and record from muscle, driving action potentials across neuromuscular junctions. Similarly, H-reflexes (Fig. 7) can be utilized to evaluate Sherrington's reflex arc (electrophysiological correlate to the knee-jerk tendon reflex), which includes dorsal and ventral roots, the motor neuron pool, sensory and motor axons, and neuromuscular junctions (46,47).

EP screen as a prelude to the (proposed) sensory EP guideline. USEPA is currently considering sensory EP guidelines which include flash evoked potentials

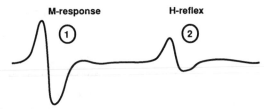

M-response   **H-reflex**

(1) M-response: Muscle action potential from direct stimulation of motor axons.

(2) H-reflex: Delayed muscle action potential from stimulation of sensory axons. Nerve action potentials ascend sensory axons, dorsal root, and synapse with motor neurons. Motor action potentials then descend through the ventral root, motor axons and neuromuscular junction to the muscle.

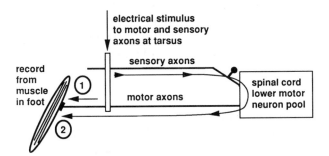

FIG. 7. Typical M-response and H-reflex. This test measures with great accuracy Sherrington's reflex arc. Instead of using a percussion hammer on the patellar tendon, the tibial nerves in the ankle (hock) are electrically stimulated. Exact measurements of response time and response amplitude are made. The H-reflex pathway is long, and includes most of the length of peripheral sensory fibers, the lower motor neurons, and the entire length of peripheral motor fibers, neuromuscular junction, and muscle. A delayed response indicates dysfunction somewhere along the pathway, and more specific motor-axon and sensory-axon tests then would be indicated. The M-response pathway is short, and focuses attention on the neuromuscular junction and muscle. The ratio of amplitudes of M:H reflects the responsiveness of the spinal cord motor neurons. A large M coupled with a small H indicates a lack of responsiveness of the lower motor neurons.

(FEPs), pattern reversal evoked potentials (PREPs), somatosensory evoked potentials (SEPs), and auditory brain stem responses (ABRs). The current draft of the guideline requires that these EPs be stimulated at multiple intensities, that PREPs use a range of pattern sizes, and ABRs use clicks and pure-tones at low, middle, and high frequencies. This detailed assessment is aimed more at initial characterization of effects, rather than being a screen to discover effects. An EP screen, in contrast, would provide less detail but more breadth of data across sensory modalities and anatomic areas of the neuraxis.

A screen, for example, could use only one stimulus intensity, a limited range of pattern sizes for the PREP, and three or four frequencies for the ABR (only one intensity), but would add measures of peripheral nerve, spinal cord, and cerebellar function. Peripheral nerve screening assessments would include two stimuli in close succession to evaluate recovery of excitability of the nerve trunk, but would not use multiple interstimulus intervals to plot recovery functions. Although stimulus variables would be minimized, they would be chosen to be challenging to normal animals, and thereby quite sensitive to toxic effects.

Use of an EP screen before a more intensive evaluation could provide several benefits. First, the absence of effects in the screen would reduce the motivation to conduct a more extensive evaluation. Resources may be better utilized elsewhere. Second, if screening data show a selective action of the xenobiotic on a single sensory modality, then more extensive tests could be limited to this modality. Last, screening data might indicate that other tests should be added to supplement the interpretation of the results. For example, knowledge of peripheral nerve function is necessary for the interpretation of SEPs. Screening data also might indicate the possibility of retinal dysfunction, and electroretinograms and more aggressive retinal histopathology would then be advisable in the follow-up study.

EPs and developmental neurotoxicity. The USEPA developmental neurotoxicity guidelines resemble classic developmental studies in that large numbers of rat pups are evaluated at several stages of life for a multiplicity of endpoints. Developmental neurotoxicity studies are triggered by evidence of adult neurotoxicity, effects on developmentally important endocrine organs, or structure-activity relationships to known developmental neurotoxicants.

Developmental neurotoxicity studies use a large number of animals; 20 dams per dose level, and eight pups per dam. To ensure a sample size of 20 dams/dose level, more than 20 dams/dose level must be bred. Thus, more than 760 animals are used in these studies. To avoid unnecessary use of large numbers of laboratory animals, we need assurance that one or more of the trigger events is real. EP evaluations in adult animal neurotoxicity studies can help in this decision process. A comprehensive set of negative EP data would weigh against proceeding with a full developmental neurotoxicity study, while positive data would assure us that adult neurotoxicity had indeed occurred. In addition, positive EP data aid considerably in the experimental design of a follow-up developmental neurotoxicity study. For example, if ototoxicity occurs in adults, then auditory function should be tested in pups, and caution would be advised in the interpretation of data from other tests such as acoustic startle habituation. High frequency auditory dysfunction in dams also might interfere with maternal-pup communication and cause alterations in pup behavior which are due to maternal dysfunction, and not due to the effects of the xenobiotic on the pups.

EPs are used often in human neonatology to evaluate the level of maturation and health status of the nervous system (48–50). EPs also have been used in animal developmental neurotoxicology (38,41,43). There are adequate numbers of pups in guideline developmental neurotoxicity studies to provide a subgroup for EP evaluations (e.g., 20 litters/dose, eight pups/litter; select one male pup/litter from ten litters, and one female pup/litter from the other ten litters). If specific EP changes were detected in the adult animal toxicity studies that lead to implementation of the developmental study, then the pups should be evaluated for similar changes. The same sensory modality, at least, should be examined in the offspring. However, a better understanding of the developmental effects would be gained with a screen of visual, somatosensory, and auditory functions.

A more common scenario would be the implementation of a developmental neurotoxicity study based on information that lacked EP data. Principal tests in the guideline are motor activity of the rat pups at days 13, 17, 21, and 60; auditory startle habituation on days 22 and 60; and simple learning in pups and a more complex cognitive task when they reach adulthood. Neuropathology is done on pups, and again when grown to adults. If interpretive confusion exists as the data are collected as the pups mature, a subset of pups can be assigned to an EP subgroup to gain more definitive information on nervous system function. On the other hand, if the developmental behavioral data is convincingly negative, one may prefer not to implement the EP portion of the study.

INTEGRATING EVOKED POTENTIALS INTO A TEST BATTERY

Subchronic neurotoxicity examples and model protocol. Several subchronic (13-week) inhalation neurotoxicity studies have included FOBs, an EP screen, and neuropathology (perfused tissues): sulfuryl fluoride (51); trichloroethane (52); dichloromethane (53); 1,1,2,2-tetrachlorethylene (62); and styrene (63). These studies provide a model for incorporating EPs into subchronic neurotoxicity studies, as shown in Table 4. Because of the large numbers of rats to be evaluated, the studies were conducted in two counterbalanced replicates, started 1 week apart. Rats were removed from the inhalation chambers for 1 day for surgical placement of chronic epidural electrodes which were embedded in acrylic. EP tests and the final FOB were conducted the week following the last exposure to minimize acute transient effects from the more slowly cleared materials, and still optimize detection of persistent neurotoxic effects.

Motor activity was not conducted during the above studies but is incorporated into Table 4 to complete the study design according to contemporary USEPA

TABLE 4. *Outline of a typical subchronic neurotoxicity study with integration of an evoked potential battery (FOB, EP screen, neuropathology)*

Week	Daily cageside	Weekly Body weight	Weekly "Mini" FOB	Monthly FOB	Monthly Motor activity	Surgery	EPs	Necropsy
0	X	X		X	X			
1	X	X	X					
2	X	X	X					
3	X	X	X					
4	X	X	X					
5	X	X		X	X			
6	X	X	X					
7	X	X	X					
8	X	X	X					
9	X	X		X	X			
10	X	X	X					
11	X	X	X			X (approx.)		
12	X	X	X					
13	X	X	X					
End of exposure								
14	X	X		X	X		X	X
15	X	X	X					
16	X	X	X			follow-up as indicated		

N = 12/sex/dose; neuropathology *N* = 5/sex/dose; The mini-FOB is the hand-held portion of the standard FOB, except that the observer is aware of treatment status. The FOB is conducted blind to treatment.

FOB, Functional Observational Battery; EP, evoked potential.

neurotoxicity test guidelines. Motor activity testing takes about 1 hour for each rat, and depending on the number of motor activity test units, usually takes up too much of the day to have inhalation exposure for those animals on that same day. Thus, extra exposure days could be added to the study to compensate for the exposure days missed for testing.

Operational advantages and problems. All techniques have both advantages and problems, and EPs are not unique in this respect. When one decides to perform neurotoxicity evaluations in greater depth and breadth than are provided for by FOBs and neuropathology, several operational factors enter into the decision process for test selection. For example, animals are euthanized for necropsy and histopathology. For the advantages of being able to examine tissues in great detail, the animals are lost to subsequent testing. EPs can be done repetitively on the same animals, which allows for direct confirmation of initial findings, or evaluation of progression or resolution of neurotoxic changes without adding more animals to the study. Repetitive testing also can be altered according to findings on the earlier EP screens, either by changing test parameters or adding other EP tests.

EP tests do not require animal training or the maintenance of training during the study. In contrast, schedule controlled operant behaviors (SCOBs), as used in subchronic neurotoxicity studies, require several weeks of training before the study begins (13). Then, to evaluate performance during the study or at the end of the study, the behavior must be maintained during the study with several test sessions per week. Test sessions last about 30 to 90 min per animal, but several operant units can be used simultaneously. Still, several hours per day are required for testing. Although operant techniques can generate very valuable data about behavior, their associated logistic requirements make it very difficult to perform operant evaluations in subchronic inhalation studies in which 6 hr of inhalation exposure already requires about 8 hr of work per day.

EP tests do not require food restriction and hunger motivation. SCOBs that use food as a positive reinforcement require food restriction so that behavior will be effectively controlled by food. The logistics of performing a diet exposure study with food as a positive reinforcement are formidable. Not only do the daily rations for each animal have to be adjusted to maintain a proper body weight (based on the amount of food obtained as a positive reinforcement), but the amount of test chemical in the daily ration also must be adjusted on an individual basis. In addition, systemic toxicity may affect appetite, and not only would the amount of food eaten change, but the effectiveness of food as a positive reinforcement also could change. Thus, EP tests have several operational advantages in diet studies over tests that require the use of food for reinforcement.

In spite of the many operational advantages of EPs, there are several reasons why toxicologists may express reservations regarding the incorporation of EPs in neurotoxicity testing protocols. First among these is the need to anesthetize the animals for electrode implantation and/or testing. There may be concern that the presence of the anesthetic or the induction of liver enzymes following anesthesia might alter the toxicity of the test compound, or that regulatory bodies will think so and refuse to accept the submitted data.

The metabolism of xenobiotics can be significantly affected by factors of species, genetics, sex, age, diet, and other synthetic chemicals (54). Probably the most important metabolic decision is the selection of the test species, a decision that seldom gets much consideration. Still, the question of drug treatment–anesthetic interaction deserves careful attention.

Fortunately, operational factors mitigate against significant treatment-anesthetic interactions. Anesthetic duration for cranial implants is short, and most anesthetics have a short half-life. In contrast, conditions that favor significant metabolic induction are repeated exposures (ethanol, phenobarbitone), or long half-lives (e.g., PCBs) that equate to repetitive exposures (54). Many anesthetics, like isoflurane, are poorly metabolized (<0.2%) (55), and are unlikely to initiate induction by being a readily available substrate. In addition, the extremely wide use of the common anesthetics in hospital patients that concurrently are treated with other drugs creates a great opportunity to discover significant drug-anesthetic interactions. The lack of such information on a specific anesthetic might, therefore, be considered indirect evidence that a particular anesthetic is not a potent inducer.

The potential for chemical-chemical interaction is always a complicated problem which is best dealt with on a case-by-case basis. If anesthetic interaction is a concern, the problem is easily dealt with in acute studies by implanting the electrodes well in advance of dosing with the test compound. For subchronic or chronic studies, the solution will depend on the nature and goals of the study. For many test compounds it will be apparent that the potential for metabolic interaction between the anesthetic and test compound is low. For these cases, implantation of electrodes and testing during the course of the experiment should provide little concern. For compounds likely to be metabolized through pathways similar to the anesthetic of choice, it may be possible to select a different anesthetic or to implant electrodes after the course of treatment is completed if the major goal of the study is to detect lasting changes. Finally, if it is unavoidable that the actions or metabolism of the anesthetic and test compounds will interact, then the best solution may be to add extra test animals to the study to check specifically for the anesthetic interaction effects.

Some may fear that valuable test animals may be inadvertently lost due to anesthetic overdose or failure of the electrode preparation. This may be a problem when first undertaking a new procedure, but with experience, laboratories should find that few animals are lost from any cause.

Reservations also may be expressed that the electrodes may themselves damage the neural tissue, and that this damage may be misinterpreted as neurotoxicity produced by the test compound. Our experience is that this is not a practical problem. Epidural electrodes are adequate for most EP work and should produce little trauma to brain tissue. Occasionally electrodes will be placed too deep and some damage will be evident in a well-prescribed area surrounding the electrode. This damage is easily recognized and should not be mistaken for neurotoxicity. In addition, such damage should serve notice that refinement of surgical technique is needed. Should the study require electrodes which penetrate the brain or other neural tissues, then damage surrounding the electrodes is to be expected, but again, this is easily distinguished from neurotoxicity on histopathological examination.

Another concern is that electrophysiological testing is excessively expensive. Setting up an electrophysiological laboratory is not cheap, but neither is it particularly expensive compared with other scientific laboratories. Off-the-shelf equipment is readily available from several manufacturers which requires little in the way of setup time. These systems are either the same as, or similar to, those used in hospitals worldwide, and purchase of a well-recognized system virtually guarantees technical support for your system. Because of the need for good laboratory practices (GLP), it is advisable to buy a good computer-based system so that GLP information management can be written into the EP management programs. Although these systems are more expensive initially, the savings in personnel time will quickly make up the difference.

Once the EP laboratory is established, then conducting electrophysiological investigations is not expensive because there is little in the way of recurring supply or material need. The main expense will be the labor of surgical implantation of electrodes (30–60 min/animal) and testing (roughly 5–60 min/animal, depending on the protocol), and data management and report writing (variable time, depending on efficiency).

Finally, there are few toxicologists with practical electrophysiological experience. This may reflect a lack of opportunities as much as a lack of trained individuals. The techniques and theoretical knowledge required are well within the reach of any good graduate of a neurophysiology/neuropharmacology program, or residents in veterinary clinical neurology. Once the promise of these techniques is more widely appreciated and the employment opportunities expand, it is likely that many qualified individuals will be available.

CONCLUSION

EPs should play at least two roles in neurotoxicity testing: (a) to test for sensory deficits and (b) to provide perspective for observational and histopathologic data. EPs can be efficiently integrated into neurotoxicity testing schemes to provide toxicologists with highly diagnostic information on nervous system function. The structure-function relationships inherent in EPs allow the toxicologist to pursue diagnostic problems on the same animals with other EP tests, with other types of functional tests, and to confer with pathologists about tissue sampling plans and histological techniques. If follow-up studies are necessary, EP data will provide critical information for the refinement of hypotheses so that resources can be focused.

REFERENCES

1. Anderson RJ, Richardson RJ, Schwab BW. The nervous system as a target organ for toxicity; Section III, electrophysiological aspects. In: Cohen GM, ed. *Target organ toxicity.* Boca Raton, FL: CRC Press, 1986;187–197.
2. Arezzo JC, Simson R, Brennan NE. Evoked potentials in the assessment of neurotoxicology in humans. *Neurobehav Toxicol Teratol* 1985;7:299–304.
3. Boyes WK, Dyer RS. Pattern reversal and flash evoked potentials following acute triethyltin exposure. *Neurobehav Toxicol Teratol* 1983;5:571–577.
4. Dyer RS. The use of sensory evoked potentials in toxicology. *Fund Appl Toxicol* 1985;5:24–40.
5. Dyer RS. Evoked potentials: physiological methods with human applications. In: Johnson BL, Anger WK, Durao A, Xintaras C, eds. *Advances in neurobehavioral toxicology: application in environmental and occupational health.* Chelsea, Michigan: Lewis Publishers, 1990;165–174.
6. Fox DA, Lowndes HE, Bierkamper GG. Electrophysiological techniques in neurotoxicology. In: Mitchell CL, ed. *Nervous system toxicology.* New York: Raven Press, 1982;299–335.
7. Lowndes HE, ed. *Electrophysiology in neurotoxicology,* vols. I & II. Boca Raton, FL: CRC Press, 1987.
8. Mattsson JL, Eisenbrandt DL, Albee RR. Screening for neurotoxicity: Complementarity of functional and morphologic techniques. *Toxicol Pathol* 1990a;18(Number 1, part 2):115–127.
9. Otto D, Hudnell K, Boyes W, Janssen R, Dyer R. Electrophysiological measures of visual and auditory function as indices of neurotoxicity. *Toxicology* 1988;49:205–218.

10. Rebert CS. Multisensory evoked potentials in experimental and applied neurotoxicology. *Neurobehav Toxicol Teratol* 1983;5:659–671.
11. Ross JF, Lawhorn GT. ZPT-related distal axonopathy: behavioral and electrophysiologic correlates in rats. *Neurotoxicol Teratol* 1990;12(2):153–159.
12. Seppalainen AMH. Neurophysiological approaches to the detection of early neurotoxicity in humans. *CRC Crit Rev Toxicol* 1988;18:245–298.
13. US Environmental Protection Agency, Pesticide Assessment Guidelines, Subdivision F. Hazard evaluation: human and domestic animals, addendum 10, neurotoxicity. EPA 540/09-91-123, NTIS PB 91-154617, Springfield, VA 22161.
14. Report of the SAB/SAP Joint Study Group on Cholinesterase. Review of cholinesterase inhibition and its effects. A Science Advisory Board Report, US Environmental Protection Agency, EPA-SAB-EC-90-014, Washington, DC, May, 1990.
15. Thompson T, Schuster CR. *Behavioral pharmacology.* Englewood Cliffs, NJ: Prentice-Hall, 1968;215.
16. Crofton KM. Reflex modification and the detection of toxicant-induced auditory dysfunction. *Neurotoxicol Teratol* 1990;12:461–468.
17. Grant WM. *Toxicology of the eye.* Springfield, IL: Charles C Thomas, 1986.
18. Ehyai A, Freemon FR. Progressive optic neuropathy and sensorineural hearing loss due to chronic glue sniffing. *J Neurol Neurosurg Psychiatry* 1983;46:349–351.
19. Mattsson JL, Gorzinski SJ, Albee RR, Zimmer MA. Evoked potential changes from 13 weeks of simulated toluene abuse in rats. *Pharmacol Biochem Behav* 1990b;36(3):683–689.
20. Pryor GT, Dickenson J, Howd RA, Rebert CS. Neurobehavioral effects of subchronic exposure of weanling rats to toluene or hexane. *Neurobehav Toxicol Teratol* 1983;5:47–52.
21. Pryor GT, Rebert CS, Howd RA. Hearing loss in rats caused by inhalation of mixed xylenes and styrene. *J Appl Toxicol* 1987;7(1):55–61.
22. Evans HL. Assessment of vision in behavioral toxicology. In: Mitchell CL, ed. *Nervous system toxicology.* New York: Raven Press, 1982;81–107.
23. Moody DD, Stebbins WC. Detection of the effects of toxic substances on the auditory system by behavioral methods. In: Mitchell CL, ed. *Nervous system toxicology.* New York: Raven Press, 1982;109–131.
24. Lochry EA. Concurrent use of behavioral/functional testing in existing reproductive and developmental toxicity screens: practical considerations. *J Am Coll Toxicol* 1987;6:433–439.
25. Tavares MA, Paula-Barbosa MM, Verwer RWH. Synapses of the cerebellar cortex molecular layer after chronic alcohol consumption. *Alcohol* 1987;4(2):109–116.
26. King MA, Hunter BE, Walker DW. Alterations and recovery of dendritic spine density in rat hippocampus following long-term ethanol ingestion. *Brain Res* 1988;459:381–385.
27. Galofre E, Ferrer I, Fabregues I, Lopez-Tejero D. Effects of prenatal ethanol exposure on dendritic spines of layer V pyramidal neurons in the somatosensory cortex of the rat. *J Neurol Sci* 1987;81:185–195.
28. Lorton D, Anderson WJ. Altered pyramidal cell dendritic development in the motor cortex of lead intoxicated neonatal rats. A Golgi study. *Neurobehav Toxicol Teratol* 1984;8:45–50.
29. Lorton D, Anderson WJ. The effects of postnatal lead toxicity on the development of cerebellum in rats. *Neurobehav Toxicol Teratol* 1986;8:51–59.
30. Scallet AC, Uemura E, Andrews A, Ali SF, McMillan DE, Paule MG, Brown RM, Slikker W. Morphometric studies of the rat hippocampus following chronic delta-9-tetrahydrocannabinol (THC). *Brain Res* 1987;436:193–198.
31. Uemura E, Ireland WP, Levin ED, Bowman RE. Effects of halothane on the development of rat brain: a Golgi study of dendritic growth. *Exper Neurol* 1985;89:503–519.
32. Becker LE, Armstrong DL, Chan F. Dendritic atrophy in children with Down's syndrome. *Ann Neurol* 1986;20:520–526.
33. Flood DG, Coleman PD. Failed compensatory dendritic growth as a pathophysiological process in Alzheimer's disease. *Can J Neurol Sci* 1986;13:475–479.
34. Quattrochi JJ, McBride PT, Yates AJ. Brainstem immaturity in sudden infant death syndrome: a quantitative rapid Golgi study of dendritic spines in 95 infants. *Brain Res (Netherlands)* 1985;325:39–48.
35. Ruiz-Marcos A, Ipina SL. Hypothyroidism affects preferentially the dendritic densities on the more superficial region of pyramidal neurons of the rat cerebral cortex. *Dev Brain Res* 1986;29:259–262.

36. Root EJ, Kirkpatrick JB, Rutter MK, Wigal TL, Longenecker JB. Dendritic and behavioral changes following subclinical nutrient deficiencies in the rat. *Nutr Reports Int* 1988;37(5):959–972.
37. Sima A, Persson L. The effect of pre- and postnatal undernutrition on the development of the rat cerebellar cortex. *Neurobiology* 1975;5:23–34.
38. Albee RR, Mattsson JL, Johnson KA, Kirk HD, Breslin WJ. Neurological consequences of congenital hypothyroidism in Fischer 344 rats. *Neurotoxicol Teratol* 1989;11(2):171–183.
39. Albee RR, Mattsson JL, Yano BL, Chang LW. Neurobehavioral effects of dietary restriction in rats. *Neurotoxicol Teratol* 1987;9:203–211.
40. Fox DA, Lewkowski JP, Cooper GP. Acute and chronic effects of neonatal lead exposure on development of the visual evoked response in rats. *Toxicol Appl Pharmacol* 1977;40:449–461.
41. Fox DA, Lewkowski JP, Cooper GP. Persistent visual cortex excitability alterations produced by neonatal lead exposure. *Neurobehav Toxicol* 1979;1:101–106.
42. Otto DA, Reiter LW. Developmental changes in slow cortical potentials of young children with elevated body lead burden. *Ann NY Acad Sci* 1983;425:377–383.
43. Church MW, Holloway JA. Effects of prenatal ethanol exposure on the postnatal development of the brainstem auditory evoked potential in the rat. *Alcoholism: Clin Exp Res* 1984;8:258–265.
44. Arezzo JD, Schaumburg HH, Spencer PS, Vaughan HG Jr. A novel approach in detecting toxic axonal neuropathies: the brainstem somatosensory evoked potential. *Ann Neurol* 1980;8:96–97.
45. Arezzo JD, Schaumburg HH, Spencer PS. Structure and function of the somatosensory system: a neurotoxicological perspective. *Environ Health Perspect* 1982;44:23–30.
46. Meinck H. Occurrence of the H-reflex and the F-wave in the rat. *Electroenceph Clin Neurophysiol* 1976;41:530–533.
47. Mattsson JL, Albee RR, Brandt LM. H-reflex waveform and latency variability in rats. *Fundam Appl Toxicol* 1984;4:944–948.
48. Cracco JB, Cracco RQ. Spinal, brainstem, and cerebral SEP in the pediatric age group. In: Cracco RQ, Bodis-Wollner I, eds. *Evoked potentials.* New York: Alan R Liss, 1986;471–482.
49. Stockard JE, Stockard JJ. Clinical applications of brainstem auditory evoked potentials in infants. In: Cracco RQ, Bodis-Wollner I, eds. *Evoked potentials.* New York: Alan R Liss, 1986;455–462.
50. Sokol S. Clinical applications of the ERG and VEP in the pediatric age group. In: Cracco RQ, Bodis-Wollner I, eds. *Evoked potentials.* New York: Alan R Liss, 1986;447–454.
51. Mattsson JL, Albee RR, Eisenbrandt DL, Chang LW. Subchronic neurotoxicity in rats of the structural fumigant, sulfuryl fluoride. *Neurotoxicol Teratol* 1988;10:127–133.
52. Testing consent order for 1,1,1-trichloroethane, USEPA and companies represented by the Halogenated Solvents Industry Alliance. Federal Register 54(162), Aug 23, 1989, 34991–34995.
53. Mattsson JL, Albee RR, Eisenbrandt DL. Neurotoxicologic evaluation of rats after 13 weeks of inhalation exposure to dichloromethane or carbon monoxide. *Pharmacol Biochem Behav* 1990c;36(3):671–681.
54. Gibson GG, Skett P. *Introduction to drug metabolism.* New York: Chapman and Hall, 1986.
55. Holaday DA, Fiseroua-Bergerova V, Latto IP, Zumbiel MA. Resistance of isoflurane to biotransformation in man. *Anesthesiology* 1975;43:325–332.
56. Spehlmann R. *Evoked potential primer: visual, auditory, and somatosensory evoked potentials in clinical diagnosis.* Boston: Butterworth Publishers, 1985.
57. Wiederholt WC, Iragui-Madoz VJ. Far field somatosensory potentials in the rat. *Electroenceph Clin Neurophysiol* 1977;42:456–465.
58. Dyer RS. The interactions of behavior and neurophysiology. *Neurobehav Toxicol* 1986;9:193–213.
59. Kimura J. Nerve conduction studies and electromyography. In: Dyck PJ, Thomas PK, Lambert EH, Bunge R, eds. *Peripheral neuropathy.* Philadelphia: WB Saunders Co., 1984;919–966.
60. Albee RR, Mattsson JL, Yano BL, Beckman MJ, Bradley GJ. Dow Chemical Company, unpublished report (1990).
61. National Agricultural Chemicals Association, letter to USEPA, Alternate Proposal for Neurotoxicity Testing, February 6, 1990;7.
62. HSIA, 1990. Halogenated Solvents Industry Alliance sponsored subchronic neurotoxicity evaluation of 1,1,2,2-tetrachlorethylene.
63. SIRC, 1991. Styrene Information and Research Center sponsored subchronic inhalation neurotoxicity evaluation of styrene.

Neurotoxicology, edited by Hugh Tilson
and Clifford Mitchell. Raven Press, Ltd.,
New York © 1992.

8

Assessment of Neurotoxicant-Induced Effects on Motor Function

Beverly M. Kulig and J. H. C. M. Lammers

*Department of Pharmacology, TNO Medical Biological Laboratory,
Rijswijk, The Netherlands*

The potential of environmental neurotoxins to affect motor function has been recognized since antiquity and the first regulatory action purportedly taken to control the effects of neurotoxic exposure in the workplace was aimed at reducing the motor dysfunction produced by mercury exposure in the mining industry in 1665 (63). Of the 167 chemicals currently regulated for their effects on the nervous system, evidence of motor dysfunction has served as a basis for recommendations of allowable workplace levels by the American Conference of Industrial Hygienists (ACIH) in over half the cases (3).

Exposure to many different classes of substances including metals, solvents, pesticides, gases, and drugs has been associated with toxicant-induced motor disturbances (4). In some cases, functional effects are the result of damage to peripheral nerve, while in others, structures within the central nervous system are involved. Not surprisingly, the types of motor effects which have been reported in the literature for different compounds are equally heterogeneous, ranging from specific deficits such as ataxia, paresis, tremor, and other types of dyskinesias, to more generalized effects such as activity changes and changes in psychomotor performance.

Observational assessments comprise an essential part of hazard evaluation studies and several observational batteries designed to document the occurrence of clinical signs in chemically exposed animals have been developed. There can be little doubt that clinical motor signs such as altered body posture, gait abnormalities, and tremor can provide compelling evidence of a given chemical's potential to adversely affect motor function. However, observational techniques are based on subjective assessments which typically yield quantal or ranked data and their sensitivity in detecting the onset of progressive neuromotor impairment depends, at least in part, on the experience and acumen of the experimenter. In relatively high-dose screening studies designed to measure the presence or absence of effects, these limitations do not necessarily present formidable difficulties.

However, in studies designed to characterize the time-course of specific effects, evaluate the least effective dose, or study mechanisms of action, more definite and precise endpoints are usually called for (147).

A number of techniques have been developed for measuring the effects of different aspects of motor function in small laboratory animals. These tests differ widely with respect to the types of motor deficits which they are designed to measure and the degree to which they are automated. Further, there are considerable differences in the psychometric properties of different tests of motor function and their suitability for application in long-term experiments. This chapter surveys some of the behavioral approaches currently in use in neurotoxicology which have been developed in recent years to quantify the severity of specific clinical motor signs, as well as paradigms for examining changes in the speed, amount, and pattern of motor behaviors in different test environments.

TESTS FOR QUANTIFYING CLINICAL SIGNS OF MOTOR IMPAIRMENT

Measures of Neuromuscular Function

Various behavioral tests including the measurement of hindlimb thrust (18), suspension from a horizontal rod (45), swimming endurance (9,104), and a noninvasive behavioral technique for obtaining continuous recording of muscle tone (66) have been used for quantifying the effects of neurotoxicants and psychopharmacological agents on different aspects of neuromuscular function.

One method employed in a number of laboratories at present utilizes mechanical strain gauges to obtain separate estimates of fore- and hindlimb grip strength in the same testing apparatus (102). Validation of this technique was carried out by Pryor et al. (115) in a study in which the effects of eight chemicals were assessed using a battery of neurobehavioral tests. Their data demonstrated low coefficients of variation for control animals (11–15% for forelimb grip strength; 10–15% for hindlimb grip strength), stability of control grip strength values during long-term exposure, and sensitivity in detecting neurotoxic effects in a manner consistent with the known neuropathology of specific test chemicals.

The general principle of the Meyer et al. technique (102) is depicted in Fig. 1. The apparatus consists of a push-pull strain gauge attached to a t-bar positioned at the end of a specially built platform. To measure forelimb grip strength, the rat is placed with its forepaws on the t-bar and gently pulled backwards by the tail until it engages the bar. It is then pulled back further until its grip is broken. To measure hindlimb grip strength, the rat is placed on the platform facing away from the t-bar and a similar procedure with the strain gauge in the push mode is employed. In order for a grip strength measurement to be valid, animals must actually grip the bar with both paws during a given trial in a manner similar to that shown in Fig. 1. If the animal fails to grip the bar on a given trial, the

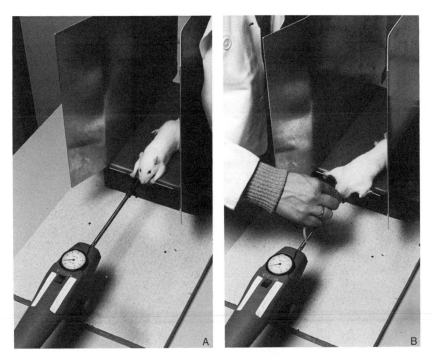

FIG. 1. The measurement of (**A**) forelimb and (**B**) hindlimb grip strength using mechanical strain gauges.

measurement is discarded and the trial is repeated. Typically, data from three valid forelimb trials and three valid hindlimb trials are averaged separately to provide estimates of fore- and hindlimb grip strength.

Since its introduction, grip strength measurements have been adopted by numerous investigators to assess the effects of a wide variety of chemical compounds on neuromuscular function (e.g., 14,79,97,107,137,148). Further, grip strength testing has been included in the Neurotoxicity Testing Guidelines published by the US Environmental Protection Agency (154).

In our laboratory, measurements of grip strength have been incorporated into a battery of neurofunctional tests to examine the effects of long-term exposure to chemicals (79), particularly organic solvents (77,78,80,83). Figure 2 demonstrates the changes in hindlimb grip strength measured during and following long-term inhalational exposure to the organic solvents carbon disulfide (80) or *n*-hexane (83) compared to those induced by subchronic exposure to the prototypic neurotoxic agent, acrylamide. As expected, acrylamide produced a progressive decrease in hindlimb grip strength during the course of the 12-week exposure period, and it persisted into the postexposure period with evidence for considerable recovery of function in acrylamide-treated animals during the first 6 weeks postexposure. These findings are in keeping with those reported by other

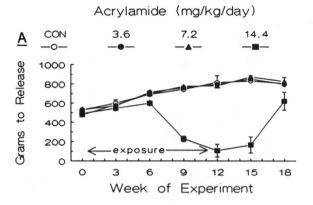

A — Acrylamide (mg/kg/day)

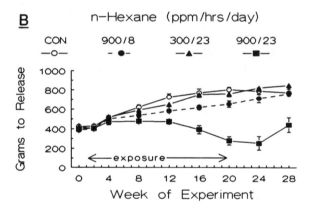

B — n-Hexane (ppm/hrs/day)

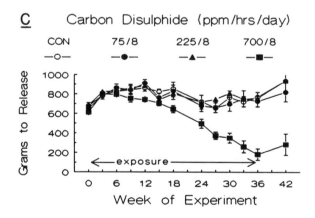

C — Carbon Disulphide (ppm/hrs/day)

investigators (115). Exposure to carbon disulfide at 700 ppm for 8 hr/day, 5 days/week for 36 weeks also resulted in the development of hindlimb weakness progressing to marked gait disturbances, with no significant improvement in grip strength detected during the postexposure period. Exposure to *n*-hexane at 900 ppm for 23 hr/day, 5 days/week for 20 weeks also produced marked and persistent changes in hindlimb grip strength, while exposure to 900 ppm on an 8 hr/day schedule produced only mild, nonpersistent changes.

Taken together, data such as these, as well as that of other investigators, demonstrate the sensitivity of grip strength measurements in detecting the effects of chemical agents which are toxic to peripheral nerve as well as the utility of this method in extended exposure studies. These qualities, together with low cost and ease of administration, make grip strength measurements suitable for both screening and mechanistic studies.

Measurement of Ataxic and Paretic Gait Disturbances

A simple and direct approach to quantifying ataxic and paretic gait disturbances is the analysis of the temporal and spatial characteristics of limb movement during walking using either cinematography (e.g., 20,54) and/or the visualization and analysis of hard-copy records of successive footprints made during locomotion (e.g., 68). Although a number of methods for obtaining footprint records have been described (29,68,92), in most studies the hindpaws are either greased or inked and the animal is allowed to walk on a nonmoving, paper-covered surface. In order to facilitate consistent walking in one direction, testing is usually conducted by placing the animal at one end of a narrow walkway covered with paper and allowing it to traverse the length of the walkway to a darkened enclosure at the other end.

Studies of gait topography in normal rats have demonstrated that the spatial and temporal characteristics of locomotion are highly consistent. In the normal rat, the most commonly employed gait up to velocities of 80 cm/sec is the walk with a gallop occasionally used at higher speeds (61). In order to achieve stability during walking, the rat utilizes a fixed sequence of limb movements characterized as a lateral sequence, diagonal couplet which consists of the following limb sequence: left front, right hind, right front, left hind, left front, etc. (Fig. 3). Ap-

---

FIG. 2. The development of persistent hindlimb weakness in rats during and following long-term exposure to chemicals producing central-peripheral axonopathies. **A:** *Acrylamide (ACR):* Groups of rats (N = 8/group) were exposed to ACR at 0 (Controls), 3.6, 7.2, or 14.4 mg/kg ip, 5 days/week for 12 weeks (from Kulig, ref. 79, with permission). **B:** n-*Hexane (HX):* Groups of rats (N = 8/group) were exposed by inhalation to HX at 0 ppm (controls), 900 ppm for 8 hr/day, 300 ppm for 23 hr/day, or 900 ppm for 23 hr/day, 5 days/week for 20 weeks (from Kulig, ref. 80, with permission). **C:** *Carbon disulphide (CS2):* Groups of rats (N = 8/group) were exposed by inhalation to CS2 at 0 (controls), 75, 225, or 700 ppm for 8 hr/day, 5 days/week for 36 weeks (from Kulig, ref. 80, with permission).

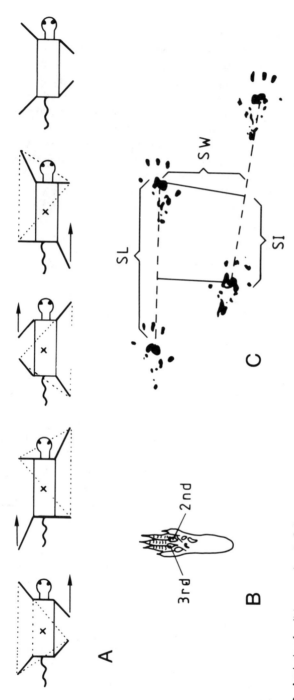

FIG. 3. Analysis of gait topography. **A:** Schematic diagram of sequential limb movements made by a quadruped walking in such a way as to be stable at all times. X marks the center of mass and the broken lines show the triangle of support. Each diagram represents the instant when the foot marked with an arrow is lifted. In each case, the center of mass will move forward before the next change of supporting feet (redrawn from Alexander, ref. 2, with permission). **B:** The sole of the right hind paw of a rat showing the second and third interdigital pads used as reference points in stride analysis (redrawn from Parker and Clarke, ref. 113, with permission). **C:** Example of parameters used in stride analysis showing stride length (SL), stride width (SW), and interpedal distance (SI) (redrawn from Parker and Clarke, ref. 113, with permission).

parently this sequence of limb movements is employed by all quadrupeds, even at fast walks involving stages in which only two feet are on the ground (2).

Analysis of successive footfalls indicates that stride length, stride width, and measures of gait symmetry for normal adult rats (Fig. 3) are highly consistent among animals of the same strain and sex over a reasonably wide range of body weights, with coefficients of variation ranging from 10 to 21% (for review, see 61,113). For male and female Wistar rats, for example, Parker and Clarke (113) reported a mean (±SEM) stride length of 11.8 (±0.06) cm and a stride width of 2.7 (±0.02) cm for females ranging in weight from 180 to 218 g, and a stride length of 11.9 (±0.08) cm and a stride width of 3.3 (±0.02) cm for males in the weight range from 220 to 270 g.

The analysis of gait topography has been applied to evaluate the effects of a number of different types of experimental treatments. These include studies of the effects of drugs such as barbiturates (122), mescaline (136), analogs of thyrotropin-releasing hormone (21); ethanol and diazepam (8); x-irradiation (109); vitamin B_1 (68) and B_6 deficiency (129); experimental models of peripheral nerve damage and disease including mechanically induced sciatic nerve damage and experimental allergic neuritis (29,162) and neurotoxic exposures including intrastriatal kainic acid injections (62), atropine administration in 6-hydroxydopamine-treated rats (130), 3-acetylpyridine, and acrylamide (68); and prenatal and neonatal ethanol exposure (55,101).

The apparent reliability and stability of gait parameters obtained from untrained animals and the sensitivity of this method in detecting the effects of a wide variety of treatments suggest that measurements of footfall patterns may provide a technique for quantifying altered gait which is suitable for screening purposes. There are, however, some considerations in the use of this technique in its present stage of development for general application in neurotoxicology assessment. First, there has been little standardization among laboratories with respect to the exact procedures used to collect footfall patterns in terms of the test apparatus used, the techniques employed for obtaining visual records of footprints, or the minimum number of successive footprints necessary for reliable results. Further, although both Hruska et al. (61) and Parker and Clarke (113) did not find significant correlations between gait parameters and body weight in adult rats of the same strain, comparisons of strains with different average body weights as well as studies in young, developing rats (e.g., 113,129) indicate that reduced body size may influence gait parameters and probably should be taken into account, especially in studies examining gait parameters in developing animals. Further, because of the demonstrated relationship between stride length and velocity of walking (61), the measurement of transit times in the walkway would also appear to warrant consideration. One of the practical drawbacks of this method is that the measurements of different gait parameters by hand are quite laborious and time-consuming for routine studies, which probably accounts for the typically small number of footprints analyzed for each rat in the earlier work. More recent studies, however, have utilized computer-supported measurement strategies with substantial savings in the time necessary to analyze hard-copy footprint records (28,140).

In addition, videorecording methods have also been applied to gait analysis. Clarke and Parker (20), for example, described a computer-supported method in which successive video images of footfall patterns were taken from a camera positioned below the floor of a transparent nonmoving walkway. Using frame-by-frame visual inspection of the video images and a computer cursor, both spatial characteristics of footfall patterns and the temporal characteristics of the step cycle could be obtained. An analogous approach using a fully automated system has also been described to examine footfall patterns of animals walking on a moving surface (144,163). However, since gait topography on moving surfaces may be different from that of terrestrial locomotion [see Parker and Clarke (113)], it may be more appropriate to consider footfall patterns collected in forced locomotion situations as specialized tests of motor coordination.

Tests of Motor Coordination

Tests of motor coordination are based on the premise that deficits in coordination may become more apparent in situations in which the capacity of the animal to maintain equilibrium or execute coordinated movements is challenged to a degree greater than that required by an animal walking at its own pace on a level, nonmoving surface. Many different techniques have been used for this purpose, and they vary considerably in the complexity of the apparatus and experimental procedures used, the degree of pretraining necessary to establish stable baseline levels of performance, the variability and stability of control performance, and whether a moving or stationary surface is employed.

Several tests which employ a nonmoving surface to measure coordination deficits have been described in the literature. In some tests, use has been made of the normal animal's tendency to reorient its head and body upwards when placed in a downwards direction as in, for example, the test of negative geotaxis in the rat (e.g., 115) and the inverted screen test for the mouse (e.g., 23,74,105), while others have exploited the normal animal's ability to walk on a grid floor without making missteps (e.g., 32,160). Further, the normal animal's ability to land from a fall with its hindpaws in a particular orientation has also been utilized as a measure of hindlimb weakness and incoordination (e.g., 14,34,107).

With respect to the test of negative geotaxis in the rat, Pryor et al. (115) reported relatively large coefficients of variation in control animals ranging from 69 to 135% which tended to increase with repeated testing. Despite the variability of control responding, however, this test was sensitive in detecting deficits in acrylamide- and tetraethyl tin-treated rats although somewhat later than detecting changes in hindlimb grip strength. With respect to the inverted screen test for mice, quantal data (e.g., the number of rats reaching the top of the screen within 60 sec) rather than latency is usually employed, suggesting considerable variability in the temporal measure of this task as well.

Relatively good psychometric properties in terms of acceptable levels of variability, stability of control performance, and sensitivity have been reported for tests of hindlimb splay. In this test, the fourth digit pad on each of the animal's hindpaws is inked in order to obtain a hard-copy record of the position of the hindpaws on landing. The animal is dropped from a height of approximately 30 cm onto a paper-covered surface and the distance between the two points of contact is measured (34). Using this technique, Broxup et al. (14) were able to detect changes in hindlimb splay in rats exposed to acrylamide at 40 mg/kg/day as early as 4 days after the start of treatment, whereas changes in hindlimb grip strength measured in the same animals were not affected until day 11. Because of its simplicity, ease of administration, and apparently good psychometric properties, tests of landing foot splay have been incorporated by some authors into functional observational batteries for screening purposes (e.g., 107).

With respect to tests of motor coordination employing moving surfaces, several investigators have used treadmill tests to study the effects of chemicals on motor coordination (e.g., 48,75,135). However, the rotarod test has been used most extensively. Rotarod testing techniques were introduced in the 1950s (33) and are still widely used for assessing motor coordination in pharmacological and toxicological experiments (e.g., 1,5,50,145). In this test, the animal is placed on a motor-driven rotating rod or drum and its ability to remain on the rod for a specified period of time is taken as a measure of motor coordination. A number of difficulties with rotarod testing techniques have been noted by several authors, including unstable baselines of control performance and the quantal nature of the data obtained. In order to minimize these problems, a number of modifications in testing procedures have been proposed, including the use of an electric shock landing platform to motivate animals to remain in motion on the rod and the use of an accelerating rod to obtain data of a continuous, rather than quantal, nature (11). In these authors' laboratory, the most difficult problems with the use of rotarod testing procedures were encountered in long-term studies involving months of exposure and repeated testing. Control animals became increasingly adept at escaping from the apparatus and affected animals would sometimes land in extremely severe and unnatural positions (i.e., on their head or back). As a result of these limitations, a test utilizing a TV-microprocessor–based system was developed (82,144,163) to quantify coordination deficits in a moving surface test environment. It was designed to avoid some of the practical problems associated with rotarod testing techniques (e.g., control performance failures, the need for shock or height to motivate stable performance, escape responding, etc.), but was still sensitive to neurotoxicant-induced motor deficits.

The test apparatus consists of a vertically positioned, transparent wheel which is slowly rotated at a constant speed of approximately 8 cm/sec. Crosspieces are attached to the floor of the wheel, forming rungs at regularly spaced intervals (Fig. 4). Just as animals learn to walk in wire mesh cages without their feet dropping through the openings, normal animals quickly learn to walk on the tops of the rungs while the wheel is in motion. For the 90 sec test trial, the left

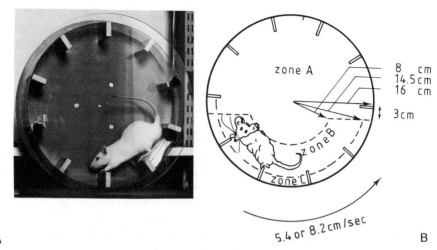

FIG. 4. Video-based automated quantitative analysis of hindlimb placement and movement. **A:** A normal rat walking on the tops of the rungs attached to the floor of a transparent motor-driven wheel rotating at a constant speed (from Kulig, ref. 76, with permission). **B:** Movements of the dyed left hindpaw are recorded by a video camera and digitized for computer analysis. When the paw enters zone A, walking failure is recorded. In zone B, correct hindlimb placement and stepping are registered. Missteps are detected when the paw falls between the rungs in zone C (from Tanger et al., ref. 144, with permission).

hindpaw of the rat is colored with a fluorescent dye and a color video camera placed in front of the wheel receives a lateral image of the wheel, the rat, and the dyed hindpaw. Video signals are processed electronically and translated into a series of X-Y coordinates with reference to the spatial characteristics of the wheel. These data are then fed into a personal computer for offline analysis of different components of correct and incorrect hindlimb movements. These include the time spent with the hindpaw correctly placed on top of the rungs (time on rung), the time the hindlimb was moving from one rung to the adjacent rung (small steps) or the subsequent rung (large steps) without dropping between the rungs, the time spent making missteps (hindpaw falling >1.5 cm below the top of the rungs) and walking failure [the amount of time the rat failed to stay in motion and was carried outside the lower quadrants of the wheel (zone B in Fig. 4)]. The Coordinated Movement Score, a composite score defined as the total time the hindpaw was either correctly placed on the rung or was moving from one rung to the next without dropping between the rungs, was used as a general index of coordination.

Using this technique, we have been able to detect significant coordination deficits in animals exposed to acrylamide as early as 1 day following a single exposure to either 50 or 100 mg/kg. As Fig. 5 demonstrates, acrylamide produced a dose-related decrease in the Coordinated Movement Score (denoted as "correct" on the left side of the graph) when measured 20 hr postexposure. Further, ex-

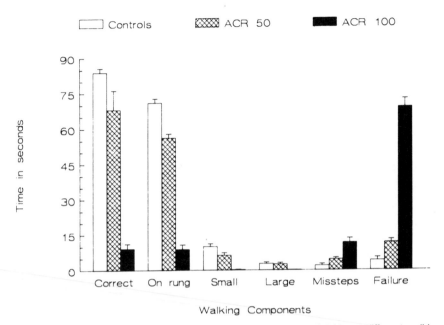

FIG. 5. Effects of a single injection of 0, 50, or 100 mg/kg ip acrylamide on different walking components measured 20 hr postinjection. (From Kulig et al., ref. 82, with permission).

amination of which parameters were affected indicated that the most pronounced effects were on walking failure. In addition, we have also applied the technique to examine changes in coordinated movement during exposure to lower levels of acrylamide for 12 weeks followed by a 6-week recovery. As Fig. 6A demonstrates, acrylamide at 14.4 mg/kg/day for 5 days/week produced progressive and significant decreases in the composite measure of correct stepping, which paralleled the effects seen in hindlimb grip strength measured in the same rats (see Fig. 2). Further, control performance was quite stable throughout the 18 weeks of the experiment, indicating that the technique could be applied successfully to the chronic testing situation. Compared to the effects of acute, high-dose acrylamide administration, however, low-dose chronic administration produced a markedly different profile of effects on the different components comprising correct and incorrect stepping. As Fig. 6B demonstrates, the effects of chronic low-level acrylamide measured during the last week of exposure indicated that incoordination was the result of a marked increase in missteps rather than walking failure.

The reason for the discrepancy in the pattern of coordinated motor effects produced by acute high-dose exposure compared to subchronic low-dose exposure is unknown. In humans, however, acute high-dose acrylamide poisoning does not produce signs of peripheral nerve damage, but rather a clinical picture in which signs of CNS involvement, including cerebellar ataxia, predominate (87).

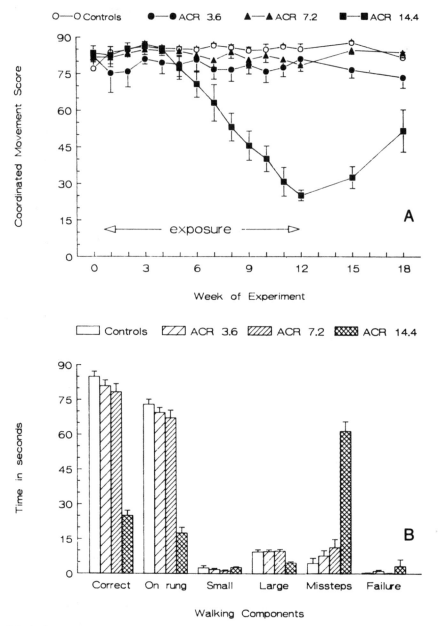

FIG. 6. Changes in coordinated hindlimb movement during and following 12 weeks of exposure to low levels of acrylamide. **A:** Effect of exposure on the composite score of coordinated movement across 18 weeks. **B:** Effects on different walking components measured during the last week of exposure. N = 8/group. (From Kulig, ref. 81, with permission).

Thus, it may be that the motor effects of acrylamide are differential in nature in the rat as well and that different types of motor impairments are manifested at different time points during the course of exposure. If such a hypothesis is correct, it may also be relevant to the use of acrylamide as a prototypic neurotoxic agent in studies comparing the relative sensitivities of different neurobehavioral test methods (e.g., 14) since some tests may be measuring the onset of different types of motor impairment with different underlying mechanisms.

The Quantification of Tremor

Tremor refers to rhythmic, involuntary oscillations of the whole body or a particular body part and, as a clinical sign, is seen in various neurological disease states such as Parkinson's disease and multiple sclerosis, and as a consequence of exposure to particular therapeutic drugs or chemicals (96,139). In addition, tremor can also result from environmental overexposure to a variety of compounds including metals, pesticides, and organic solvents (4). Some of the most dramatic instances of human neurotoxic disease have involved occupational exposure to tremorogens. Mercury-induced tremor was so widespread in the felt hatting industry at the turn of the century, for example, that the term "hatter's shakes" (or, in the case of the Connecticut hatting industry, "Danbury shakes") became part of the general lexicon (161). In a more recent outbreak, pesticide-induced tremor, referred to as "Kepone shakes," was described in industrial workers involved in the production of the organochlorine insecticide chlordecone (69).

The need to characterize chemically induced tremor in laboratory animals has led to the development of a number of qualitative and quantitative approaches. Techniques employing visual observation and simple rating scales have been employed by a number of investigators, either alone or in combination with polygraphic recordings, to study the effects of tremorogenic drugs (24,70,73,93,120).

There are a number of different types of transducers which have been employed to convert the mechanical energy of tremor to electrical energy for the purpose of visual display and/or quantification. Dill (30), for example, described techniques for the detection of tremor in the forelimb of the rat by use of an implanted magnet and an electric field. Knowles and Philips (73) used the analog output from a commercially available electronic activity monitor (Stoelting) which detects movement on the basis of changes in an electromagnetic field in order to obtain polygraphic recordings. As more accurate and sensitive transducers have become more widely available, other techniques for tremor measurement have been described, including the use of force-displacement transducers (22,93); load-cell transducers (e.g., 47); accelerometers (86); and strain gauges (67). As pointed out by Newland (110) in his excellent review of tremor detection and quantification, many of the test systems described in the literature measure not only the movement and oscillations of the rat, but also those of the test platform or cage as well. Thus, whichever transducer is selected, it is important to determine the sensitivity of the method and the extent to which the physical properties of the test system in the form of resonance and frequency-dependent damping contribute to the overall measurement (for an example of system-related resonance, see ref. 67).

In addition to a variety of techniques for the detection of tremor, different approaches have also been utilized for quantifying tremorogenic activity. Clement and Dyck (22), for example, used a single measure based on the number of signals occurring within a specified frequency range and greater in amplitude than a predetermined threshold value in order to determine an index of tremor severity in terms of "tremorogenic activity counts." Although such an index may be efficient for screening purposes, a considerable amount of information is also lost with such an approach. In human and nonhuman primate studies of tremor, spectral analytic techniques have been used extensively to provide a more comprehensive description of the characteristics of tremor in terms of tremor magnitude, dominant frequencies, and the shape of the spectral profile (for discussions of the mathematical basis and use of spectral analysis, see refs. 53,110). Similar techniques aimed at describing tremor intensity as a function of tremor frequency are being used increasingly at the rodent level as well (e.g., 43,47,67).

Such an approach has been used recently in an elegant series of studies designed to examine the tremor-inducing effects of different organochlorine pesticides and their possible sites and mechanisms of action (46,47,57,58,150,151). Figure 7 illustrates the changes in spectral analysis functions obtained from rats exposed to the pesticide chlordecone and examined 12 hr postexposure following either an acute dose (Fig. 7A) or 10 days of dosing at lower dose levels (Fig. 7B). As the figure demonstrates, animals repeatedly exposed to chlordecone displayed tremor having a peak frequency similar to that of rats after a single exposure. It is interesting to note that exposure to an acute dose of chlordecone at 10 mg/kg did not produce tremor, while 10-day repeated dosing at 5 mg/kg produced significant tremorogenesis, demonstrating the cumulative toxic nature of this compound.

In addition to methods based on whole body tremor, an ingenious technique employing operant procedures to examine the spectral profile of tremor in the forelimb alone has also been described (43,44). In this paradigm, food-deprived rats were trained with sweetened milk as the reward in an operant chamber equipped with an isometric force transducer as the operandum and a recessed dipper. The force transducer was located in a recessed aperture in such a way that the animal had access to both the operandum and the reinforcement dipper as long as the animal exerted a force of more than 20 g on the force transducer (Fig. 8). During response execution, oscillations of the rat's right forelimb detected by the force transducer were recorded and analyzed to obtain spectral density functions of tremor. Using this technique to study the effects of haloperidol,

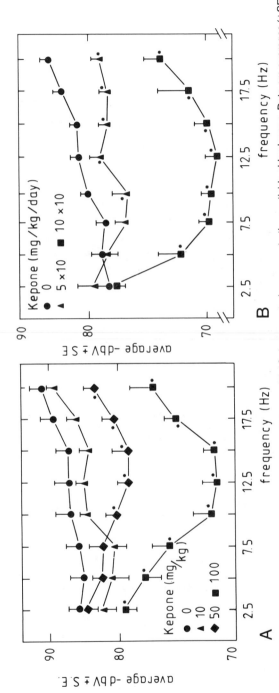

FIG. 7. Spectral analysis of tremorogenic activity following a (**A**) single or (**B**) 10-day repeated exposure to the pesticide chlordecone. Data are averages (±SEM) of 10 rats per group. High doses of chlordecone produce significant tremorogenesis as indicated by less negative −dBVs. (From Gerhart et al., ref. 47, with permission).

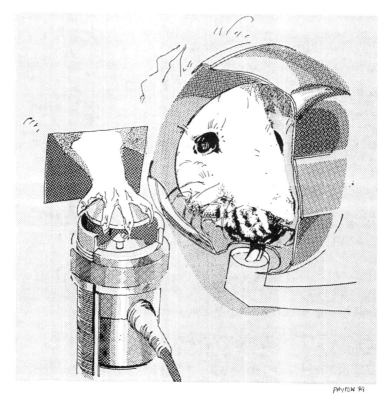

PAYTON 87

FIG. 8. The use of operant techniques to measure limb tremor. The figure shows an outside-the-chamber view of a rat performing the task of drinking milk from a dipper and keeping the dipper within reach by using the right forelimb to exert force on an isometric force transducer. (From Fowler et al., ref. 43, with permission).

Fowler and his colleagues reported dose-related increases in the amplitude of force oscillations in the 0 to 4.9 Hz and 10.0 to 25.0 Hz frequency bands, and an increase in the band width at 7 Hz, indicating frequency-dependent changes in force oscillations in haloperidol-treated rats.

The Measurement of Stereotypies, Catalepsy, and Other Dyskinesias

Research on schizophrenia and neurodegenerative diseases such as Parkinson's disease and Huntington's chorea has produced a voluminous literature on chemically induced abnormal movement and posture. Stereotypies produced by the dopamine agonists amphetamine and apomorphine, as well as catalepsy produced by reserpine and neuroleptic drugs blocking central dopamine receptors, for example, are widely recognized phenomena which have been studied for more than 20 years (see ref. 65 for review). In addition, animal models of basal ganglia movement disorders involving chemical-lesioning techniques with 6-hydroxy-dopamine (152), kainic acid (62), or quinolinic acid (132), as well as systemic administration of neurotoxins such as 1-methyl-4-phenyl-1,2,5,6-tetrahydro-pyridine (MPTP) (56) or iminodipropionitrile (IDPN) (19), produce a spectrum of dyskinesias ranging from hypo- and hyperkinesias, ataxic gait, circling behaviors, barrel rotations, lateral head weaving, backward head tilting, and retropulsion. With such a wide array of qualitatively different effects, it is not surprising that most studies have typically used observational techniques to document the presence of specific signs and, in some cases, to provide estimates of the severity of these behavioral phenomena in terms of either the frequency of occurrence, duration, or ranked severity scores (e.g., 31,35,89).

Although most of the work on stereotypy has utilized CNS-active drugs or experimental neurotoxins as tools for studying neurotransmitter functions, some environmentally relevant compounds may also be capable of producing similar types of effects. Using a computer-supported observational method, for example, Walker et al. (157) recently characterized the profile of stereotyped behaviors induced by exposure to triadimefon, a triazole fungicide used to protect cereal and fruit crops. As Fig. 9 demonstrates, triadimefon exposure produced a non-monotonic (inverted U-shape) dose-response function for forward locomotion, rearing, and grooming, and dose-related increases in licking, nose poking, and gnawing. At the highest dose tested (200 mg/kg), retropulsion, circling, and head weaving were also increased. In addition, biochemical studies indicated the involvement of monoaminergic neurons in triadimefon-induced stereotypies with the pattern of neurochemical effects resembling those produced by indirect-acting CNS stimulants, especially *d*-amphetamine.

In addition to observational methods, several tests of specific postural and motor abnormalities have been developed, including tests of catalepsy, rigidity, and circling behavior. The most common behavioral test to quantify the intensity of drug-induced catalepsy, for example, is the "bar test" originally described by Kuschinsky and Hornykiewicz (84) to quantify morphine-induced cataleptic reactions. The general procedure to test for catalepsy is to place the forepaws of the animal on a round bar at a height appropriate to the animal's size and to measure the time taken for the rat to descend from this posture. The catalepsy test has been used extensively in neuropharmacology to assess the antipsychotic properties of neuroleptics and as a tool for the study of the interactions between dopaminergic and other neurotransmitter systems. The procedural variables affecting measures on this test and the application of catalepsy testing in neuropharmacological studies of dopaminergic function have recently been reviewed (123).

With respect to the measurement of circling, several techniques for quantification of rotational behavior employing a "rotometer" have been described (7,13,52,133,152,158). The "rotometer" test chamber consists of a hemispherical or spherical bowl or a flat-bottomed cylinder or rectangular box. In most experiments, the rat is placed in a harness that is connected to a pivot of a vane

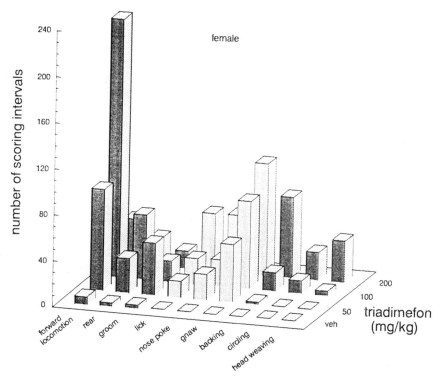

FIG. 9. Effects of acute exposure to the fungicide triadimefon on stereotyped behaviors in the rat. Starting 4 hr after dosing, each animal was observed for four consecutive 15-sec intervals every 5 min over the course of a 40-min session. The number of instances in which a particular behavior was observed during the 15-sec observation period was summed for each animal and each treatment group (N = 10/group). (From Walker et al., ref. 157, with permission).

which activates a sensing device such as a photocell or microswitch when the rat turns. Such techniques have been extensively used to quantify drug-induced turning in the unilateral striatal 6-hydroxydopamine model, both alone and in combination with an analysis of the laterality of stepping (164).

In addition to these relatively simple approaches, there is also a growing use of video-based computer methods for quantifying stereotyped behaviors. In a recent study aimed at examining the effects of dopamine D_1 and D_2 antagonists on oral behavior, for example, rats were confined to a restraining tube and small spots were painted on the upper and lower jaws using a fluorescent dye (88). A video camera equipped with a close-up lens was positioned in front of the rat's mouth and the output of the camera was fed to a computer with a movement detection circuit. The computer calculated the vertical distance between the upper and lower spots at a sampling rate of 60 times per second. By computing the amplitude and frequency of mouth openings based on the distance between the upper and lower spots, oral movements corresponding to stereotyped, vacuous chewing movements could be quantified and differentiated from orofacial tremor.

Further, video and photocell-based techniques for quantifying whole body and part-of-body stereotypies have also been described (e.g., 12,124,138). However, since these techniques are usually used in conjunction with the measurement of other types of locomotor activity, they have been included in the discussion of activity measurement presented below.

QUANTIFYING THE AMOUNT AND PATTERN OF MOVEMENT: MEASURES OF LOCOMOTION AND OTHER MOTOR ACTIVITIES

Of all the tests of motor function, measures of spontaneous activity have become the behavioral parameters most extensively used to examine the effects of neurotoxicant exposures. Although some of the recent focus on motor activity may be due in part, to its proposed use as a regulatory endpoint for neurotoxicity screening (154), efforts to study motor activity date back almost a century. A number of features make motor activity an attractive behavioral endpoint for examining the effects of chemical exposures. Motility is an inherent feature of all animals and motor activity occurs spontaneously without the need for deprivation or pretraining of the animals. Further, measures of motor activity have been shown to be sensitive to treatments known to affect central nervous function, including brain damage and drugs (118). Finally, the quantification of motor activity lends itself relatively easily to automation, which is an important consideration in a field of study where the number of compounds which are potentially worthy candidates for evaluation can be expected to be substantial.

Table 1 presents examples from the recent literature of some of the different types of studies being conducted to examine the effects of chemical exposures on motor activity. Although far from exhaustive, Table 1 illustrates that motor activity assessment is a very active area in many laboratories involved in neurotoxicology research and that the approaches being used to study toxicant-induced activity changes differ widely along a number of dimensions including the method of detection; the shape, size, and complexity of the test environments; the length of the measurement period; and the number and types of measures which are being used to describe motor activity effects.

In their excellent survey of motor activity assessment techniques, Reiter and MacPhail (118) presented a detailed discussion of the different detection methods which had up until that time been used, particularly in the field of behavioral pharmacology, for automatically detecting animal movement. With the exception of mechanical devices such as stabilimeters and jiggle cages, almost all of the different technical approaches for quantifying motor activity discussed in their review, including running wheels, photocell-based systems, and field detectors, are currently being used for neurotoxicity assessment; several new techniques,

TABLE 1. *Examples of motor activity studies in the recent literature using different types of measurement strategies*

Test environ.	Session length	Movement parameters	Compounds examined	References
I. Studies using photocell detection methods				
A. Photocell Detection in Simple Test Environments				
Rect	25 min	Hor/Ver	chlordimeform	85
Rect	25 min	Hor/Ver	amitraz	106
Rect	25 min	Hor	triethyltin	103
Rect	25 min	Hor	p-xylene	15
Rect	25 min	Hor	endosulfan	94
Cir	5 min	Hor	acrylamide, IDPN	131
Rect	20 min	Multivar	d-amphetamine, chlorpromazine, physostigmine, scopolamine	40
Rect	10 min	Multivar	iminodipropionitrile (IDPN)	64
Rect	60 min	Multivar	intratympanic sodium arsanilate	112
Rect	12 hr	Multivar	quinolinic acid lesions, MK801	49
HoCage	24 hr/day	Hor/Ver	trimethyltin	16
HoCage	24 hr/day	Hor/Ver	triethyltin	17
HoCage	24 hr/day	Hor	lead/disulfiram	111
HoCage	24 hr/day	Hor	haloperidol	146
B. Photocell Detection and Complex Maze-type Test Environments				
Fig-8	60 min	Hor	carbaryl, propoxur	121
Fig-8	60 min	Hor	cismethrin, deltamethrin	26
Fig-8	60 min	Hor	permethrin, cypermethrin, RU26607, fenvalerate, cyfluthrin, flucylthrinate, fluvalinate, p,p'-DDT	27
Fig-8	1–2 hr	Hor	triadimefon with drug challenge	25
Residential	23 hr	Hor/patterns	methyl mercury	36,38
Residential	23 hr	Hor/patterns	methylazoxymethanol	6
Residential	4 hr	Hor/patterns	lindane	91
II. Studies using field detectors and related devices				
A. Radiofrequency and Capacitance-based Devices				
Rect	10 min	Undifferentiated	toluene, n-hexane	114
Rect	10 min	Undifferentiated	acrylamide, tetraethyltin, arsenic trioxide, methyl mercury, chlordecone, lead acetate, monosodium salicylate	115
Rect	2–3 hr	Undifferentiated	toluene	59
B. Doppler radar				
HoCage	12 hr/day	Undifferentiated	methylene chloride, perchloroethylene, toluene, trichloroethylene, 1,1,1-trichloroethane	72
C. Infrared sensors				
HoCage	24 hr/day	Hor	acetanilid, phenylmercuric acid	142
III. Video-based systems for quantifying activity				
Rect	10 min	Multivar	acrylamide, carbon disulfide	79,80
Complex	10 min	Hor/Ver	styrene, tichloroethylene	77,78
Rect	5 min	Multivar	substantia nigra 6-hydroxydopamine	12

TABLE 1. *Continued.*

Test environ.	Session length	Movement parameters	Compounds examined	References
IV. Running wheels equipped with home cage monitors				
Wheel	24 hr/day	Running	diisopropyl phosphorofluoridate	117
Wheel	24 hr/day	Running	trimethyltin, triethyltin	99
Wheel	24 hr/day	Running	caffeine	100
Wheel	24 hr/day	Running	methoxychlor, progesterone	51

such as infrared sensors and video pattern recognition systems, recently have been added to the list.

With respect to the method of detection, the most widely used technique for measuring changes in motor activity is based on the use of photocells. Typically, the photocells are positioned in such a way that they are primarily sensitive to horizontally directed locomotor movement which is measured in either simple test environments, such as a rectangular box or circular enclosure, or in complex maze-type test environments, such as the figure-8 maze and the residential maze (see Table 1). In addition to photocell-based methods, field detection systems such as that used by Pryor et al. (115) in their validation studies have also been employed to provide estimates of chemically induced changes in activity levels. Video-based systems are also becoming increasingly applied in motor activity measurement.

The detection system which has been employed in the authors' laboratory is an automated video-based system capable of tracking horizontally directed locomotor activity. In this system, a camera positioned above the test apparatus receives a video image of the white rat on a black background which is digitized by computer into a series of X-Y coordinates, providing information on the amount of locomotor activity in terms of "total meters run," the distribution of movement at different speeds, and the amount of locomotor activity in different locations within the test chamber (143). Because video systems are, in general, poorly suited for detecting vertical movements when the camera is located above the animal, a specially designed capacitance system specifically sensitive to vertical movements is utilized to detect rearing (155). Figure 10 demonstrates the results of studies examining the effects of subchronic acrylamide (Fig. 10A) and methyl mercury (Fig. 10B) exposure on ambulation and rearing measured in this system. In the case of acrylamide, rearing and ambulation were depressed in the highest dose group by week 3 of exposure, but these effects did not reach statistical significance until week 6 and 9, respectively. A similar progressive reduction in rearing and ambulation was also seen during the course of subchronic methyl mercury exposure. For both compounds, rearing was virtually abolished in the final stages of exposure. For acrylamide, there was some evidence of recovery of function during the postexposure period which paralleled the recovery of

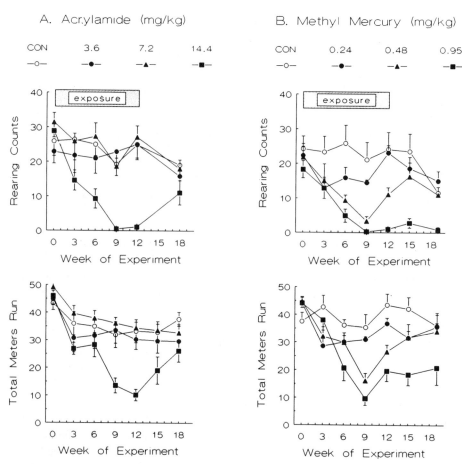

FIG. 10. Changes in spontaneous rearing and ambulation during and following subchronic exposure of rats to **(A)** 0, 3.6, 7.2, or 14.4 mg/kg acrylamide (*ACR*) or to **(B)** 0, 0.24, 0.48, or 0.95 mg/kg methyl mercury (MM). Activity was measured for 10 min in a 1 m² simple test environment using an automated video- and capacitance-based system. N = 8/group. (From Kulig, ref. 81, with permission).

function seen in measures of grip strength and coordinated movement measured in the same animals (see Figs. 2 and 6A). Effects of methyl mercury treatment, however, persisted to a greater degree and were most apparent for the rearing measure. These effects of acrylamide and methyl mercury on horizontal and vertical locomotor activity are in contrast to those reported in the Pryor et al. study (115). In that study, no significant effects of subchronic acrylamide or methyl mercury exposure could be found on measures of undifferentiated motor activity when activity was measured using a radio frequency-based field device. The device used by Pryor et al. detects both whole-body and part-of-body movements to provide a single index of undifferentiated activity. Such a measure is obviously quite different in nature from that obtained with methods designed to track specific locomotor movements such as ambulation and rearing. The extent to which the differences in measurement techniques contributed to the discrepancy in the effects seen with acrylamide and methyl mercury found in our laboratory compared to those reported in the Pryor et al. study is unknown. However, there is some evidence that measures of undifferentiated motor activity obtained with field devices do not necessarily correlate with locomotor movement (90).

Although cross-comparison studies of different devices to measure motor activity are relatively rare, the available data indicate that, at least for photocell-based devices, measures of activity obtained in both simple and complex test environments yield comparable results, in general (94,95). In a cross-comparison study using a commercially available photodetection system from Motron Products with a rectangular enclosure as the simple test environment and a figure-8 maze as the complex environment, for example, MacPhail (94) compared the effects of the insecticide endosulfan on activity levels measured in these two devices. As Fig. 11 demonstrates, the dose-response curves found using these two devices were virtually identical up to a dose of 10 mg/kg (94), indicating that the two devices were, in fact, measuring the same phenomenon with the same degree of sensitivity. Only at very high doses, which induce nonlocomotor activity changes, did the results obtained with these two devices show divergence, in that locomotor activity showed a further decline in the figure-8 maze and a stabilization in the Motron device. Although the reason for such divergence is uncertain, the Motron device uses a much shorter distance between photocells in order to detect movement, a feature which makes the detection of nonlocomotor activity more likely than in the figure-8 maze. Thus, it may be that photocell density rather than environmental complexity per se accounts for the

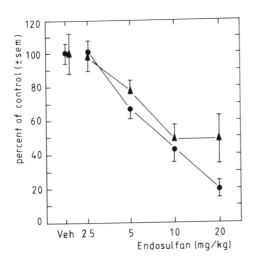

FIG. 11. Effects of acute endosulfan administration on the motor activity of adult rats measured in a complex figure-8 maze test environment (*circles*) or a simple rectangular Motron activity chamber (*triangles*). N = 12/group. Activity was measured in the figure-8 maze for 60 min beginning 60 min after dosing. Test duration in the Motron chamber was 25 min beginning 120 min after dosing. (From MacPhail, ref. 94, with permission).

differences in the degree of motor activity depression that were seen at very high doses of this compound.

In all of the studies discussed above, relatively short test sessions were employed to examine chemically induced changes in motor activity. In most cases, the choice of session length is made on the basis of the rate of habituation which animals display in a particular test environment (119). In simple test environments, the test session length chosen is usually on the order of 10 to 30 min, while typically longer times have been employed for complex environments such as the figure-8 maze. However, the rate of habituation is not the only consideration in the choice of testing times. Several studies have indicated that the effects of some treatments on motor activity are apparent only during the nocturnal phase of the light-dark cycle. In studies examining prenatal exposure to polychlorinated biphenyls, for example, effects on motor activity could be detected only during the dark phase (149), and similar findings have been reported for other compounds as well (6,49). Because of the possible differential effects of treatments on motor activity as a function of diurnal rhythms, some investigators, (e.g., 6,36,38) have employed session lengths of up to 23 hr using the residential maze to study the effects of prenatal chemical exposures on both temporal and spatial patterns of activity. In addition, there is also evidence that chemical exposures may disrupt diurnal activity patterns, and some investigators have developed large-scale systems which can be used to measure daily activity patterns in the animal's home cage on a 24-hr basis (39). Both photocell-based systems (16,17,111,146) and wheel running devices have been utilized for this purpose (e.g., 51,99) and an infrared sensor system designed for 24-hr motor activity monitoring has also recently been described (142).

Although motor activity studies differ widely in terms of the techniques and paradigms employed, the majority of studies tend to be similar in that they employ only one or two measures of motor activity to assess neurotoxicant-induced effects. This has been due in part to a tendency to conceptualize motor activity as a unitary phenomenon (98,108,116), but it also reflects, to some degree, what has been technically feasible at different points in time during the long history of automated motor activity methods.

For the human observer, classes of whole-body locomotor movements such as forward progression and rearing are relatively easy to recognize and to distinguish from each other as well as from part-of-body movements, such as sniffing and grooming. As a result, observational methods typically make distinctions among different classes of motor behavior as an integral part of the measurement system. For example, nine parameters have been used for characterizing whole-body movements and eight parameters for part-of-body movements in observational techniques using the open field (159), and seven categories of whole-body movement (locomotion, rearing, circling, patterned locomotion, retropulsion, jumping, and inactive) and eleven part-of-body movement categories have been delineated in observational techniques for quantifying stereotyped behaviors (89). With automated tests, however, the discriminatory ability of the human eye is not easily duplicated. As a result, the approach typically used in the automated assessment of motor activity has been to reduce motor activity to a single variable in one of two ways: either by employing combinations of sensing devices and environmental test chambers which favor the detection of a single or several aspects of motor activity (e.g., horizontal locomotion and rearing measured with photocells or video systems) to the exclusion (or partial exclusion) of other types of movement, or to use test systems which detect, but do not differentiate, different types of whole- and part-of-body movements (e.g., undifferentiated motor activity measured with field detection devices).

Although this has been the general approach to automated motor activity assessment, increasing efforts have been directed in recent years toward the development of systems which provide multivariate information regarding whole-body and part-of-body movement patterns using refined capacitance (141), and photocell (126) and video-based (12,60,71) detection systems. Recently, for example, a commercially available photocell-based system, the Digiscan (Omnitech Electronics), employed very dense spacing of photocells in a simple test environment in order to derive 23 motor activity measures which provide information on both the total amount of horizontally and vertically directed locomotor movements as well as movement patterns indicative of whole-body (e.g., circling) and part-of-body movements. This device has been used most extensively in neuropharmacological studies aimed at examining effects on different activity parameters of treatments known to produce hyperactivity and stereotypies (10,49,125,126,127,128) and has been recently examined in studies aimed at validating this device for the purpose of neurotoxicity assessment (40,64). Similarly, a number of recent reports have described complex video-based pattern recognition systems capable of detecting whole-body and part-of-body stereotypies as well as the temporal sequencing of normal motor acts (12,60,71,138). Because these techniques have not, as yet, been widely used to examine the effects of environmentally relevant neurotoxic agents, it is not possible to judge their usefulness for neurotoxicity assessment vis à vis other, more simple systems. However, the successful application of these techniques in behavioral pharmacology and the neurosciences indicates that they will also become increasingly used for examining the effects of chemical exposures.

QUANTIFICATION OF PSYCHOMOTOR PERFORMANCE USING OPERANT TECHNIQUES

The method for measuring forelimb tremor in rats performing on a learned operant task described earlier is an example of the use of operant methods to study fine motor control in laboratory animals. Although operant studies typically use response rate as the primary dependent variable, fine motor control experiments exploit the highly structured operant test situation to measure multiple biophysical parameters of motor response execution. A number of different paradigms have been described in the literature for examining different aspects of motor performance using operant techniques (41). Some experiments employ standard operant levers to examine response duration (156) or response initiation (134), while other experiments employ force-sensing devices in order to examine changes in applied force during response execution (42). The biophysical parameters of response execution are sensitive to a variety of drugs. For example, multivariate techniques for making distinctions between major drug classes such as neuroleptics, benzodiazepines, and barbiturates have been described (156).

The extent to which operant methodologies for studying fine motor control can successfully be applied to the study of the motor effects of low-level exposure to environmental chemicals is uncertain since few studies have addressed this point. Bushnell (15) demonstrated that differences in response force requirements could modulate p-xylene–induced changes in operant responding, but exposure and parameters of response execution were not fully explored. Further, Elsner et al. (37) described a force titration paradigm for examining the effects of 2,5-hexanedione (2,5HD) on multiple biophysical aspects of response execution, but their data did not provide convincing evidence that response force parameters were superior to measures of grip strength in detecting the onset of 2,5HD-induced motor impairment. It was noted in passing, however, that measures of response speed, rather than response force, were significantly affected before effects on grip strength and response force could be measured. We have also found measures of response speed in the performance of discrete-trial operant tasks to be sensitive to the effects of chemical exposure, particularly to the acute effects of organic solvents (77,80). To what extent techniques designed to study fine motor control are, in general, sensitive to the acute effects of low-level chemical exposures on psychomotor performance has not been systematically explored. However, it may be that chemicals which produce motor effects through central mechanisms are better suited for study with these techniques than chemicals which produce peripheral neuropathies.

SUMMARY

Neurotoxic exposures can affect motor functions in a variety of ways, ranging from specific clinical signs such as tremor and paralysis to more general effects such as changes in spontaneous activity. A number of behavioral techniques have been developed or refined in recent years to quantify these different types of chemically induced motor changes. Methods for measuring the severity of clinical signs such as neuromuscular weakness, gait disturbances, coordination deficits, and tremor have been described in the literature and are being increasingly applied in experiments employing repeated measures as designs for tracking the development of chemically induced motor impairments. Innovative, multivariate approaches utilizing motor activity assessment techniques and operant methodologies have been described which use novel technologies for studying the expression of motor impairments in different types of test environments. The techniques described in this chapter differ widely in terms of their complexity, the degree to which they are automated, and the number and types of measurements which they provide. Some techniques—grip strength measurements, for example—are simple and inexpensive to perform and can be easily incorporated into neurotoxicity screening batteries. Others require a considerable amount of technical expertise and support. Despite the diversity of the technical approaches involved, however, the general trend in the quantification of motor function seems to be toward the development of precise methods which show not only sensitivity, but specificity as well. This is particularly apparent in the development and use of computerized, video-based techniques which are being increasingly used to study not only the amount of spontaneous locomotion, but the temporal and spatial patterns of whole-body and part-of-body movements as well. As detection methods become increasingly sensitive and pattern recognition techniques more sophisticated, further developments in multivariate approaches to examining the effects of chemical exposures on different aspects of motor function can be expected.

REFERENCES

1. Alder S, Zbinden G. Neurobehavioral tests in single- and repeated-dose toxicity studies in small rodents. *Arch Toxicol* 1983;54:1–23.
2. Alexander R McN. Terrestrial locomotion. In: Alexander R McN, Goldspink G, eds. *Mechanics and energetics of animal locomotion.* London: Chapman and Hall, 1977;169–203.
3. Anger WK. Neurobehavioral testing of chemicals: impact on recommended standards. *Neurobehav Toxicol Teratol* 1984;6:147–153.
4. Anger WK, Johnson BL. Chemicals affecting behavior. In: O'Donaghue JL, ed. *Neurotoxicity of industrial and commercial chemicals,* vol. I. Boca Raton, FL: CRC Press, 1985;51–148.
5. Atchison WD. Reduced safety factor for neuromuscular transmission and abnormal quantal secretion precede neuromuscular weakness induced by dithiobiuret. *Toxicol Appl Pharmacol* 1990;106:234–244.
6. Balduini W, Lombardelli G, Peruzzi G, Cattabeni F, Elsner J. Nocturnal hyperactivity induced by prenatal methylazoxymethanol administration as measured in a computerized residential maze. *Neurotoxicol Teratol* 1989;11:339–343.
7. Barber DL, Blackburn TP, Greenwood DT. An automatic apparatus for recording rotational behaviour in rats with brain lesions. *Physiol Behav* 1973;11:117–120.
8. Becker A, Palissa A, Grimm M. Gait analysis—a useful method for quantitatively measuring ataxia in mice. *Z Versuchstierkd* 1988;31:89–94.
9. Bhagat B, Wheeler M. Effect of nicotine on the swimming endurance of rats. *Neuropharmacology* 1973;12:1161–1165.
10. Boast CA, Pastor G. Characterization of motor activity patterns induced by N-methyl-D-aspartate antagonists in gerbils. *Pharmacol Biochem Behav* 1987;27:553–557.
11. Bogo V, Hill TA, Young RW. Comparison of accelerod and rotarod sensitivity in detecting ethanol- and acrylamide-induced performance decrement in rats: review of experimental considerations of rotating rod systems. *Neurotoxicology* 1981;2:765–787.
12. Bonatz AR, Steiner H, Huston JP. Video image analysis of behavior by microcomputer: categorization of turning and locomotion after 6-OHDA injection into the substantia nigra. *J Neurosci Meth* 1987;22:13–26.
13. Brann MR, Hacker M, Finnerty M, Ellis J, Lenox RH, Erlich YH. Automated analysis of stereotypic behavior induced by psychomotor stimulants. *Pharmacol Biochem Behav* 1983;19:57–62.

14. Broxup B, Robinson K, Losos G, Beyrouty P. Correlation between behavioral and pathological changes in the evaluation of neurotoxicity. *Toxicol Appl Pharmacol* 1989;101:510–520.
15. Bushnell PJ. Behavioral effects of acute p-xylene inhalation in rats: autoshaping, motor activity, and reversal learning. *Neurotoxicol Teratol* 1988;10:569–577.
16. Bushnell PJ, Evans HL. Effects of trimethyltin on homecage behavior in rats. *Toxicol Appl Pharmacol* 1985;79:134–142.
17. Bushnell PJ, Evans HL. Diurnal patterns in homecage behavior of rats after acute exposure to triethyltin. *Toxicol Appl Pharmacol* 1986;85:346–354.
18. Cabe PA, Tilson HA. The hindlimb extensor response: a method of assessing motor dysfunction in rats. *Pharmacol Biochem Behav* 1978;9:133–136.
19. Cadet JL. The iminodipropionitrile (IDPN)-induced dyskinetic syndrome: behavioral and biochemical pharmacology. *Neurosci Biobehav Rev* 1989;13:39–45.
20. Clarke KA, Parker AJ. Locomotion in the rat-an underview. *Med Sci Res* 1988;16:901–902.
21. Clarke KA, Steadman P. Abnormal locomotion in the rat after administration of a TRH analogue. *Neuropeptides* 1989;14:65–70.
22. Clement JG, Dyck WR. Device for quantifying tremor activity in mice: antitremor activity of atropine versus soman- and oxotremorine-induced tremors. *J Pharmacol Meth* 1989;22:25–36.
23. Coughenour LL, McClean JR, Parker RB. A new device for the rapid measurement of impaired motor function in mice. *Pharmacol Biochem Behav* 1977;6:351–353.
24. Coward OM, Doggett NS. Potentiation of LON 954 tremor by typical and atypical neuroleptics—an indication of striatal dopamine antagonism. *Psychopharmacology* 1980;67:177–180.
25. Crofton KM, Boncek VM, MacPhail RC. Evidence for monoaminergic involvement in triadimefon-induced hyperactivity. *Psychopharmacology* 1989;97:326–330.
26. Crofton KM, Reiter LW. Effects of two pyrethroid insecticides on motor activity and the acoustic startle response in the rat. *Toxicol Appl Pharmacol* 1984;75:318–328.
27. Crofton KM, Reiter LW. The effects of type I and II pyrethroids on motor activity and the acoustic startle response in the rat. *Fund Appl Toxicol* 1988;10:624–634.
28. De Medinaceli L, De Renzo E, Wyatt RJ. Rat sciatic functional index data management system with digitized input. *Comput Biomed Res* 1984;17:184–192.
29. De Medinaceli L, Freed WJ, Wyatt RJ. An index of the functional condition of rat sciatic nerve based on the measurements made from walking tracks. *Exp Neurol* 1982;77:634–643.
30. Dill RE. Induction and measurement of tremor and other dyskinesias. In: Myers RD, ed. *Methods in psychobiology,* vol. 3. New York: Academic Press, 1977;241–257.
31. Dourish CT. A pharmacological analysis of the hyperactivity syndrome induced by β-phenylethylamine in the mouse. *Br J Pharmacol* 1982;77:129–139.
32. Dudek BC, Philips TJ. Distinctions among sedative, disinhibitory, and ataxic properties of ethanol in inbred and selectively bred mice. *Psychopharmacology* 1990;101:93–99.
33. Dunham NW, Miya TS. A note on a simple apparatus for detecting neurological deficits in rats and mice. *J Am Pharmacol Assoc* 1957;46:208–209.
34. Edwards PM, Parker VH. A simple, sensitive, and objective method for early assessment of acrylamide neuropathy in rats. *Toxicol Appl Pharmacol* 1977;40:589–591.
35. Eilam D, Szechtman H. Biphasic effect of D-2 agonist quinpirole on locomotion and movements. *Eur J Pharmacol* 1989;161:151–157.
36. Elsner J. Testing strategies in behavioral teratology: III. Microanalysis of behavior. *Neurobehav Toxicol Teratol* 1986;8:573–584.
37. Elsner J, Fellmann C, and Zbinden G. Response force titration for the assessment of the neuromuscular toxicity of 2,5-hexanedione in rats. *Neurotoxicol Teratol* 1988;10:3–13.
38. Elsner J, Looser R, Zbinden G. Qualitative analysis of rat behavior patterns in a residential maze. *Neurobehav Toxicol* 1979;1(Suppl. 1):163–179.
39. Evans HL. Behaviors in the home cage reveal toxicity: recent findings and proposals for the future. *J Am Coll Toxicol* 1989;8:35–52.
40. Fitzgerald RE, Berres M, Schaeppi U. Validation of a photobeam system for assessment of motor activity in rats. *Toxicology* 1988;49:433–439.
41. Fowler SC. Force and duration of operant response and dependent variables in behavioral pharmacology. In: Thompson T, Dews PB, eds. *Advances in behavioral pharmacology* New York: Erlbaum, 1987.
42. Fowler SC, LaCerra MM, Ettenberg A. Effects of haloperidol on the biophysical characteristics of operant responding: implications for motor and reinforcement processes. *Pharmacol Biochem Behav* 1986;25:791–796.

43. Fowler SC, Liao RM, Skjoldager P. A new rodent model for neuroleptic-induced pseudo-parkinsonism: low doses of haloperidol increase forelimb tremor in the rat. *Behav Neurosci* 1990;104:449–456.

44. Fowler SC, Morgenstern C, Notterman JM. Spectral analysis of variations of force during a bar pressing time discrimination. *Science* 1972;176:1126–1127.

45. Frischer RE, King JA, Rose KJ, Strand FL. Maturational changes in neonatal rat motor system with early postnatal administration of nicotine. *Int J Dev Neurosci* 1988;6:149–154.

46. Gerhart JM, Hong JS, Tilson HA. Studies on the possible sites of chlordecone-induced tremor in rats. *Toxicol Appl Pharmacol* 1983;70:382–389.

47. Gerhart JM, Hong JS, Uphouse LL, Tilson HA. Chlordecone-induced tremor: quantification and pharmacological analysis. *Toxicol Appl Pharmacol* 1982;66:234–243.

48. Gibbins, RJ, Kalant H, LeBlanc AE. A technique for accurate measurement of moderate degrees of alcohol intoxication in small animals. *J Pharmacol Exp Ther* 1968;159;236–242.

49. Giordano M, Ford LM, Brauckmann JL, Norman AB, Sanberg PR. MK801 prevents quinolinic acid-induced behavioral deficits and neurotoxicity in the striatum. *Brian Res Bull* 1990;24: 313–319.

50. Goodlett CR, Thomas JD, West JR. Long-term deficits in cerebellar growth and rotarod performance of rats following "binge-like" alcohol exposure during the neonatal brain growth spurt. *Neurotoxicol Teratol* 1991;13:69–74.

51. Gray LE, Ostby JS, Ferrell JM, Sigmon ER, Goldman JM. Methoxychlor induces estrogen-like alterations of behavior and the reproductive tract in the female rat and hamster: effects on sex behavior, running wheel activity, and uterine morphology. *Toxicol Appl Pharmacol* 1988;96: 525–540.

52. Greenstein S, Glick SD. Improved automated apparatus for recording rotation (circling behavior) in rats or mice. *Pharmacol Biochem Behav* 1975;3:507–510.

53. Gresty M, Buckwell D. Spectral analysis of tremor: understanding the results. *J Neurol Neurosurg Psychiatr* 1990;53:976–981.

54. Gruner JA, Altman J, Spivack N. Effects of arrested cerebellar development on locomotion in the rat: cinematographic and electromyographic analysis. *Exp Brain Res* 1980;40:361–373.

55. Hannigan JH, Riley EP. Prenatal ethanol alters gait in rats. *Alcohol* 1989;5:451–454.

56. Heikkila RE, Hess A, Duvoisin RC. Dopaminergic neurotoxicity of 1-methyl-4-phenyl-1,2,5,6-tetrahydropyridine (MPTP) in mice. *Science* 1984;224:1451–1453.

57. Herr DW, Mailman RB, Tilson HA. Blockade of only spinal α_1 adrenoceptors is insufficient to attenuate DDT-induced alterations in motor function. *Toxicol Appl Pharmacol* 1989;101: 11–26.

58. Herr DW, Tilson HA. Modulation of p,p'-DDT-induced tremor by catecholaminergic agents. *Toxicol Appl Pharmacol* 1987;91:149–158.

59. Hinman DJ. Biphasic dose-response relationship for effects of toluene inhalation on locomotor activity. *Pharmacol Biochem Behav* 1987;26:65–69.

60. Hopper DL, Kernan WJ, Wright JR. Computer pattern recognition: an automated method for evaluating motor activity and testing for neurotoxicity. *Neurotoxicol Teratol* 1990;12:419–428.

61. Hruska RE, Kennedy S, Silbergeld EK. Quantitative aspects of normal locomotion in rats. *Life Sci* 1979;25:171–180.

62. Hruska RE, Silbergeld EK. Abnormal locomotion in rats after bilateral injection of kainic acid. *Life Sci* 1979;25:181–194.

63. Hunter D. *The diseases of occupations,* 5th ed. London: English Universities Press, 1974;240–297.

64. Ivens I. Neurotoxicity testing during long-term studies. *Neurotoxicol Teratol* 1990;12:637–641.

65. Jackson DM. Drug-induced behavioral models of central disorders. In: Boulton AA, Baker GB, Juorio AV, eds. *Drugs as tools in neurotransmitter research, neuromethods,* vol. 12. Clifton, NJ: Humana Press, 1989;385–441.

66. Johnels B, Steg G, Ungerstedt U. A method for mechanographical recording of muscle tone in the rat: the effects of some antiparkinsonian drugs on rigidity induced by reserpine. *Brain Res* 1978;140:177–181.

67. Johnson JD, Meisenheimer TL, Isom GE. A new method for quantification of tremors in mice. *J Pharmacol Meth* 1986;16:329–337.

68. Jolicoeur FB, Rondeau DB, Hamel E, Butterworth RF, Barbeau A. Measurement of ataxia and related neurological signs in the laboratory rat. *Can J Neurol Sci* 1979;6:209–215.

69. Joy RM. The chlorinated hydrocarbon insecticides. In: Ecobichon DJ, Joy RM, eds. *Pesticides and neurological diseases.* Boca Raton, FL: CRC Press, 1982;91–150.
70. Kelly DM, Naylor RJ. Mechanisms of tremor induction by harmaline. *Eur J Pharmacol* 1974;27: 14–24.
71. Kernan WJ, Mullenix PJ, Hopper DL. Pattern recognition of rat behavior. *Pharmacol Biochem Behav* 1987;27:559–564.
72. Kjellstrand P, Holmquist B, Jonsson I, Romare S, Månsson L. Effects of organic solvents on motor activity in mice. *Toxicology* 1985;35:35–46.
73. Knowles WD, Philips MI. Neurophysiological and behavioral maturation of cerebellar function studied with tremorogenic drugs. *Neuropharmacology* 1980;19:745–756.
74. Koplovitz I, Romano JA, Stewart JR. Assessment of motor performance decrement following soman poisoning in mice. *Drug Chem Toxicol* 1989;12:221–235.
75. Kulig BM. Some behavioral effects of antiepileptic drugs during acute and prolonged treatment in the rat. *Acta Neurol Scand* 1981;64:(Suppl. 89):59–66.
76. Kulig BM. An automated technique for the evaluation of coordinated movement deficits in rats. *Zbl Bakt Hyg B* 1987;185:28–31.
77. Kulig BM. The effects of chronic trichloroethylene exposure on neurobehavioral functioning in the rat. *Neurotoxicol Teratol* 1987;9:171–178.
78. Kulig BM. The neurobehavioral effects of chronic styrene exposure in the rat. *Neurotoxicol Teratol* 1988;10:511–517.
79. Kulig BM. A neurofunctional test battery for evaluating the neurotoxic effects of long-term exposure to chemicals. *J Am Coll Toxicol* 1989;8:71–83.
80. Kulig BM. Methods and issues in evaluating the effects of organic solvents. In: Russell RW, Flattau PE, Pope AM, eds. *Behavioral measures of neurotoxicity.* National Academy Press, Washington, D.C., 1991;159–183.
81. Kulig BM. Evaluating the effects of chemical exposures on adaptive functioning. *Comp Biochem Physiol* 1991;100C:263–268.
82. Kulig BM, Vanwersch RAP, Wolthuis OL. The automated analysis of coordinated hindlimb movement in rats during acute and prolonged exposure to toxic agents. *Toxicol Appl Pharmacol* 1985;80:1–10.
83. Kulig BM, Wolthuis OL. The development of neurological and behavioral impairments in rats exposed subchronically to n-hexane. *Arbete Och Halsa* 1984;29:83.
84. Kuschinsky K, Hornykiewicz O. Morphine catalepsy in the rat. Relation to striatal dopamine metabolism. *Eur J Pharmacol* 1972;19:119–122.
85. Landauer MR, Tomlinson WT, Balster RL, MacPhail RC. Some effects of the formamidine pesticide chlordimeform on the behavior of mice. *Neurotoxicology* 1984;5:91–100.
86. Lehtinen MS, Gothoni PR. A system for measuring tremor intensity in rats. *IEEE Trans Biomed Eng* 1985;8:BME–32.
87. Le Quesne PM. Acrylamide. In: Spencer PS, Schaumberg HH, eds. *Experimental and clinical neurotoxicology.* Baltimore: Williams and Wilkins, 1980;309–325.
88. Levin ED, See RE, South D. Effects of dopamine D_1 and D_2 receptor antagonists on oral activity in rats. *Pharmacol Biochem Behav* 1989;34:43–48.
89. Lewis MH, Baumeister AA, McCorkle DL, Mailman RB. A computer-supported method for analyzing behavioral observations: studies with stereotypy. *Psychopharmacology* 1985;85:204–209.
90. Ljungberg T. Reliability of two activity boxes commonly used to assess drug induced behavioral changes. *Pharmacol Biochem Behav* 1978;8:191–195.
91. Llorens J, Tusell JM, Suñol C, Rodriguez-Farré E. Effects of lindane on spontaneous behavior of rats analyzed by multivariate statistics. *Neurotoxicol Teratol* 1989;11:145–151.
92. Lowdon IMR, Seaber AV, Urbaniak JR. An improved method of recording rat tracks for measurement of the sciatic functional index of Medinaceli. *J Neurosci Meth* 1988;24:279–281.
93. Lutes J, Lorden JF, Beales M, Oltmans GA. Tolerance to the tremorogenic effects of harmalin: evidence for altered olivo-cerebellar function. *Neuropharmacology* 1988;27:849–855.
94. MacPhail RC. Observational batteries and motor activity. *Zbl Bakt Hyg B* 1987;185:21–27.
95. MacPhail RC, Peele DB, Crofton KM. Motor activity and screening for neurotoxicity. *J Am Coll Toxicol* 1989;8:117–125.
96. Martinelli P. Tremor: a clinical and pharmacological survey. *J Neural Transm* 1986;22(Suppl.): 141–148.

97. Mattsson JL, Albee RR, Johnson KA, Quast JF. Neurotoxicologic examination of rats dermally exposed to 2,4-D amine for three weeks. *Neurobehav Toxicol Teratol* 1986;8:255–263.

98. Maurissen JPJ, Mattsson JL. Critical assessment of motor activity as a screen for neurotoxicity. *Toxicol Ind Health* 1989;5:195–202.

99. McMillan DE, Chang LW, Idemudia SO, Wenger GR. Effects of trimethyltin and triethyltin on lever pressing, water drinking and running in an activity wheel: associated neuropathology. *Neurobehav Toxicol Teratol* 1986;8:499–507.

100. Meliska CJ, Landrum RE, Landrum TA. Tolerance and sensitization to chronic and subchronic oral caffeine: effects on wheelrunning in rats. *Pharmacol Biochem Behav* 1990;35:477–479.

101. Meyer LS, Kotch LE, Riley EP. Alterations in gait following ethanol exposure during the brain growth spurt in rats. *Alcohol Clin Exp Res* 1990;14:23–27.

102. Meyer OA, Tilson HA, Byrd WC, Riley MT. A method for the routine assessment of fore- and hindlimb grip strength of rats and mice. *Neurobehav Toxicol* 1979;1:233–236.

103. Miller DB. Pre- and postweaning indices of neurotoxicity in rats: effects of triethyltin (TET). *Toxicol Appl Pharmacol* 1984;72:557–565.

104. Molinengo L, Orsetti M. Drug action on the "grasping" reflex and on swimming endurance: an attempt to characterize experimentally antidepressant drugs. *Neuropharmacology* 1976;15:257–260.

105. Moser VC, Balster RL. Acute motor and lethal effects of inhaled toluene, 1,1,1-trichloroethane, halothane, and ethanol in mice: effects of exposure duration. *Toxicol Appl Pharmacol* 1985;77:285–291.

106. Moser VC, MacPhail RC. Investigations of amitraz neurotoxicity in rats. III. Effects on motor activity and inhibition of monoamine oxidase. *Fund Appl Toxicol* 1989;12:12–22.

107. Moser VC, McCormick JP, Creason JP, MacPhail RC. Comparison of chlordimeform and carbaryl using a functional observational battery. *Fund Appl Toxicol* 1988;11:189–206.

108. Mullenix PJ. Evolution of motor activity tests into a screening reality. *Toxicol Ind Health* 1989;5:203–219.

109. Mullenix P, Norton S, Culver B. Locomotor damage in rats after X-irradiation in utero. *Exp Neurol* 1975;48:310–324.

110. Newland MC. Quantification of motor function in toxicology. *Toxicol Lett* 1988;43:295–319.

111. Oskarsson A, Ljungberg T, Ståhle L, Tossman U, Ungerstedt U. Behavioral and neurochemical effects after combined perinatal treatment of rats with lead and disulfiram. *Neurobehav Toxicol Teratol* 1986;8:591–599.

112. Ossenkopp K-P, Prkacin A, Hargreaves EL. Sodium arsanilate-induced vestibular dysfunction in rats: effects on open-field behavior and spontaneous activity in the automated Digiscan monitoring system. *Pharmacol Biochem Behav* 1990;36:875–881.

113. Parker AJ, Clarke KA. Gait topography in rat locomotion. *Physiol Behav* 1990;48:41–47.

114. Pryor GT, Dickinson J, Howd RA, Rebert CS. Neurobehavioral effects of subchronic exposure of weanling rats to toluene or hexane. *Neurobehav Toxicol Teratol* 1983;5:47–52.

115. Pryor GT, Uyeno ET, Tilson HA, Mitchell CL. Assessment of chemicals using a battery of neurobehavioral tests: a comparative study. *Neurobehav Toxicol Teratol* 1983;5:91–117.

116. Rafales LS. Assessment of locomotor activity. In: Annau Z, ed. *Neurobehavioral toxicology.* London: Edward Arnold Ltd., 1987;54–68.

117. Raslear TG, Kaufman LW. Diisopropyl phosphorofluoridate (DFP) disrupts circadian activity patterns. *Neurobehav Toxicol Teratol* 1983;5:407–411.

118. Reiter LW, MacPhail RC. Motor activity: a survey of methods with potential use in toxicity testing. *Neurobehav Toxicol* 1979;1(Suppl. 1):53–66.

119. Reiter LW, MacPhail RC. Factors influencing motor activity measurements in neurotoxicology. In: Mitchell CL, ed. *Nervous system toxicology.* New York: Raven Press, 1982;45–65.

120. Robertson HA. Harmaline-induced tremor: the benzodiazepine receptor as a site of action. *Eur J Pharmacol* 1980;67:129–132.

121. Ruppert PH, Cook LL, Dean KF, Reiter LW. Acute behavioral toxicity of carbaryl and propoxur in adult rats. *Pharmacol Biochem Behav* 1983;18:579–584.

122. Rushton R, Steinberg H. Mutual potentiation of amphetamine and amylobarbitone measured by activity in rats. *Br J Pharmacol* 1963;21:295–305.

123. Sanberg PR, Bunsey MD, Giordano M, Norman AB. The catalepsy test: its ups and downs. *Behav Neurosci* 1988;102:748–759.

124. Sanberg PR, Calderon SF, Giordano M, Tew JM, Norman AB. The quinolinic acid model of Huntington's disease: locomotor abnormalities. *Exp Neurol* 1989;105:45–53.
125. Sanberg PR, Hagenmeyer SH, Henault MA. Automated measurement of multivariate locomotor behavior in rodents. *Neurobehav Toxicol Teratol* 1985;7:87–94.
126. Sanberg PR, Moran TH, Kubos KL, Coyle JT. Automated measurement of stereotypic behavior in rats. *Behav Neurosci* 1983;97:830–832.
127. Sanberg PR, Moran TH, Kubos KL, Coyle JT. Automated measurement of rearing behavior in adult and neonatal rats. *Behav Neurosci* 1984;98:743–746.
128. Sanberg PR, Zoloty SA, Willis R, Ticarich CD, Rhoads K, Nagy RP, Mitchell SG, Laforest AR, Jenks JA, Harkabus LJ, Gurson DB, Finnefrock JA, Bednarik EJ. Digiscan activity: automated measurement of thigmotactic and stereotypic behavior in rats. *Pharmacol Biochem Behav* 1987;27:569–572.
129. Schaeffer MC, Cochary EF, Sadowski JA. Subtle abnormalities of gait detected in vitamin B_6 deficiency in aged and weanling rats with hind leg gait analysis. *J Am Coll Nutr* 1990;9:120–127.
130. Schallert T, Whishaw IQ, Ramirez VD, Teitelbaum P. Compulsive, abnormal walking caused by anticholinergics in akinetic, 6-hydroxydopamine-treated rats. *Science* 1978;199:1461–1463.
131. Schulze GE. Large-scale assessment of motor activity in rodents: procedures for routine use in toxicology studies. *J Am Coll Toxicol* 1990;9:455–463.
132. Schwarcz R, Foster HC, French ED, Whetsell WO, Kohler C. Excitotoxic models for neurodegenerative disorders. *Life Sci* 1984;35:19–32.
133. Schwarz RD, Stein JW, Bernard P. Rotometer for recording rotation in chemically or electrically stimulated rats. *Physiol Behav* 1978;20:351–354.
134. Skjoldager P, Fowler SC. Effects of pimozide, across doses and within sessions, on discriminated lever release performance in rats. *Psychopharmacology* 1988;96:21–28.
135. Smith SS, Woodward DJ, Chapin JK. Sex steroids modulate motor-correlated increases in cerebellar discharge. *Brain Res* 1989;476:307–316.
136. Sykes EA. Mescaline induced motor impairment in rats, assessed by two different methods. *Life Sci* 1986;39:1051–1058.
137. Somkuti SG, Tilson HA, Brown HR, Campbell GA, Lapadula DM, Abou-Donia MB. Lack of delayed neurotoxic effect after tri-o-cresyl phosphate treatment in male Fischer 344 rats: biochemical, neurobehavioral, and neuropathological studies. *Fund Appl Toxicol* 1988;10:199–205.
138. Spruijt BM, Gispen WH. Prolonged animal observation by use of digitized videodisplays. *Pharmacol Biochem Behav* 1983;19:765–769.
139. Stein RB, Lee RG. Tremor and clonus. In: Brooks VB, ed. *Handbook of physiology. Section I. The nervous system,* vol. 2. Motor control, part 1. Bethesda, MD: American Physiological Society, 1981;325–343.
140. Steinberg H, Sykes EA, McBride A, Terry P, Robinson K, Tillotson H. Computer analysis, using a digitizer, of ataxic mouse gait due to drugs. *J Pharmacol Methods* 1989;21:103–113.
141. Stoff DM, Stauderman K, Wyatt RJ. The time and space machine: continuous measurement of drug-induced behavior patterns in the rat. *Psychopharmacology* 1983;80:319–324.
142. Tamborini P, Sigg H, Zbinden G. Quantitative analysis of rat activity in the home cage by infrared monitoring. Application to the acute toxicity testing of acetanilide and phenylmercuric acetate. *Arch Toxicol* 1989;63:85–96.
143. Tanger HJ, Vanwersch RAP, Wolthuis OL. Automated TV-based system for open field studies: effects of metamphetamine. *Pharmacol Biochem Behav* 1978;9:555–557.
144. Tanger HJ, Vanwersch RAP, Wolthuis OL. Automated quantitative analysis of coordinated locomotor behaviour in rats. *J Neurosci Meth* 1984;10:237–245.
145. Tanii H, Hashimoto K. Neurotoxicity of acrylamide and related compounds in rats: effects on rotarod performance, morphology of nerves and neurotubulin. *Arch Toxicol* 1983;54:203–213.
146. Thiel R, Chahoud I, Schwabe R, Neubert D. Device for monitoring locomotor activity of 120 animals: motility of offspring of dams exposed to haloperidol. *Neurotoxicology* 1989;10:621–628.
147. Tilson HA. Behavioral indices of neurotoxicity. *Toxicol Pathol* 1990;18:96–104.
148. Tilson HA, Cabe PA. Assessment of chemically-induced changes in the neuromuscular function of rats using a new recording grip meter. *Life Sci* 1978;23:1365–1370.
149. Tilson HA, Davis GJ, McLachlan JA, Lucier GW. The effects of polychlorinated biphenyls given prenatally on the neurobehavioral development of mice. *Environ Res* 1979;18:466–474.

150. Tilson HA, Hong JS, Mactutus CF. Effects of 5,5-diphenylhydantoin (phenytoin) on neurobehavioral toxicity of organochlorine insecticides and permethrin. *J Pharmacol Exp Ther* 1985;233:285–289.
151. Tilson HA, Peterson NJ, Mactutus CF. Comparative effects of three organochlorine pesticides on hyperreactivity and tremor in rats. *Toxicologist* 1984;4:24.
152. Ungerstedt U. 6-Hydroxydopamine induced degeneration of central monoamine neurons. *Eur J Pharmacol* 1968;5:107–110.
153. Ungerstedt U, Arbuthnott GW. Quantitative recording of rotational behavior in rats after 6-hydroxydopamine lesions of the nigrostriatal dopamine system. *Brain Res* 1970;24:485–493.
154. US Environmental Protection Agency. Toxic Substances Control Act Test Guidelines: Final Rules. Subpart G-Neurotoxicity *Federal Register.* 1985;50:39458–39469.
155. Van den Steen L, Vanwersch RAP. Improved capacitive transducer for animal movements. *Med Biol Eng Comp* 1981;19:479–482.
156. Walker CH, Faustman WO, Fowler SC, Kazar DB. A multivariate analysis of some operant variables used in behavioral pharmacology. *Psychopharmacology* 1981;74:182–186.
157. Walker QD, Lewis MH, Crofton KM, Mailman RB. Triadimefon, a triazole fungicide, induces stereotyped behavior and alters monoamine metabolism in rats. *Toxicol Appl Pharmacol* 1990;102:474–485.
158. Walsh MJ, Silbergeld EK. Rat rotation monitoring for pharmacology research. *Pharmacol Biochem Behav* 1979;10:433–436.
159. Walsh RN, Cummins RA. The open-field test: a critical review. *Psychol Bull* 1976;83:482–504.
160. Watson M, McElligott JG. Cerebellar norepinephrine depletion and impaired acquisition of specific locomotor tasks in rats. *Brain Res* 1984;296:129–138.
161. Weiss B. Behavioral toxicology and environmental health science: opportunity and challenge for psychology. *Amer Psychol* 1983;38:1174–1187.
162. Wiethölter H, Eckert S, Stevens A. Measurement of atactic and paretic gait in neuropathies of rats based on analysis of walking tracks. *J Neurosci Meth* 1990;32:199–205.
163. Wolthuis OL, Vanwersch RAP. Behavioral changes in the rat after low doses of cholinesterase inhibitors. *Fund Appl Toxicol* 1984;4:S195–S208.
164. Ziegler MGM, Szechtman H. Relation between motor asymmetry and direction of rotational behaviour under amphetamine and apomorphine in rats with unilateral degeneration of the nigrostriatal dopamine system. *Behav Brain Res* 1990;39:123–133.

Neurotoxicology, edited by Hugh Tilson
and Clifford Mitchell. Published by
Raven Press, Ltd., New York 1992.

9

Reflex Modification and the Assessment of Sensory Dysfunction

K. M. Crofton

*Neurotoxicology Division, Health Effects Research Laboratory, US Environmental
Protection Agency, Research Triangle Park, North Carolina 27711*

Sensation is a fundamental part of the interaction of an organism with its environment. A wide variety of environmental and pharmaceutical chemicals are known to alter the structure and function of sensory systems (50,74,76,78). In fact, alterations in sensory function are frequently reported as the first signs of chemical exposure in humans (29).

Determining the magnitude of the potential for chemicals to adversely impact sensory system function is difficult at best. There are no precise accounts of the number of chemicals that disrupt sensory processes. Recently, Crofton and Sheets (19) estimated that 44% of the chemicals reported to be neurotoxic affect some aspect of sensory function (Fig. 1). Estimates of the percentage of all known chemicals with neurotoxic effects range from 3% (81) to 25% (85) and 28% (4). From these admittedly crude estimates, one can presume that 1.5 to 16% of all chemicals may be sensory toxicants. Although the potential magnitude of concern is justification enough, other reasons for the importance of assessing the effects of toxicants on sensory function include (a) a potentially greater susceptibility to toxicants of sensory receptors due to their location in the periphery, and (b) the need to eliminate sensory alterations as confounds in tests of cognitive and motor function. Therefore, it would appear crucial for proper assessment of health risks associated with chemical exposure that rapid, objective, and cost-effective methods be made available to screen chemicals for sensory system toxicity.

The assessment of sensory functions in animals has been a significant research tradition in comparative psychology, and many of the methods developed in that field have been applied in neurotoxicology. These methods vary from rel-

This manuscript has been reviewed by the Health Effects Research Laboratory, US Environmental Protection Agency, and approved for publication. Mention of trade names or commercial products does not constitute endorsement or recommendation for use.

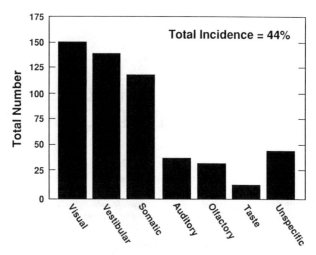

FIG. 1. Number of chemicals that affect sensory function. Data are expressed as the number of chemicals, by sensory modality, for which there are reported effects on sensory function. One chemical may affect more than one sensory modality. Note that visual system effects include reports of eye irritation. Total Incidence is the percentage of the total number of chemicals in the list that affect sensory function, with each chemical counted once. Data were derived from Anger and Johnson (5). (From Crofton and Sheets, ref. 19, with permission.)

atively simple and subjective sensory reflex tests, such as elicitation of the pinna reflex and pupil constriction, to more complicated operant-discrimination paradigms and evoked potential procedures. In addition to the traditional concerns for validity and sensitivity, the application of methods in neurotoxicology also raises concerns regarding generality and economy. Table 1 compares some of the more common tests used today. A cursory view of this table implies that some of the techniques may be better than others for screening or hazard identification. Even though tests of simple reflexes (e.g., pinna reflex, orienting response) may be inexpensive and easy to perform, they may lack precision in estimating psychophysical thresholds, and they are subject to experimenter bias, are hard to automate, and generally result in large interlaboratory variability (130). These tests are also thought to be relatively insensitive to toxicant-induced alterations in sensory function (70). Operant and conditioned discrimination procedures, on the other hand, represent some of the more sensitive and specific methods used for assessing sensory system dysfunction. Such tests have been used to characterize various sensory deficits following exposure to a wide variety of agents, including auditory deficits produced by aminoglycoside antibiotics (48,80), visual deficits produced by methyl mercury (76) or acrylamide (77) ammonia-induced disruption of olfaction (131); and somatosensory dysfunction produced by acrylamide (75). Although efforts have been made to improve the efficiency of tests in this class (57), these tests generally require extensive training, are time consuming, and are therefore limited to small numbers of subjects

TABLE 1. *Comparison of psychophysical techniques*

Test characteristic	Simple reflex	Reflex modification	Conditioning techniques		
			Classical	Operant	Suppression
Control of stimulus	0	+	+	+	+
Control of stimulus locus	0	+	+	+	+
Precision of estimates	−	+	+	+	+
Sensory-motor separation	−	0	+	0	0
Response bias	+	+	+	0	0
Motivational factors	+	+	+	−	−
Amount of response training	+	+	+	−	−
Amount of testing time	−	+	+	0	−
Replicability	+	+	+	+	+
Applicability to neonates	+	+	+	0	0
Labor intensiveness	−	0	0	+	+
Automatability	−	+	+	+	+
Equipment cost, complexity	+	0	−	−	−

Rating: +, advantage; −, disadvantage; 0, no advanage/disadvantage.
Adapted and modified from Cabe, ref. 10.

(116). One alternative to the extremes presented here is a methodology based on reflex modification.

REFLEX MODIFICATION

In rodents, reflex modification makes use of the whole body startle response, a characteristic sequence of reflexive muscle movements elicited by a sudden intense sensory stimulus. Reflex modification refers to a change in the reflex response due to a perceptible, and antecedent, change in the sensory environment. There are a number of reviews on the psychophysiology (56), anatomy and pharmacology (31,32,86), and use in toxicology (19,38,61,62,126) of the startle response and reflex modification.

Many characteristics of this procedure make it applicable for toxicological studies (Table 2). Reflex modification requires no prior training because a reflex reaction is used as the baseline. The phenomenon can be demonstrated in as few as two trials (58). There are no necessary invasive procedures, as opposed

TABLE 2. *Advantages of reflex modification*

1. No prior training requirement.
2. No invasive surgical or pharmacological procedures.
3. Rapid and objective measurements of sensory function.
4. Multiple sensory modalities can be tested.
5. High extrapolation potential.
6. Large neurobiological database.
7. Differentiation of the sensory and nonsensory effects of toxicants.
8. Ease of integration into multidisciplinary studies.

to electrophysiological procedures that require both anesthetics and surgery, either prior to (35) or during testing (73,92). A number of different sensory modalities can be tested. As shown in Fig. 2, reflex modification has been demonstrated using auditory, visual, and tactile stimuli (56,61). Reflex modification procedures have provided very rapid and objective assessments of toxicant-induced alterations in auditory function (24,43,135). Reflex modification testing yields data useful for extrapolation of animal data to humans because of the involvement of homologous neurobiological processes in all mammalian species. Figure 3 illustrates similar reflex inhibition processes in mice, rats, rabbits, and humans (61). Moreover, humans and animals have been tested using analogous procedures following exposures to solvents (106,107). Furthermore, a possible animal model of schizophrenia has been developed using reflex modification (7,8,45). There is also a growing database on the anatomy, pharmacology, and toxicology of the startle reflex and the reflex modification process to aid in interpretation of toxicant-induced effects (31,32,67,86,108,109,132,134). Table 3 lists chemicals that have been studied with automated equipment for testing the startle response and/or reflex modification. This table is evidence for the widespread use of the startle response and reflex modification procedures, as well as the sensitivity of these procedures to a wide variety of chemicals and other physical agents.

Reflex modification also has an ability to discriminate between the sensory

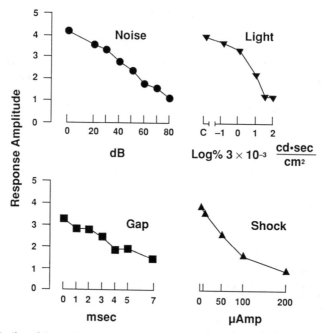

FIG. 2. Illustration of the similarity in the relationship between response inhibition and the intensity of S1 using qualitatively different stimuli. Response is in arbitrary units. (From Ison, ref. 61, with permission).

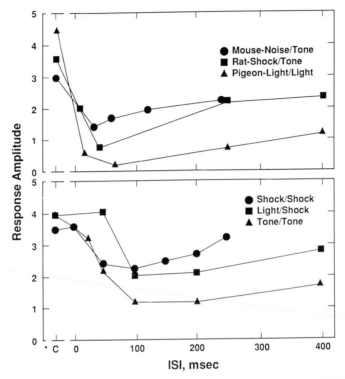

FIG. 3. Similarities in the relationship between the interstimulus interval and the inhibition of the startle response across different sensory modalities and different species. *Upper panel*: Data from the mouse, rat, and pigeon. *Lower panel*: Data collected using the eyeblink reflex in humans. Response is in arbitrary units. (From Ison, ref. 61, with permission.)

and nonsensory effects of toxicants (37,127). Another advantage of this procedure, due mainly to its noninvasive and rapid nature, is the ease with which it can be integrated into multidisciplinary studies. Startle response testing has been integrated as part of a neurobehavioral test battery in the National Toxicology Program (122). Recent work has demonstrated the usefulness of reflex modification in multidisciplinary assessments of the ototoxicity of trimethyltin (24,43,55) and the potentiation of noise-induced hearing loss by carbon monoxide (41,137).

The intent of the rest of this chapter is three-fold. The first is to review the basic methodology of reflex modification and to describe some of its characteristics. The second is to illustrate, using examples of chemical, genetic, and physical-trauma induced changes, the utility of reflex modification in detecting visual, somatosensory, and auditory dysfunction. The third purpose is to discuss how careful interpretation of reflex modification data can provide evidence for the differentiation of sensory from nonsensory (e.g., motor) effects of chemicals.

TABLE 3. Chemicals studied using the startle response

Chemical	Species	Age	SR	RM	References
METALS					
Arsenic trioxide	r	a	n	nt	89
Lead acetate	r	a	n	nt	89
Methyl mercury	r	d	in	nt	9,46,101
	r	a	n	nt	89
	r	a	in	de-i	133
Manganese	r	d	n	nt	66
cis-Platinum	r	a	de	nt	93
Triethyltin	r	d	de	nt	49
	r	a	de	ne	37
	r	a	de	nt	95,111
Trimethyltin	r	d	de	nt	103
	r	a	de	nt	104
	r	a	de	in-t	24,36,39,136
	gp	a	de	in-t	43
(bis)Tributyltin oxide	r	d	de	nt	22
Tetraethyltin	r	a	de	nt	89
Triethyl lead	r	a	n	nt	89
	r	a	in	nt	120,124
Trimethyl lead	r	a	de	nt	124
PESTICIDES					
Amitraz	r	a	de	nt	21
Chlordecone	r	d	de	nt	71
	r	d	n	nt	113
	r	a	in	nt	51,89,114,121
Lindane	r	a	in	in-i	54
	r	a	n	nt	121
p,p'-DDT	r	a	in	de-i	108
	r	a	in	nt	18,51,121
PCBs	r	a	de	nt	118
Pyrethroids					
Cismethrin	r	a	in	nt	16,17
Permethrin	r	a	in	nt	18,52,122
RU 11697	r	a	in	nt	18
NAK 1901	r	a	in	ne	53
Deltamethrin	r	a	de	nt	16,17,52
Cypermethrin (cis)	r	a	de	nt	18
(cis, trans)	r	a	n	ne	53
(cis)	r	a	in	ne	52
Cyfluthrin	r	a	de	nt	18
Fenvalerate	r	a	in	nt	18
Flucythrinate	r	a	de	nt	18
Fluvalinate	r	a	n	nt	18
SOLVENTS					
Benzene	r	d	n	nt	119
Ethylbenzene	r	a	de	ne	106
Ethanol	r	d	in	nt	1
	r	d	in	ne	88
	r	a	de	ne	127
	r	a	de	nt	47,87
n-Hexane	r	a	de	nt	91
Methanol	r	a	de	ne	127
	r	d	n	ne	115

TABLE 3. *Continued.*

Chemical	Species	Age	SR	RM	References
Methylbromide	rb	a	de	nt	3,105
Methylethylketone	h	a	n	de-i	107
Toluene	r	d	n	nt	110
	h	a	n	ne	107
	r	a	n	nt	91
Xylene	r	a	de	de-i	106
	r	d	n	nt	100
MISCELLANEOUS					
Acrylamide	r	a	de	nt	89
	r	a	de	ne	28
Aspertame	r	a	n	ne	123
Butyl nitrite	r	a	de	in-t	42
Carbon monoxide +					
noise	r	a	de	in-t	41,137
Cocaine	h	d	in	in-i	2
	r	a	in	nt	33
Hyperbaric pressure	r	a	de	nt	30
Kanamycin	r	a	n	de-i	126
Iminodipropionitrile	r	d	n	in-t	25
	r	a	de	in-t	26,69
n-Methyloacrylamide	r	a	de	nt	122
Monosodium					
glutamate	r	d	de	nt	112
	r	d	n	in-t	20
Monosodium					
salicylate	r	a	n	nt	89
Neomycin	r	a	n	in-t	135
Misonidazole	r	a	nr	in-t	15
Metronidazole	r	a	nr	in-t	15
Nitrous oxide	r	d	de	nt	97
Picrotoxin	r	a	de	nt	17,44
	r	d	in	nt	44
Posttraumatic stress					
disorder	h	d	de	de-i	84
Quinine	r	a	ne	ne	14
Strychnine	r	a	in	nt	65
	r	a	in	ne	34
Viral infection					
(LCMV)	r	d	de	in-t	12
X-irradiation	r	d	in	nt	79
GENETIC DISORDERS					
Hypertension	r	a	de	nt	117
Retinal dystrophy	r	a	n	de-i	128

Species: r, rat; rb, rabbit; h, human.
Age (at exposure): a = adult; d = development (prenatal/postnatal).
SR (effect on response amplitude): in = increase; de = decrease; n = no effect,
RM (reflex modification testing): in-t = increase in sensory threshold; de-i = decrease in inhibition;
nt = not tested, ne = no effect; nr = not reported.

METHODOLOGY

Reflex modification of the acoustic startle response is illustrated in Fig. 4. The left panel depicts the response to the eliciting stimulus (S_2: in this case, a 120 dB, 40 msec white noise burst). The response is recorded in grams as the output from a force transducer affixed to the platform under the rat, from which the latency and peak amplitude are derived (24). The right panel depicts the effects of a reflex-modifying "prepulse" stimulus (S_1: in this case, a 60 dB, 5 kHz, 20 msec tone) delivered 90 msec prior to S_2. Presenting S_1 prior to S_2 results in a decrease in the peak amplitude of the startle response (prepulse inhibition or reflex modification) when compared to the response produced by S_2 alone (Figure 4, *left panel*, blank trial). The psychophysical details of this procedure have been reviewed by Hoffman and Ison (56).

A basic principle of the phenomenon is the proportional relationship between the intensity of S_1 and the reduction in the response to S_2. If it is assumed that the preliminary stimulus must be perceived in order to reduce the response to S_2, then an appropriate definition of threshold would be a stimulus intensity level above which inhibition occurs. Figure 5 illustrates the determination of a threshold for one animal using a nonlinear segmented line regression model (24). Other methods for deriving thresholds from similar data have been published (40,135). The data in Fig. 5 were collected using a 5 kHz tone as S_1, at intensities ranging from 0 to 90 dB sound pressure levels (SPL). In this analysis, the y-intercept is an estimate of the baseline amplitude (baseline startle amplitude), corresponding to the amplitude on blank trials. The analysis also calculates the

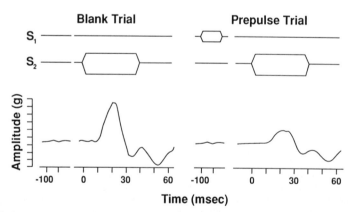

FIG. 4. Reflex modification of the startle response. *Left panel*: The response output (amplitude) in a trial in which only the reflex-eliciting stimulus (S_2) is presented (Blank Trial). *Right panel*: Reflex modification, i.e., the amplitude reduction in a trial when the prepulse (S_1) is presented 90 msec prior to S_2 (Prepulse Trial). [Note that S_1 is subthreshold for eliciting a response, and can be almost any sensory modality (56)]. Response data are actual output amplitudes from a force transducer using S_2 = 120 dB, 40 msec white noise burst and S_1 = 75 dB, 5 kHz tone. (From Crofton, ref. 23, with permission.)

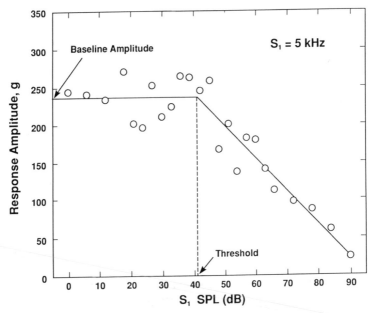

FIG. 5. An example of reflex modification demonstrating the relationship between the intensity of S_1 and the inhibition of the response to S_2. S_1 was a 5 kHz, 40 msec pure tone; S_2 was a 120 dB (A), 40 msec, white noise burst; the interstimulus interval was 90 msec, with 10 trials at each of 24 S_1-intensity levels. Data are from one rat and each point represents the mean of ten trials. The lines were fit using a nonlinear regression procedure. The threshold is defined as the intersection between the two lines. SPL, sound pressure level (re. 0.0002 μbar). (From Crofton, ref. 23, with permission.)

slope of the descending function that results from the inhibition of the response by S_1, and the threshold is defined as the x-value where the two lines intersect. Sensory thresholds can be quickly determined; for example the data in Fig. 5 were collected in 50 min. Computer automation allows for simultaneous control of multiple chambers and therefore the potential to test relatively large numbers of subjects in short periods of time.

Detailed audiometric functions can also be obtained relatively rapidly with this technique (40,135). Figure 6 presents data collected from 16 animals tested over 13 daily sessions. Also presented are the data of Kelly and Masterton (64), which were collected from three animals using a conditioned lick suppression paradigm over approximately 150 days. Comparison of these data sets illustrates two points. First, auditory thresholds derived from reflex modification procedures approximate those obtained with conditioned operant suppression paradigms (40,135). Second, compared to conditioned suppression and other operant procedures, reflex modification procedures can be used to rapidly and efficiently determine the effects of chemicals on auditory thresholds.

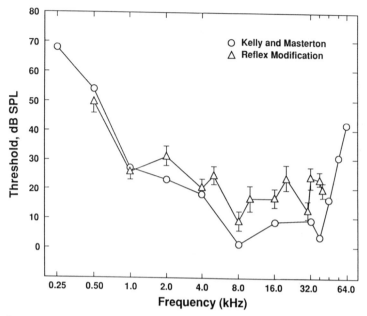

FIG. 6. A comparison of audiograms for separate groups of rats tested with either reflex modification of the startle response or a conditioned suppression procedure. Reflex modification data represent the means (±SE) of 16 animals (Crofton, unpublished). The condition suppression data are the means of three animals (data adapted from Kelly and Masterton, ref. 64). SPL, sound pressure level. (From Crofton, ref. 23, with permission.)

Other procedures have been proposed which do not involve a threshold assessment, and which may be shorter because they examine a more limited number of prestimulus intensities. These procedures usually employ only one or two S_1 intensity levels in addition to blank control trials. Comparisons are made on the shifts in the amounts of inhibition due to experimental treatments relative to control groups (72,98). Although these procedures do not provide estimates of sensory thresholds, they are still very useful in determining the effects of toxicants on the inhibition process and/or sensory function (45,123,127,133).

TOXICANT-INDUCED SENSORY DYSFUNCTION

Conroy et al. (15) were the first to document toxicant-induced sensory dysfunction using reflex modification procedures (Fig. 7). These authors demonstrated a complete loss of inhibition to a high frequency auditory stimulus ($S_1 = 32$ kHz) in rats exposed to nitroimidazoles, a class of radiosensitizers. Subsequently, Young and Fechter (135) demonstrated a toxicant-induced high frequency selective hearing loss in rats exposed to the ototoxicant, neomycin.

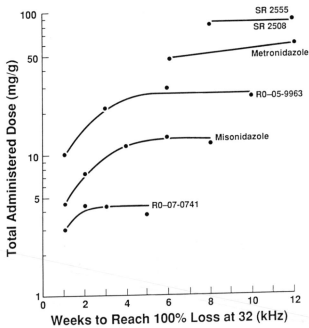

FIG. 7. Data from mice exposed to nitroimidazoles. Data are presented as the number of treatment weeks required to produce a loss of inhibition to a 32 kHz tone vs. the total administered dose. (From Conroy et al., ref. 15, with permission.)

Figure 8 illustrates the high frequency selective nature of the hearing loss induced by neomycin in that the inhibition is normal at low frequencies (less than 10 kHz), whereas higher frequencies are either less effective or ineffective in inhibiting the response. Loss of auditory function has also been demonstrated in humans. Reiter and Ison (94) demonstrated an impaired ability of auditory stimuli to inhibit the blink reflex (elicited by a cutaneous stimulus) in human subjects with etiologically known cases of hearing loss (e.g., noise-induced and sensory-neural) (Fig. 9).

There are also a number of examples of somatosensory dysfunction detected by reflex modification procedures. Ison (61) presented data showing that cutaneous S_1 stimulus (to the hand) failed to inhibit the eyeblink response in a human subject whose hand had been denervated (Fig. 10). Wu et al. (133) demonstrated a progressive loss of inhibition to an auditory-induced startle response using an electrodermal S_1 stimulus in rats exposed to methyl mercury (Fig. 11). More recently, Leitner (68) demonstrated that morphine decreased footshock-induced (S_1) inhibition of the auditory startle reflex in rats (Fig. 12).

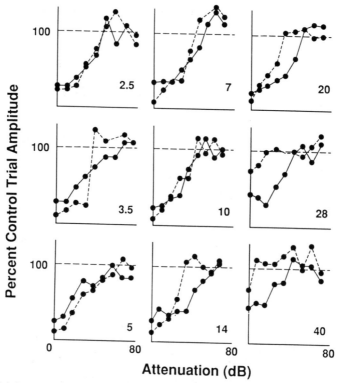

FIG. 8. High frequency hearing loss in a rat treated with the ototoxic aminoglycoside antibiotic, neomycin. Animals were tested before exposure (*solid line*), exposed to 100 mg/kg/day neomycin for 5 days, and tested 7 days later (*dotted line*). The number within each panel indicates the frequency of S₁. An upward shift during postexposure testing indicates a hearing loss. Data demonstrate high frequency selective hearing loss that ranged from 19 to 38 dB at 40 kHz. (From Young and Fechter, ref. 135, with permission.)

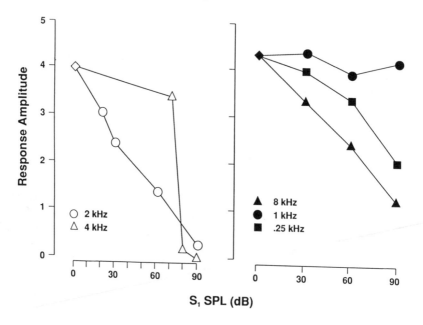

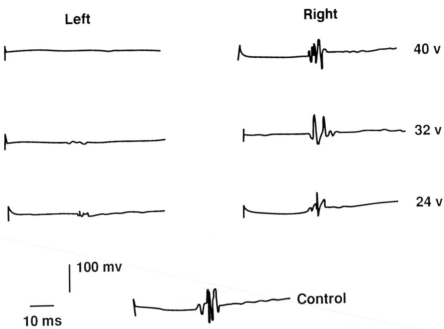

Left　　　　　　　　**Right**

40 v

32 v

24 v

100 mv

10 ms

Control

FIG. 10. Tracing of individual trials of the EMG responses taken from the eyelid muscles in reaction to a mild shock to the forehead, and when that shock was preceded by shocks of various intensities to the ring fingers of the left hand (intact) and the right hand (surgically repaired). Note the lack of inhibition to S_1 stimuli presented to the right hand. (From Ison, ref. 61, with permission.)

Visual system dysfunction has also been demonstrated using reflex modification procedures. Rats with inherited retinal dystrophy showed a loss of inhibition of the acoustic startle reflex by a visual prestimulus, but not an auditory prestimulus, indicating normal auditory function but impaired visual function (128) (Fig. 13). Campo et al. (12) demonstrated functional impairment following viral-induced retinal degeneration using visual S_1 stimuli in rats.

Lastly, there are also data indicating that reflex modification procedures may also be useful in the elucidation of dysfunction in processes other than the simple perception of sensory stimuli. Inhibition to gaps in background noise has been

FIG. 9. Relative size of the eyeblink response to a cutaneous stimulus when it was preceded by tones of various intensities and frequencies. *Left:* The subject was a mechanic with a high frequency hearing loss and loudness recruitment. The detection threshold at 2 kHz was 20 dB, and at 40 kHz was 70 dB. *Right:* The subject was a 9-year-old girl who had a moderate sensory neural hearing loss in the mid-frequencies. Detection thresholds were 40, 88, and 20 dB for 0.25, 1, and 8 kHz, respectively. SPL, sound pressure level. (From Ison, ref. 61, with permission.)

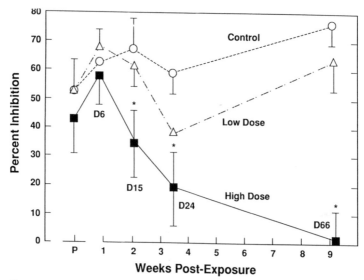

FIG. 11. Changes in the amount of startle inhibition by a leading 140 μA electrodermal prestimulus as a function of dose and time over a 9-week period following methyl mercury treatment. Data are percent inhibition (decreased inhibition relates to loss of sensory function) vs. weeks post-exposure (group means ±SE). (Control: 0 mg/kg, N = 6; low dose: 13.3 mg/kg, N = 6; high dose: 40 mg/kg, N = 5). (From Wu et al., ref. 133, with permission.)

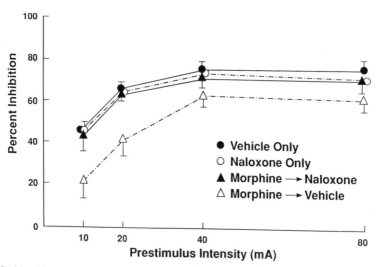

FIG. 12. Morphine-induced alterations in acoustic startle response inhibited by footshock. S_1 stimuli blocked by naloxone pretreatment. Decreased inhibition relates to loss of sensory function. Data are percent inhibition as a function of S_1 intensity (group mean ±SE). Vehicle: saline; naloxone: 2 mg/kg alone or prior to morphine; morphine: 5 mg/kg. (From Leitner, ref. 68, with permission.)

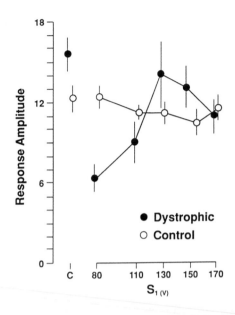

FIG. 13. Mean normalized acoustic startle response amplitude for the control (*closed circles*) and dystrophic (*open circles*) rats following the presentation of S_2 (auditory) alone (C) or preceded by a visual. A decrease in the normalized response relative to the S_2 alone condition shows sensory perception-induced inhibition. The dystrophic animals failed to show any such inhibition. Data are presented as group means (±SE). (From Wecker and Ison, ref. 128, with permission.)

used to assess auditory acuity, a process necessary for human speech perception, and possibly involving temporal resolution of auditory transients (59,60). Figure 14 shows that both ethanol and methanol impair the ability of rats to detect gaps in background noise (127). Methyl mercury has also been shown to impair this function (133) (Fig. 15).

These examples provide evidence that reflex modification procedures can successfully detect toxicant-induced deficits in sensory dysfunction in both laboratory rats and humans. Furthermore, these procedures may be used to access deficits in a number of different sensory modalities (i.e., visual, somatosensory, auditory).

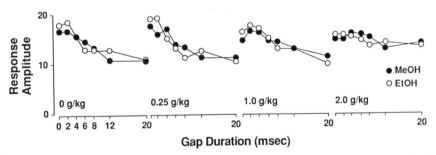

FIG. 14. Decreased auditory acuity following exposure to either ethanol (*open circles*) or methanol (*closed circles*). Data are presented as the mean normalized acoustic startle response amplitude as a function of a preceding gap in background noise S_1 stimulus. Increasing doses of both ethanol and methanol block the inhibition seen in the control group. A decrease in the normalized response amplitude relative to the 0 msec gap condition shows sensory perception-induced inhibition of the response. Data are presented as group means. (From Wecker and Ison, ref. 127, with permission.)

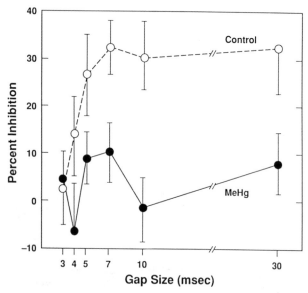

FIG. 15. Decreased inhibition to gap in background noise as a function of dose and time following exposure to methyl mercury. S_2 was an acoustic noise burst, S_1 was a 7 msec gap in the background noise. Decreased inhibition relative to the control group suggests an impairment in auditory acuity. Data are presented as group means (±SE). (From Wu et al., ref. 133, with permission.)

SENSORY VS. MOTOR INVOLVEMENT IN THE STARTLE RESPONSE

The startle response is a sensory-evoked motor reflex that is measured by a change in motor output. Consequently, when a toxicant alters the reflex (i.e., changes the measured motor output), the only conclusion that can be made is that the toxicant has altered some aspect(s) of the sensory-motor reflex pathway. This may involve anything from damage to sensory receptors (e.g., hair cells) to musculature. Reflex modification audiometry can be used to discriminate sensory from motor involvement in the effect (27,37,127). This is depicted in Fig. 16 where three theoretical outcomes are drawn. Outcome A represents an elevation in auditory threshold with no concurrent alteration in the motor component of the response. Outcome B represents an alteration in the motor component with no alteration in threshold. Outcome C represents a change in both the motor and sensory components. Only decreases in the motor component are considered here, although increases in the motor component can also occur. The ability to differentiate motor and sensory effects is dependent on the independence of these two components of the response. This requires careful data interpretation in light of concomitant measurements of sensory and motor function with other

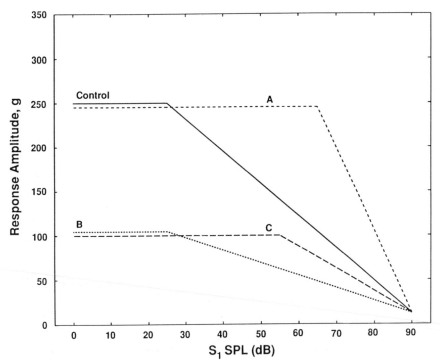

FIG. 16. Theoretical chemical-induced alterations in reflex modification of the startle response. *A–C:* Possible effects on the sensory and/or motor components. SPL, sound pressure level. (From Crofton, ref. 23, with permission.)

endpoints (e.g., motor function with grip strength). The extent to which these measures are independent can best be evaluated through studying the effects of physical- or chemical-induced lesions. This testing should utilize toxicants as tools through the use of chemicals with well-defined sites or mechanisms of action. The discussion that follows summarizes data that can be used to begin to address these issues.

Fechter and Young (37) were the first to suggest an independence of sensory and motor effects by studying the effects of the toxicant triethlytin using reflex modification. Triethyltin is an organotin compound that induces demyelination in the nervous system (13,125) resulting in a reversible decrement in motor function (95,96,111). Figure 17 illustrates the data of Fechter and Young (37). The *left panel* shows the decreases in the baseline startle response amplitude (i.e., the motor component), and the *right panel* demonstrates the lack of effect or triethyltin on auditory thresholds. Data from the ototoxic aminoglycoside neomycin were included as a positive control for high frequency hearing loss. Overall, these data illustrate two important points: (a) the technique can measure normal auditory thresholds in animals with greater than 70% depression in the

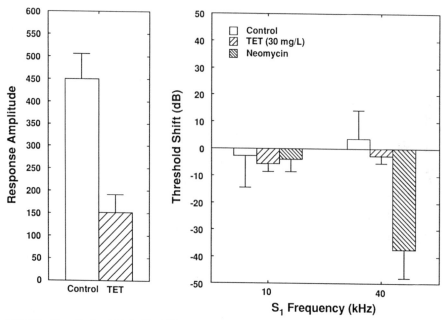

FIG. 17. The effects of triethyltin (*TET*) on reflex modification of the startle response. Rats were tested during the third week of exposure to 30 mg/l TET in the drinking water. *Left panel*: A TET-induced depression in the baseline startle amplitude. *Right panel*: There was no shift (from preexposure levels) in the thresholds for either a 10 or 40 kHz tone. The positive control (neomycin) produced a high frequency selective threshold shift indicative of a hearing impairment. (From Fechter and Young, ref. 37, with permission.)

baseline amplitude; and (b) thresholds can be quantified independent of the level of the startle response amplitude.

Similar data have been obtained from animals exposed to 2,4-dithiobiuret (27). This chemical produces a reversible motor impairment through a depletion of acetylcholine at the neuromuscular junction (6,129). The effects of 2,4-dithiobiuret on reflex modification of the startle response are shown in Fig. 18. The *left panel* documents a dosage-dependent depression in the baseline startle amplitude, and the *right panel* shows a lack of effect on the auditory thresholds for either a 5- or 40-kHz tone. Other independent measurements of motor function, including grip strength and motor activity, also demonstrate decreases in dithiobiuret-exposed animals (Fig. 19). Moreover, other tests of sensory function (e.g., pain, visual) have also failed to reveal sensory dysfunction (6,27). The effects of these two chemicals also elucidate the ability of reflex modification to measure normal auditory thresholds even though the motor output has been depressed by greater than 70%. These two examples document the ability of the reflex modification procedure to detect motor dysfunction (and document normal auditory function) resulting from two entirely different mech-

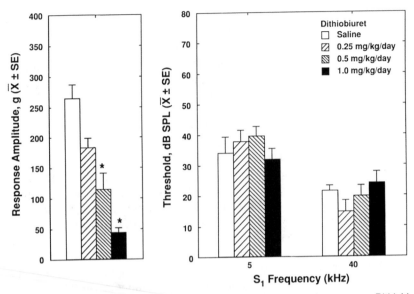

FIG. 18. The effects of dithiobiuret on reflex modification of the startle response. Dithiobiuret was administered ip to rats for 5 days. Testing occurred 48 hr postexposure. Test conditions were similar to those for Fig. 5. *Left panel:* A dosage-dependent decrease in the baseline startle amplitude. *Right panel:* There were no effects on the thresholds for either 5 or 40 kHz tones. Data represent group means (±SE) for 9–12/group. (From Crofton et al., ref. 27, with permission.)

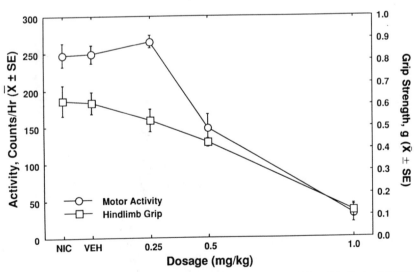

FIG. 19. Effects of 2,4-dithiobiuret on motor activity (*circles*, left y-axis) and hindlimb grip strength (*squares*, right y-axis). Data were collected 24 hr following a 1-week exposure. Data are presented as group means (±SE). (From Crofton et al., ref. 27, with permission.)

anisms, one involving demyelination of axons and the other a depletion of acetylcholine at the neuromuscular junction (example B in Fig. 16).

Exposure to the neurotoxicant trimethyltin produces a different pattern of results on sensory and motor function. This pattern involves a reduction in the baseline startle amplitude as well as an increase in high frequency hearing thresholds (104,136). Trimethyltin induces histopathological damage to a wide variety of neural structures, most notably the limbic system (13), and recent evidence also has linked trimethyltin-induced loss of high frequency hearing to the destruction of hair cells in the cochlea (24,43). Figure 20 illustrates data from rats exposed to trimethyltin. The *left panel* shows a dosage-dependent decrease in the baseline startle amplitude. The *right panel* indicates a high frequency selective effect of trimethyltin on auditory thresholds. These data replicate the previous work of Young and Fechter (136), and again illustrate the independence of the two measures (i.e., sensory and motor) in that although the startle response (i.e., motor response) is depressed almost 80%, auditory thresholds are either normal or elevated depending on the frequency (examples A and C from Fig. 16, respectively). Moreover, these data appear to indicate the ability of reflex modification procedures to detect alterations in auditory function due to pathological damage to auditory sensory receptors (i.e., hair cells).

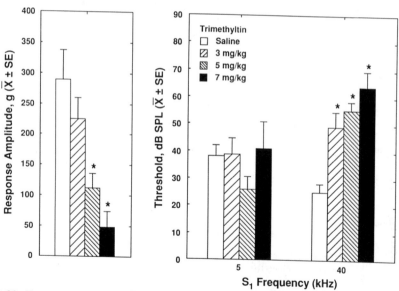

FIG. 20. The effects of acute exposure to trimethyltin on reflex modification of the startle response 9 weeks postexposure. Rats were tested under conditions similar to those for Fig. 2. *Left panel:* A dosage-dependent decrease in the baseline startle amplitude. *Right panel:* Trimethyltin produced an elevation in the threshold for the 40 kHz tone, but was without effect on the 5 kHz tone. Data represent group means (±SE) for 9–12/group. SPL, sound pressure level. (From Crofton et al., ref. 24, with permission.)

Exposure to the neurotoxicant iminodipropionitrile results in an even different pattern of effects compared to trimethyltin or dithiobiuret. Iminodipropionitrile exposure causes a distinct hyperkinetic syndrome consisting of spontaneous head movements, abnormal circling, and backwards locomotion (11). Figure 21 shows data from adult animals 9 weeks after a 3-day exposure to iminodipropionitrile. The *left panel* shows a decrease in the baseline startle amplitude. The *right panel*, in contrast to the effect of trimethyltin, shows an increase in auditory thresholds for both the low and high frequencies. Further work has demonstrated an elevation in auditory thresholds for all frequencies between 0.5 and 40 kHz in rats exposed to iminodipropionitrile (Fig. 22). Although the mechanism for this effect is not known, these data again demonstrate that reflex modification procedures can detect auditory dysfunction in the presence of a depressed baseline startle amplitude. Data from other endpoints would also argue that the depressed amplitude is not due to a decrease in motor function since the animals are hyperactive and show no alteration in hindlimb grip strength (Fig. 23).

Recent work has investigated the effects of early postnatal exposure to toxicants in rats, and subsequent testing as adults. Rats exposed to iminodipropionitrile on postnatal days 5 to 7 demonstrate effects that are distinct from adult exposure in two respects (Fig. 24). The first difference is that early postnatal exposure to iminodipropionitrile has no effect on the baseline startle amplitude when animals

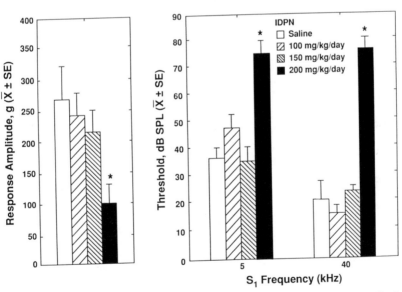

FIG. 21. The effects of 3-day exposure to iminodipropionitrile (IDPN) on reflex modification of the startle response 9 weeks postexposure. Rats were tested under conditions similar to those for Fig. 5. *Left panel*: A dosage-dependent decrease in the baseline startle amplitude. *Right panel*: IDPN produced an elevation in the thresholds for both the 5 and 40 kHz tones. Data represent group means (±SE) for 9–10/group. SPL, sound pressure level. (From Crofton and Knight, ref. 26, with permission.)

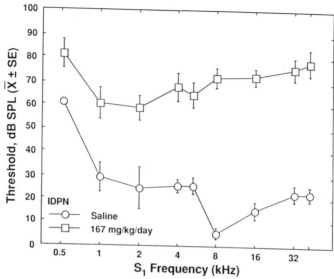

FIG. 22. The effect of a 3-day exposure to iminodipropionitrile (IDPN) on the audiogram of the rat. Rats were tested 6–7 weeks postexposure. Test conditions were similar to those for Fig. 2. Each rat was tested for one frequency once per day. IDPN produced an elevation in thresholds at all frequencies tested. Data represent group means (±SE) for 8–10/group. SPL, sound pressure levels. (From Crofton, ref. 23, with permission.)

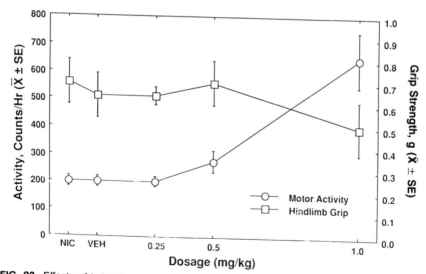

FIG. 23. Effects of iminodipropionitrile on motor activity (*circles*, left y-axis) and hindlimb grip strength, (*squares*, right y-axis). Data were collected 3 weeks following a 3-day exposure. Data are presented as group means (±SE). (From Crofton and Knight, ref. 26, with permission.)

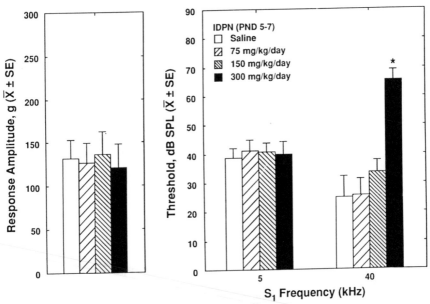

FIG. 24. The effects of early postnatal exposure to iminodipropionitrile (IDPN) on reflex modification of the startle response. IDPN was administered (ip) to rat pups on postnatal days 5–7 (PND 5–7). Testing was conducted on postnatal days 64 and 65 under conditions similar to those for Fig. 5. *Right panel*: Postnatal exposure to IDPN had no effect on the baseline startle amplitude. *Left panel*: IDPN produced a high frequency selective elevation in the threshold for 40 kHz tone (contrast with data in Fig. 9). Data represent group means (±SE) for 12–16/group. SPL, sound pressure levels. (From Crofton et al., ref. 25, with permission.)

are tested in adulthood (*left panel*). The second difference is that neonatal exposure produces an elevation in auditory thresholds that is selective for the high frequency (*right panel*). Similar to adult-exposed animals, these animals also exhibit hyperactivity, suggesting no correspondence between baseline startle amplitude measurements and overall motor function. These data also illustrate the independence of sensory and motor effects in that threshold elevation was demonstrated without concurrent alteration in the baseline startle amplitude (outcome B in Fig. 16).

Last is an example of a compound that increases the baseline startle amplitude as well as alters high frequency auditory thresholds. Monosodium glutamate (MSG) is a neonatal neurotoxicant (82,83). Neonatal exposure to MSG on postnatal days 2 to 9 results in a novel pattern of effects (Fig. 25): the baseline startle amplitude was elevated (*left panel*) and the threshold to a high frequency tone was also elevated (*right panel*). Recently Janssen et al. (63) have demonstrated that the high frequency hearing loss in neonatal animals treated with glutamate is related not to hair cell loss, but instead to neuronal cell loss in the spiral ganglion. These data may be important for two reasons. First, they illustrate a condition not seen in Fig. 16, an elevation in the baseline with a shift in the

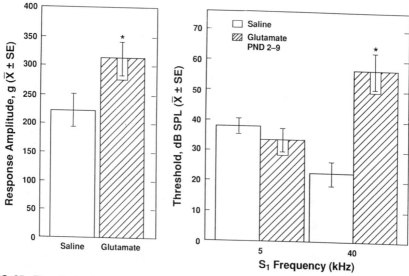

FIG. 25. The effects of early postnatal exposure to monosodium glutamate on reflex modification of the startle response. Glutamate (4 mg/g/day) was administered (ip) to rat pups on postnatal days 2–9 using the model of Rigdon et al. (99). Testing was conducted on postnatal days 80 and 81 under conditions similar to Fig. 5. *Left panel*: Postnatal exposure to glutamate resulted in an increase in the baseline startle amplitude as well as an elevation in the threshold for the 40 kHz tone. *Right panel*: Data represent group means (±SE) for 9/group. SPL, sound pressure levels. (From Crofton, ref. 23, with permission.)

threshold, indicating that elevations in the baseline startle amplitude do not preclude measurements of thresholds, although a better case for independence would be made with a chemical that elevates the baseline startle amplitude without altering thresholds. Second, these data suggest that the reflex modification procedure may be sensitive to toxicant-induced damage in other areas of the startle pathway (i.e., spiral ganglia and cochlear nucleus).

CONCLUSIONS

Table 4 summarizes these and other findings on the effects of chemicals on the sensory and motor components of reflex modification. Although the list is short and the validation process is not complete, it does document the fact that there is no predictable relationship between the effects of a chemical on the motor component of the startle response and auditory thresholds determined with reflex modification procedures. Reflex modification can give detailed information on the sensory effects (e.g., frequency dependence) of chemicals, and alterations in thresholds have been correlated with structural damage to sensory processing structures. Both are clear indications that sensory processes are being evaluated. However, alterations in the baseline startle amplitude do not always

TABLE 4. Comparison of effects of chemicals on sensory and motor components
of reflex modification of the ASR

Chemical	Motor	Sensory dysfunction	Reference
Neomycin	0	+	135
IDPN (neonatal)	0	+	25
Butyl nitrate	0	+	40
Triethyltin	↓	0	37
Dithiobiuret	↓	0	27
Ethanol	↓	0	127
Methanol	↓	0	127
Acrylamide	↓	0	28
Trimethyltin	↓	+	24,39,136
IDPN (adult)	↓	+	26
Noise + carbon monoxide	↓	+	41,137
MSG (neonatal)	↑	+	20

For motor effects; O, no change; ↓, decreased ASR baseline amplitude; ↑, increased ASR
baseline amplitude. For sensory effects; O, no change; +, sensory dysfunction.
IDPN, iminodipropionitrile; MSG, monosodium glutamate.

correlate with changes in motor function and care must be taken in interpreting the data. For example, trimethyltin causes a depression in the baseline startle amplitude (104,136), yet motor function levels (e.g., motor activity) are elevated (102). Adult exposure to iminodipropionitrile causes no alteration in grip strength, a very pronounced hyperactivity, and stereotyped head movements, yet the baseline startle amplitude is depressed (26). This depression in baseline startle amplitude may be related to a profound hearing loss, in which case the response to S_2 would be diminished regardless of the influence of S_1. On the other hand it may be due to an alteration in the neural circuitry responsible for sensorimotor integration. As suggested by this possibility, toxicants may affect the integration of sensory and motor processes, rather than affect either the sensory or motor components themselves.

It is also important to point out that the sensory and motor components of a reflex are not always entirely separable. Estimation of the sensory component (e.g., thresholds) involves measuring the inhibition of the motor output. Complete suppression of the baseline startle amplitude will not allow determination of inhibition functions. However, adequate inhibition functions can be obtained even for chemicals that cause greater than 80% depression of the baseline startle amplitude (27,37,127).

In summary, reflex modification of the startle response is a technique that can provide rapid, objective, and quantitative assessments of sensorimotor function. Advantages of this technique involve the ability to test animals rapidly, test without prior training, test without utilizing invasive procedures, and provide objective, independent estimates of both sensory and motor functioning. Reflex modification techniques hold great potential for use in neurotoxicology in both the identification and characterization of neurotoxic chemicals.

REFERENCES

1. Anandam N, Felegi W, Stern JM. In utero alcohol heights juvenile reactivity. *Pharmacol Biochem Behav* 1980;13:531–535.
2. Anday EK, Cohen ME, Kelley NE, Leitner DS. Effect of in utero cocaine exposure on startle and its modification. *Dev Pharmacol Ther* 1989;12:137–145.
3. Anger WK, Setzer JV, Russo JM, Brightwell WS, Wait RG, Johnson BL. Neurobehavioral effects of methyl bromide inhalation exposures. *Scand J Work Environ Health* 1981;7(Suppl. 4):40–47.
4. Anger WK. Neurobehavioral testing of chemicals: impact on recommended standards. *Neurobehav Toxicol Teratol* 1984;6:147–153.
5. Anger WK, Johnson BL. Chemicals affecting behavior. In: *Neurotoxicity of industrial and commercial chemicals,* vol. I, O'Donoghue JL, ed. Boca Raton, FL: CRC Press, 1985;51–148.
6. Atchison WD, Peterson RE. Potential neuromuscular toxicity of 2,4-dithiobiuret in the rat. *Toxicol Appl Pharmacol* 1981;57:63–68.
7. Braff DL, Geyer M. Acute and chronic LSD effects on rat startle: data supporting an LSD-rat model of schizophrenia. *Biol Psychiatry* 1980;15:909–916.
8. Braff DL, Stone C, Callaway E, Geyer M, Glick I, Bali L. Prestimulus effects on human startle reflex in normals and schizophrenics. *Psychophysiology* 1989;15:339–343.
9. Buelke-Sam J, Kimmel CA, Adams J, Nelson CJ, Vorhees CV, Wright DC, Omer V St, Korol BA, Butcher RE, Geyer MA, Holson JF, Kutscher CL, Wayner MJ. Collaborative behavioral teratology study: results. *Neurobehav Toxicol Teratol* 1985;7:591–624.
10. Cabe PA. Psychophysical methods for the measurement of somatosensory dysfunction in laboratory animals. In: Hayes AW, ed. *Toxicology of the eye, ear, and other senses.* New York: Raven Press, 1985;183–194.
11. Cadet JL. The iminodipropionitrile (IDPN)-induced dyskinetic syndrome: behavioral and biochemical pharmacology. *Neurosci Biobehav Rev* 1989;13:39–45.
12. Campo A, delCerro M, Foss JA, Ison JR, Orr JL, Warren PH, Monjan AA. Impairment of auditory and visual function follows perinatal viral infection in the rat. *Int J Neurosci* 1985;27:85–90.
13. Chang LW. Neuropathological changes associated with accidental or experimental exposure to organometallic compounds. In: Tilson HA, Sparber SB, eds. *Neurotoxicants and neurobiological function: effects of organoheavy metals.* New York: Wiley-Interscience, 1987;81–116.
14. Colley JC, Edwards JA, Heywood R, Purser D. Toxicity studies with quinine hydrochloride. *Toxicology* 1989;54:219–226.
15. Conroy PJ, McNeill TH, Passalacqua W, Merritt J, Reich KR, Walker S. Nitroimidazole neurotoxicity: are mouse studies predictive? *Int J Radiat Oncol Biol Phys* 1982;8:799–803.
16. Crofton KM, Reiter LW. Effects of two pyrethroid insecticides on motor activity and the acoustic startle response in the rat. *Toxicol Appl Pharmacol* 1984;75:318–328.
17. Crofton KM, Reiter LW. Pyrethroids insecticides and the γ-aminobutyric acid$_A$ receptor complex: motor activity and the acoustic startle response in the rat. *J Pharmacol Exp Therap* 1987;243:946–954.
18. Crofton KM, Reiter LW. The effects of Type I and Type II pyrethroids on motor activity and the acoustic startle response in the rat. *Fundam Appl Toxicol* 1988;10:624–634.
19. Crofton KM, Sheets LP. Evaluation of sensory system function using reflex modification of the startle response. *J Am Coll Toxicol* 1989;8:199–211.
20. Crofton KM, Gilbert ME. Data on the effects of neonatal exposure to monosodium glutamate on auditory function (*unpublished*).
21. Crofton KM, Boncek VM, Reiter LW. Acute effects of amitraz on the acoustic startle response and motor activity. *Pestic Sci* 1989;27:1–11.
22. Crofton KM, Dean KF, Boncek VM, Rosen MB, Sheets LP, Chernoff N, Reiter LW. Prenatal or postnatal exposure to bis (tri-n-butyltin) oxide in the rat: postnatal evaluation of teratology and behavior. *Toxicol Appl Pharmacol* 1989;97:113–123.
23. Crofton KM. Reflex modification and the detection of toxicant-induced auditory dysfunction. *Neurotoxicol Teratol* 1990;12:461–468.
24. Crofton KM, Dean KF, Menache MG, Janssen R. Trimethyltin effects on auditory function and cochlear pathology. *Toxicol Appl Pharmacol* 1990;105:123–132.

25. Crofton KM, Stanton ME, Peele DB. Developmental neurotoxicity following neonatal exposure to iminodipropionitrile in the rat. *Toxicologist* 1990;10:305(abstract).
26. Crofton KM, Knight T. Auditory deficits and motor dysfunction following iminodipropionitrile exposure in the rat. *Neurotoxicol Teratol (in press)*.
27. Crofton KM, Dean KF, Hamrick RC, Boyes WK. The effects of 2,4-dithiobiuret on sensory and motor function. *Fundam Appl Toxicol* 1991;16:469–481.
28. Crofton KM, Somerholter LJ, Farmer J, Padilla S, MacPhail RC. The effect of dose rate on acrylamide neurotoxicity: a preliminary report. *Toxicologist* 1991;11:307(abstract).
29. Damstra T. Environmental chemicals and nervous system dysfunction. *Yale J Biol Med* 1978;51:457–468.
30. David J-M, Risso JJ, Weiss M, Brue F, Pellet J. Hyperbaric effects on sensorimotor reactivity studied with acoustic startle in the rat. *Physiol Behav* 1988;42:237–244.
31. Davis M. Neurochemical modulation of sensory-motor reactivity: acoustic and tactile startle reflexes. *Neurosci Biobehav Rev* 1980;4:241–263.
32. Davis M, Gendelman DS, Tischler MD, Gendelman PM. A primary acoustic startle circuit: lesion and stimulation studies. *J Neurosci* 1982;2:791–805.
33. Davis M. Cocaine: excitatory effects on sensorimotor reactivity measured with acoustic startle. *Psychopharmacology* 1985;86:31–36.
34. Davis M. Apomorphine, d-amphetamine, strychnine and yohimbine do not alter prepulse inhibition of the acoustic startle reflex. *Psychopharmacology* 1988;95:151–156.
35. Dyer RS. The use of sensory evoked potentials in toxicology. *Fundam Appl Toxicol* 1985;5:24–40.
36. Eastman CL, Young JS, Fechter LD. Trimethyltin ototoxicity in albino rats. *Neurotoxicol Teratol* 1987;9:329–332.
37. Fechter LD, Young JS. Discrimination of auditory from nonauditory toxicity by reflex modulation audiometry: effects of triethyltin. *Toxicol Appl Pharmacol* 1983;70:216–227.
38. Fechter LD. Reflexive measures. In: Annau Z, ed. *Neurobehavioral toxicology.* Baltimore: Johns Hopkins Press, 1986;23–42.
39. Fechter LD, Young JS, Nuttall AL. Trimethyltin ototoxicity: evidence for a cochlear site of injury. *Hear Res* 1986;23:275–282.
40. Fechter LD, Sheppard L, Young JS, Zeger S. Sensory threshold estimation from a continuously graded response produced by reflex modification audiometry. *J Acoust Soc Am* 1988;84:179–185.
41. Fechter LD, Young JS, Carlisle L. Potentiation of noise induced threshold shifts and hair cell loss by carbon monoxide. *Hear Res* 1988;34:39–48.
42. Fechter LD, Richard CL, Mungekar M, Gomez J, Strathern D. Disruption of auditory function by acute administration of a "room odorizer" containing butyl nitrite in rats. *Fundam Appl Toxicol* 1989;12:56–61.
43. Fechter LD, Carlisle L. Auditory dysfunction and cochlear vascular injury following trimethyltin injury in the guinea pig. *Toxicol Appl Pharmacol* 1990;105:133–143.
44. Gallagher DW, Kehne JH, Wakeman EA, Davis M. Developmental changes in pharmacological responsivity of the acoustic startle reflex: effects of picrotoxin. *Psychopharmacology* 1983;79:87–93.
45. Geyer MA, Swerdlow NR, Mansbach RS, Braff DL. Startle response models of sensorimotor gating and habituation deficits in schizophrenia. *Brain Res Bull* 1990;25:485–498.
46. Geyer MA, Butcher RE, Fite K. A study of startle and locomotor activity in rats prenatally exposed to methylmercury. *Neurobehav Toxicol Teratol* 1985;7:759–765.
47. Gibbons RW, Kalant H, LeBlanc HE, Clark JW. The effects of chronic ethanol on startle thresholds in rats. *Psychopharmacologia* 1971;19:95–104.
48. Harpur ES. Ototoxicological testing. In: Gorrod JW, ed. *Testing for toxicity.* London: Taylor and Francis, 1981;219–240.
49. Harry GJ, Tilson HA. The effects of postpartum exposure to triethyl tin in the neurobehavioral functioning of rats. *Neurotoxicology* 1981;2:283–296.
50. Hayes AW, ed. *Toxicology of the eye, ear, and other special senses.* New York: Raven Press. 1985.
51. Herr DW, Gallus JA, Tilson HA. Pharmacological modification of tremor and enhanced acoustic startle by chlordecone and p,p'-DDT. *Psychopharmacology* 1987;91:320–325.
52. Hijzen TH, Slangen JL. Effects of type I and type II pyrethroids on the startle response in rats. *Toxicol Lett* 1988;40:141–152.

53. Hijzen TH, DeBeun R, Slangen JL. Effects of pyrethroids on the acoustic startle reflex in the rat. *Toxicology* 1988;49:271–276.
54. Hijzen TH, Slangen JL. Effects of midazolam, DMCM and lindane on potentiated startle in the rat. *Psychopharmacology* 1989;99:362–365.
55. Hoeffding V, Fechter LD. Trimethyltin disrupts auditory function and cochlear morphology in pigmented rats. *Neurotoxicol Teratol* 1991;13:135–146.
56. Hoffman HS, Ison JR. Reflex modification in the domain of startle: I. Some empirical findings and their implication for how the nervous system processes sensory input. *Psychol Rev* 1980;87:175–189.
57. Howd RA, Pryor GTM. Effect of chronic morphine on the response to and disposition of other drugs. *Pharmacol Biochem Behav* 1980;12:557–586.
58. Ison JR, Hammond GR, Krauter EE. Effects of experience on stimulus produced reflex inhibition in the rat. *J Comp Physiol Psychol* 1973;83:324–336.
59. Ison JR. Temporal acuity in auditory function in the rat: reflex inhibition by brief gaps in noise. *J Comp Physiol Psychol* 1982;96:945–954.
60. Ison JR, Pinckney LA. Reflex inhibition in humans: sensitivity to brief silent periods in white noise. *Percept Psychophys* 1983;34:84–88.
61. Ison JR. Reflex modification as an objective test for sensory processing following toxicant exposure. *Neurobehav Toxicol Teratol* 1984;6:437–445.
62. Ison JR. Behavioral methods applicable to laboratory animals and humans. In: Johnson BL, ed. *Advances in neurobehavioral toxicology: applications in environmental and occupational health.* Chelsea, MI: Lewis Publishers, 1990;389–402.
63. Janssen R, Schweitzer L, Jensen KF. Glutamate neurotoxicity in the developing rat cochlea: physiological and morphological approaches. Glutamate excitotoxicity in rat auditory system. *Brain Res* 1991;522:255–264.
64. Kelly JB, Masterton B. Auditory sensitivity of the albino rat. *J Comp Physiol Psychol* 1977;91:930–936.
65. Kehne JH, Gallagher DW, Davis M. Strychnine: brainstem and spinal mediation of excitatory effects on acoustic startle. *Eur J Pharmacol* 1981;76:177–186.
66. Kontur PJ, Fechter LD. Brain manganese, catecholamine turnover, and the development of startle in rats prenatally exposed to manganese. *Teratology* 1985;32:1–11.
67. Leitner DS, Cohen ME. Role of the inferior colliculus in the inhibition of acoustic startle in the rat. *Physiol Behav* 1985;34:65–70.
68. Leitner DS. Inhibition of acoustic startle in the rat by a footshock prestimulus: effects of morphine and naloxone. *Behav Neurosci* 1988;102:526–533.
69. Llorens J, Crofton KM. Enhanced neurotoxicity of 3,3'-iminodipropionitrile following pretreatment with carbon tetrachloride in the rat. *Neurotoxicology,* 1991;12:000–000 *(in press).*
70. Lochry EA. Concurrent use of behavioral/functional testing in existing reproductive and developmental toxicity screens: practical considerations. *J Am Coll Toxicol* 1987;6:433–439.
71. Mactutus CF, Unger KL, Tilson HA. Evaluation of neonatal chlordecone neurotoxicity during early development: initial characterization. *Neurobehav Toxicol Teratol* 1984;6:67–73.
72. Mansbach RS, Geyer MA, Braff DL. Dopaminergic stimulation disrupts sensorimotor gaiting in the rat. *Psychopharmacology* 1988;94:507–514.
73. Mattsson JL, Albee RR. Sensory evoked potentials in neurotoxicology. *Neurotoxicol Teratol* 1988;10:435–443.
74. Maurissen JPJ. Effects of toxicants on the somatosensory system. *Neurobehav Toxicol* 1979;1(Suppl. 1):15–23.
75. Maurissen JPJ, Weiss B, Davis HT. Somatosensory thresholds in monkeys exposed to acrylamide. *Toxicol Appl Pharmacol* 1983;71:266–279.
76. Merigan WH. Effects of toxicants on visual systems. *Neurobehav Toxicol* 1979;1(Suppl. 1):15–23.
77. Merigan WH, Barkdoll E, Maurissen JPJ. Acrylamide-induced visual impairment in primates. *Toxicol Appl Pharmacol* 1982;64:342–345.
78. Merigan WH, Weiss B, eds. *Neurotoxicity of the visual system.* New York: Raven Press, 1980.
79. Mickley GA, Ferguson JL. Enhanced acoustic startle responding in rats with radiation-induced hippocampal granule cell hypoplasia. *Exp Brain Res* 1989;75:28–34.
80. Moody DB, Stebbins WC. Detection of the effects of toxic substances on the auditory system by behavioral methods. In: Mitchell CL, ed. *Nervous system toxicology.* New York: Raven Press, 1982;109–132.

81. O'Donoghue JL. Neurotoxicity and risk assessment. AAAS Symposium, Evaluating the neurotoxic risk posed by chemicals in the workplace and environment. Annual Meeting of AAAS, Washington, DC, 1987.

82. Olney JW. Toxic effects of glutamate and related amino acids on the developing central nervous system. In: Nynan WL, ed. *Heritable disorders of amino acid metabolism.* New York: MacMillan, 1974;501–512.

83. Olney JW. Excitotoxic food additives—relevance of animal studies to human safety. *Neurobehav Toxicol Teratol* 1984;6:455–462.

84. Ornitz EM, Pynoos RS. Startle modulation in children with posttraumatic stress disorder. *Am J Psychiat* 1989;146:866–869.

85. Office of Technological Assessment, *Impacts of neuroscience: background paper.* Washington, DC: US GPO, OTA-BP-BA-24; 1984.

86. Pellet J. Neural organization in the brainstem circuit mediating the primary acoustic head startle: an electrophysiological study in the rat. *Physiol Behav* 1990;48:727–739.

87. Pohorecky LA, Cagan M, Brick J, Jaffe LS. The startle response in rats: effects of ethanol. *Pharmacol Behav* 1976;4:311–316.

88. Potter BM, Berntson GG. Prenatal alcohol exposure: effects on acoustic startle and prepulse inhibition. *Neurotox Teratol* 1987;9:17–21.

89. Pryor GT, Uyeno ET, Tilson HA, Mitchell CL. Assessment of chemicals using a battery of neurobehavioral tests: a comparative study. *Neurobehav Toxicol Teratol* 1983a;5:91–117.

90. Pryor GT, Dickinson J, Howd RA, Rebert CS. Transient cognitive deficits and high-frequency hearing loss in weanling rats exposed to toluene. *Neurobehav Toxicol Teratol* 1983b;5:53–57.

91. Pryor GT, Dickinson J, Howd RA, Rebert CS. Neurobehavioral effects of subchronic exposure of weanling rats to toluene or hexane. *Neurobehav Toxicol Teratol* 1983c;5:47–52.

92. Rebert CS. Multisensory evoked potentials in experimental and applied neurotoxicology. *Neurobehav Toxicol Teratol* 1983;5:659–671.

93. Rebert CS, Pryor GT, Frick MS. Effects of vincristine, maytansine and cis-platinum on behavioral and electrophysiological indicies of neurotoxicity in the rat. *J Appl Toxicol* 1984;4:330–338.

94. Reiter LA, Ison JR. Reflex modulation and loudness recruitment. *J Aud Res* 1973;19:201–207.

95. Reiter LW, Kidd K, Heavner G, Ruppert P. Behavioral toxicity of acute and subacute exposure to triethyltin in the rat. *Neurotoxicology* 1980;2:97–112.

96. Reiter LW, Ruppert PH. Behavioral toxicity of trialkyltin compounds: a review. *Neurotoxicology* 1984;5:177–186.

97. Rice SA. Effect of prenatal N$_2$O exposure on startle reflex reactivity. *Teratology* 1990;42:373–381.

98. Rigdon GC. Differential effects of apomorphine on prepulse inhibition of acoustic startle reflex in two rat strains. *Psychopharmacology* 1990;102:419–421.

99. Rigdon GC, Boyes WK, Dyer RS. Effect of perinatal monosodium glutamate administration on visual evoked potentials of juvenile and adult rats. *Neurotoxicol Teratol* 1989;11:121–128.

100. Rosen MB, Crofton KM, Chernoff N. Postnatal evaluation of prenatal exposure to p-xylene in the rat. *Toxicol Lett* 1986;34:223–229.

101. Royalty J, Taylor GT, Korol BA. The effects of prenatal exposure to methylmercury on aggressive behavior in the rat. *Neurotoxicol Teratol* 1987;9:87–93.

102. Ruppert PH, Walsh TJ, Reiter LW, Dyer RS. Trimethyltin-induced hyperactivity: time course and pattern. *Neurobehav Toxicol Teratol* 1982;4:135–139.

103. Ruppert PH, Dean KF, Reiter LW. Developmental and behavioral toxicity following acute postnatal exposure of rat pups to trimethyltin. *Neurobehav Toxicol Teratol* 1983;5:421–429.

104. Ruppert PH, Dean KF, Reiter LW. Trimethyltin disrupts acoustic startle responding in adult rats. *Toxicol Lett* 1984;22:33–38.

105. Russo JM, Anger WK, Setzer JV, Brightwell WS. Neurobehavioral assessment of chronic low-level methylbromide exposure in the rabbit. *J Toxicol Environ Hlth* 1984;14:247–255.

106. Russo JM, Junnila M. Behavioral assessment of xylene and ethylbenzene using the startle reflex of the rat. *Toxicologist* 1988;8:77(abstract).

107. Russo JM, Dick RB, Taylor BJ, Putz-Anderson V. Prestimulus inhibition of the human blink reflex during exposure to solvents. *Toxicologist* 1990;10:302(abstract).

108. Saitoh K, Shaw S, Tilson HA. Noradrenergic influence on the prepulse inhibition of acoustic startle. *Toxicol Lett* 1986;34:209–216.

109. Saitoh K, Tilson HA, Shaw S, Dyer RS. Possible role of the brainstem in the mediation of prepulse inhibition in the rat. *Neurosci Lett* 1987;75:216–222.

110. daSilva VA, Malherious LR, Bueno FMR. Effects of toluene exposure during gestation on neurobehavioral development of rats and hamsters. *Brazil J Med Biol Res* 1990;23:533–537.
111. Squibb RE, Carmichael NG, Tilson HA. Behavioral and neuromorphological effects of triethyl tin bromide in adult rats. *Toxicol Appl Pharmacol* 1980;55:188–197.
112. Squibb RE, Tilson HA, Meyer OA, Lamartinere CA. Neonatal exposure to monosodium glutamate alters the neurobehavioral performance of adult rats. *Neurotoxicology* 1981;2:471–484.
113. Squibb RE, Tilson HA. Effects of gestational and perinatal exposure to chlordecone (Kepone®) on the neurobehavioral development of Fisher 344 rats. *Neurotoxicology* 1982;3:17–26.
114. Squibb RE, Tilson HA. Neurobehavioral changes in adult Fisher 344 rats exposed to dietary levels of chlordecone (Kepone®): a 90-day chronic dosing study. *Neurotoxicology* 1982;3:59–66.
115. Stanton ME, Crofton KM, Gray LE, Gordon CM, Bushnell PJ, Mole ML, Peele DB. Assessment of offspring development and behavior following gestational exposure to inhaled methanol in the rat. *Toxicologist* 1991;11:118(abstract).
116. Stebbins WC. Concerning the need for more sophisticated animal models in sensory behavioral toxicology. *Environ Hlth Perspect* 1982;44:77–85.
117. Sutterer JR, McSparren J, Ingerman B. Auditory startle in normotensive and hypertensive rats. *Behav Neural Biol* 1988;49:310–314.
118. Tilson HA, Cabe PA. Studies on the neurobehavioral effects of polychlorinated biphenyls in rats. *Ann NY Acad Sci* 1979;320:325–336.
119. Tilson HA, Squibb RE, Meyer OA, Sparber SB. Postnatal exposure to benzene alters the neurobehavioral functioning of rats when tested during adulthood. *Neurobehav Toxicol* 1980;2:101–106.
120. Tilson HA, Mactutus CF, McLamb R, Burne TA. Characterization of triethyllead chloride neurotoxicity in adult rats. *Neurobehav Toxicol Teratol* 1982;4:671–681.
121. Tilson HA, Hong JS, Mactutus CF. Effects of 5,5-diphenylhydantoin (Phenytoin) on neurobehavioral toxicity of organochlorine insecticides and permethrin. *J Pharmacol Exp Therap* 1985;233:285–289.
122. Tilson HA. Animal neurobehavioral test battery in NTP assessment. In: Johnson BL, ed. *Advances in neurobehavioral toxicology: applications in environmental and occupational health.* Chelsea, MI: Lewis Publishers, 1990;403–418.
123. Tilson HA, Hong JS, Sobotka TJ. High doses of aspartame have no effects on sensorimotor function or learning and memory in rats. *Neurotoxicol Teratol* 1991;13:27–35.
124. Walsh TJ, McLamb RL, Bondy SC, Tilson HA, Chang LW. Triethyl and trimethyl lead: effects on behavior, CNS morphology and concentration of lead in blood and brain of rat. *Neurotoxicology* 1986;7:21–34.
125. Watanabe I. Organotins (triethyltin). In: Spencer PS, Schaumburg HH, eds. *Experimental and clinical neurotoxicology.* Baltimore: Williams and Wilkins, 1980;545–558.
126. Wecker JR, Ison JR, Foss JA. Reflex modification as a test for sensory function. *Neurobehav Toxicol Teratol* 1985;7:733–738.
127. Wecker JR, Ison JR. Acute exposure to methyl or ethyl alcohol alters auditory function in the rat. *Toxicol Appl Pharmacol* 1984;74:258–266.
128. Wecker JR, Ison JR. Visual function measured by reflex modification in rats with inherited retinal dystrophy. *Behav Neurosci* 1986;100:679–684.
129. Weiler MH, Williams KD, Peterson RE. Effects of 2,4-dithiobiuret treatment in rats on cholinergic function and metabolism of the extensor digitorium longus muscle. *Toxicol Appl Pharmacol* 1986;84:220–231.
130. World Health Organization, Geneva. *Environmental health criteria 60: principles and methods for the assessment of neurotoxicity associated with exposure to chemicals.* Helsinki, 1986.
131. Wood RW. Behavioral evaluation of sensory irritation evoked by ammonia. *Toxicol Appl Pharmacol* 1979;50:157–162.
132. Wu M-F, Suzuki SS, Siegel JM. Anatomical distribution and response patterns of reticular neurons active in relation to acoustic startle. *Brain Res* 1988;457:399–406.
133. Wu M-F, Ison JR, Wecker JR, Lapham LW. Cutaneous and auditory function in rats following methyl mercury poisoning. *Toxicol Appl Pharmacol* 1985;79:377–388.
134. Wu M-F, Siegel JM, Shouse MN, Schenkel E. Lesions producing REM sleep without atonia disinhibit the acoustic startle reflex without affecting prepulse inhibition. *Brain Res* 1990;528:330–334.

135. Young JS, Fechter LD. Reflex inhibition procedures for animal audiometry: a technique for assessing ototoxicity. *J Acoust Soc Am* 1983;73:1686–1693.
136. Young JS, Fechter LD. Trimethyltin exposure produces an unusual form of toxic auditory damage in rats. *Toxicol Appl Pharmacol* 1986;82:87–93.
137. Young JS, Upchurch MB, Kaufman MJ, Fechter LD. Carbon monoxide exposure potentiates high-frequency auditory threshold shifts induced by noise. *Hear Res* 1987;26:37–43.

Neurotoxicology, edited by Hugh Tilson
and Clifford Mitchell. Published by
Raven Press, Ltd., New York 1992.

10

The Neurotoxicology of Cognition: Attention, Learning, and Memory

David A. Eckerman* and Philip J. Bushnell†

*Department of Psychology, University of North Carolina at Chapel Hill, Chapel Hill, North Carolina 27514; †Neurotoxicology Division, Health Effects Research Laboratory US Environmental Protection Agency, Research Triangle Park, North Carolina 27711

We use the term cognition in the chapter title to identify a subfield within the larger domain of behavioral neurotoxicology—specifically, that subfield concerned with the study of the effects of toxic substances on attention, learning, and memory. Since much of the current research on these three topics is carried out under the banners of cognitive psychology and cognitive neuroscience, the term cognitive provides a link to this important work. In addition to connoting a set of topics, however, the term often implies deliberate use of mental models to summarize observed effects. By using the term we do not intend to restrict our approach to particular theories, nor do we imply an opposition to behaviorism.

The contrast between behavioral and cognitive theoretical approaches represents an appropriate tension we do not seek to resolve. Behaviorists have for many years usefully cautioned the field against the conceits of facile mental modeling. By focusing attention inside the individual, mental models almost by necessity ignore the complex web of environmental factors that continuously modulate performance. Behaviorists call us back to the importance of these influences and correctly note that a focus on mind misses the point of evolution: behavior, like form, has been selected for optimal alignment with environmental contingencies. We would also note, however, that cognitivists appropriately accuse behaviorists of denying the use of metaphor as an important scientific tool. To focus exclusively on the datum reduces behavior to uninterpretable, molecular events.

We favor a middle course, between stark induction and etheric mentalism. In this middle approach, our talent for metaphor should permit us to construct a set of heuristic processes that emphasize salient features of a systematic analysis of behavior. These processes should generalize across procedural detail and guide research toward even more accurate scientific theories of behavior. Developing

The research described in this article has been reviewed by the Health Effects Research Laboratory, U.S. Environmental Protection Agency, and approved for publication. Approval does not signify that the contents necessarily reflect the views and policies of the Agency nor does mention of trade names or commercial products constitute endorsement or recommendation for use.

such a "theory of processes" has been proposed as a guiding principle for behavioral toxicology (98). It would seem especially appropriate for the complex field of cognitive toxicology.

In its archaic meaning, "cognition" was one of three "faculties of the mind": cognition, conation, and affection (for a brief history, see ref. 69). These faculties included the following functions: cognition included apprehending and knowing; conation included willing (i.e., volition); and affection included feeling (i.e., emotion). Miller and Gazzaniga (88) enumerated the topics subsumed under the heading of cognition as "perception, memory, language, intelligence, and reasoning," noting specifically that "motivation, intention, feeling, and emotion" should be excluded from its purview. Bruner et al. (18) defined cognition as "the means whereby organisms achieve, retain, and transform information." Miller and Gazzaniga take care to note that "attention" is difficult to classify within these faculties of the mind, since it straddles all three. We, however, contend that cognition is incomplete without attention, and will therefore consider it a critical component of the field. We propose that the topics of attention, learning, and memory encompass much of the study of cognition, especially as it can be applied to the study of animal behavior. Our review will therefore consider recent developments in these three topic areas as they have extended and enlightened the field of neuroscience and, in particular, neurotoxicology.

There are now several good resources for someone interested in undertaking the study of the neurotoxicology of cognition. A review of the neurotoxicity of organometals included chapters on learning and memory (147), a general overview of behavioral toxicology can be found in Johnson (75), and tutorials on behavioral method and results can be found in Cory-Slechta (36) and Rice (115). The neurotoxicology of human occupational exposure has recently been reviewed as well (2). Two prior reviews form the specific base for the present chapter: Cabe and Eckerman (26) and Miller and Eckerman (89). The former emphasized assessment of learning dysfunction, while the latter emphasized assessment of memory dysfunction. To orient the present review, it may help to enumerate the topics addressed in these two chapters.

Cabe and Eckerman (26) advocated a "battery" approach for the screening of toxic effects on learning and memory. They proposed that a multitask screen is required to distinguish these effects from effects on other functions (e.g., performance deficits arising from sensory, motor, or motivational changes). Further, they emphasized that since learning is not considered a unitary process, it is important to assess it multiply. They reviewed methods used to assess nonassociative learning (e.g., habituation), associative learning (e.g., reflex modification and Pavlovian conditioning), and selective learning—i.e., the reinforcement of operant behavior. In this latter topic, a review was offered for measures of the selection of response (acquisition), of extinction of this selection, and of the acquisition of discriminated responding and its transfer to new situations (e.g., repeated acquisition procedures). The review used examples in and out of mazes, and of conditional as well as direct discriminative control. The assessment of memory was distinguished from the assessment of learning primarily as a matter of emphasis, not kind, that is, an emphasis on the persistence of a change in behavior over time.

Miller and Eckerman (89) extended the review of learning measures by providing additional examples. They also expanded considerably the review of memory measures and provided a fuller discussion of theoretical and procedural distinctions that can be drawn between different types of memory assessments. Though again primarily focused on work involving animal subjects, they attempted to link more closely the measures of animal memory to those of human memory. Special emphasis was given to measures of memory provided by the radial-arm maze and other recently developed spatial-memory procedures.

The present chapter will expand on these prior analyses, primarily by reviewing progress made using several exemplary learning and memory procedures that have seen increased use over the last decade. Passive avoidance is considered an apical test of learning and memory, and several repeated acquisition procedures are presented as more specific tests of learning. We then update progress that has been made in measuring neurochemical effects on working memory in rodents. We will expand these prior reviews by addressing an additional topic: measures and dysfunctions of attention. We hope that this discussion, taken with the other reviews cited, will serve as a continuing guide to the field of cognitive neurotoxicology.

LEARNING AND MEMORY

Learning is well defined as "an enduring change in the mechanisms of behavior that results from experience with environmental events" (44). The reason for the word mechanisms in this definition is that behavior is multiply determined. The effect of prior experience (i.e., learning) is to change the influence of other factors rather than to change behavior directly.[1] As we cannot see changes in these mechanisms, we must infer that learning has occurred from a change in behavior. As an aside it might be noted that we follow Domjan and Burkhard (44) in distinguishing learning (an enduring change) from memory (a change that can be either short-lasting or long-lasting) on the basis of the duration rather than type of influence of past experience. Thus learning involves the development of enduring changes while memory need not endure. Temporary changes involve memory but not learning.

It must be emphasized, however, that capacity for learning and memory is limited by sensory capacity, overall arousal, motivation, motor ability, etc. Designating a neurotoxic effect as an effect on learning or memory, then, requires that direct effects of the toxicant on these other functions be evaluated and set aside. To evaluate these other factors adequately requires that a number of control

---

[1] This distinction is often described as the "learning-performance" distinction.

conditions be included in the research design. It may also be the case that a toxicant will have multiple effects, for example, affecting learning ability *as well as* motivation. Since assessment of learning ability is so very dependent on setting aside these other factors, proving a learning deficit in this case will be difficult if not impossible unless a way is found to "see through" the motivational effect. We present a rather full account of these issues below when addressing the use of a passive avoidance task to assess neurotoxic effects on rapidly acquired reference memory.

To say that a neurotoxicant affects learning implies that the *acquisition* of a behavioral change is impaired. We would prefer to describe the effect as a change in *attention* rather than a change in learning, however, if we found that despite normal sensory function, behavior was unaffected because the critical environmental factors were not "noticed." Further, to describe the effect as resulting from impaired learning requires that we distinguish the effect from one on impaired memory. If a neurotoxicant affects the *persistence* of a behavioral change (or its retrieval) *after the behavioral change has been acquired,* then it is more appropriate to describe the effect as resulting from a change in *memory* rather than a change in learning. Only rarely are these distinctions carefully drawn. As a result, neurotoxic effects on attention, learning, and memory are usually comingled. Though the differences may be subtle, their analysis seems worthwhile. As will be seen below, many of the effects that have been called "memory" effects might be better described as "learning" effects. Also, many of the effects that at first appear to be effects on learning might turn out to be better described as "attention" effects.

An analysis of what can be said about learning and memory from results obtained with a commonly used task might help to clarify some of the distinctions we have been seeking to make. The following section provides an analysis of what can be learned from the passive avoidance task that has been a widely used measure of rapid learning and its persistence (103).

Assessing Learning With Rapidly Acquired Reference Memory Tasks

If a sufficiently aversive stimulus follows a high probability response, an animal commonly suppresses that response thereafter. The acquisition and retention of this "passive avoidance" have been assessed by a number of procedures that at first appear to be quite similar. In one version of this procedure, for example, rats or mice learn to remain in a larger, lighted chamber rather than enter a (preferred) dark area where footshock has been received (called step-through passive avoidance or inhibitory avoidance; for recent examples, see refs. 39,121). Performance can be tested after a delay (e.g., 24 to 72 hr) if the goal is to assess memory for this brief training. When shock levels are moderate to high, this learning is accomplished in one trial and the effects are quite persistent. Shock levels can also be lowered (e.g., ref. 40) in order to study the learning process across several trials; with lowered shock levels, the rate of initial learning can be estimated.

The single-trial passive avoidance procedure is perhaps the animal model most widely used to assess dysfunction of learning and memory (103). Further, amnesia for passive avoidance produced by cholinergic blockade provides a very popular animal model of recent memory, and reversal of this amnesia is widely used to assess the potential memory-enhancing actions of nootropic drugs (but see ref. 61 for concerns regarding this use). Impairment of passive avoidance retention is easily produced, and this sensitivity is a strong argument for its use in screening. Yet, the interpretation of changes in passive avoidance is a subtle enterprise; several control conditions must be included in order to firmly establish that impaired performance can be ascribed to impaired learning or memory, and these control conditions are regularly omitted from experimental protocols. Additionally, learning and memory are complex functions that can show many types of impairment. Interpretation of passive avoidance effects is therefore suspect. A brief summary of the current understanding of the effects of cholinergic blockade on this performance is offered below. This review will demonstrate the kinds of distinctions that can be drawn with this procedure and will indicate the cautions that are required.

Dissociating Cognitive from Other Kinds of Effects

Though passive avoidance procedures have been widely used to assess learning and memory, many investigators (including 89) have noted that results obtained using these procedures are very difficult to interpret. The confusion derives from many sources. First, investigators have used a great diversity of protocols to assess rapidly learned avoidance. For example, the length of the observation period would at first seem a trivial element in the protocol. Yet, effects of posttrial scopolamine are seen only if a long (e.g., 600 sec) retention test observation period is used. Investigators who adopted a commonly used shorter retention period (180 sec) found no effect (121). Slight shifts in protocol of this sort may not even be specified in the report. Levels of illumination and details of animal care and handling can also be crucial to the kind of result obtained. It is no wonder, then, that results differ between studies that appear on first reading to be quite similar.

A second source of confusion derives from individual differences in effect for different subjects. Individuals tend to be affected in an all-or-none fashion in rapidly learned avoidance tasks (freeze or move rapidly). At "sensitive" levels of the procedural parameters, these individual differences in latency of response are often very large, thus reliability is low across subjects. What may seem to be a graded effect of an agent may, in fact, merely represent a shift in the number of animals affected. Even a high dose having a major effect on the group mean may still leave several animals that show no apparent effect at all, and differences in group means can be greatly affected by a change in just a few subjects. How such changes in the incidence of impairment relate to the more graded impairments obtained from other procedures is not always clear. In addition, treatments typically affect both mean and variance of the measures, complicating statistical evaluation of the mean effects.

By far the most common source of confusion in interpreting changes in passive avoidance derives from the many factors critical to passive avoidance performance. The passive avoidance procedure is apical in that changes in many different kinds of functioning lead to changes in performance. If more than one of these functions is affected by an agent, the difficulty in interpreting the effect can be considerable since it becomes very difficult to set aside these other influences successfully and identify the source of the change as an alteration in learning and/or memory. For example, to set aside differences in sensory capacity, different stimuli should be tested and similar effects on learning should be observed with more than one set of stimuli. To set aside differences in overall arousal, initial performance prior to the learning experience should be evaluated for exposed and nonexposed subjects. Initial performances must be comparable if arousal is to be set aside. To set aside differences in motivation, different types of reinforcers should be used to support the performance (e.g., aversive as well as appetitive reinforcers). To set aside differences in motor ability, different performances should be required.

Several years ago Bammer (5) provided a helpful metaanalysis of studies showing the effects of cholinergic, noradrenergic, dopaminergic, and serotoninergic manipulations on passive avoidance. As Bammer indicates, if an agent's effect on this performance is to be "attributed to learning and memory processes, other factors which can change performance must be ruled out [by control experiments]." Further, "in most studies, these control experiments have not been carried out, so that the results are difficult to interpret" (5). A metaanalysis of the literature is required to sort through these issues. Though additional data have accumulated since his review, Bammer's careful account of the methodological issues that surround the passive avoidance procedure and the controversies that arise regarding interpretation of effects has not been improved and is well worth reviewing. He identified four factors that must be ruled out in order to designate a change in performance as specifically indicating a change in learning or memory: (a) a change in shock sensitivity, (b) a change in locomotor activity, (c) state-dependent effects of the manipulation, and (d) a change in the biochemical response to stress. We suggest an additional factor to evaluate: (e) a change in exploration or response bias (e.g., preference for dark). These factors will be noted below to demonstrate the difficulties that arise when interpreting the effects of cholinergic blockade on passive avoidance.

1. *Changes in shock sensitivity.* Since small changes in shock intensity produce very large changes in performance, any change in shock sensitivity (e.g., through changes in sweating) will affect performance even though learning and memory are intact. Assessment of shock-threshold must therefore be included, and change in this threshold makes it difficult to interpret an alteration in passive avoidance as due to changes in learning or memory. Bammer (5) noted that since cholinergic

agonists and antagonists alter shock sensitivity, interpretation of changes in passive avoidance brought about by these agents as due to changes in learning or memory is suspect.

2. *Changes in locomotor activity.* Passive avoidance is easily altered by any change in locomotor activity or "response inhibition." Such changes may well be independent of learning or memory and therefore must be set aside before an effect can be correctly attributed to changes in learning or memory. Analysis of initial response latency (latency to respond prior to the aversive event) offers some insight into this factor. Again, Bammer indicated that cholinergic agonists and antagonists do alter response inhibition and thus again, the effects of these agents on passive avoidance become more difficult to ascribe to changes in learning or memory.

3. *State-dependent effects.* Decreased passive avoidance at the time of the test might result if the stimuli present during the test differ from those present during training (generalization decrement). For subjects given a drug, the drugged state itself is a salient part of the training environment. If testing is done in an undrugged state, a decrease in performance may result from this stimulus change rather than from an impairment of learning or memory. To assess such state dependency, results for tests carried out without the agent should be compared to those for tests carried out in the presence of the agent. If the performance is improved when the agent is present during the test, a dependence on state is confirmed and it is difficult to interpret the changes as resulting from changes in learning or memory. Again, Bammer presented evidence confirming that cholinergic manipulations produce a state-dependency. This is yet another reason why interpretation of cholinergic manipulations as producing an effect on learning or memory is problematic.

4. *Changes in response to stress.* Passive avoidance might be indirectly disrupted by changes in the biological reaction to stress (e.g., sequelae following presentation of electric shock). If an agent alters biological sequelae to stress, it would be inappropriate to interpret the resulting change in passive avoidance as due to a direct change in learning or memory processes. Bammer again confirms that cholinergic manipulations affect this factor.

5. *Changes in exploration or response bias.* Similarly, if an agent alters a subject's sensitivity to a stimulus or its pattern of exploration of its environment, those changes might at first appear to be changes in learning or memory. For example, a drugged rat might be more sensitive to light and therefore show an increased preference for the dark chamber in a step-through passive avoidance apparatus. Since cholinergic manipulations would likely produce changes in pupil dilation, a change in dark preference seems quite plausible. Reduced latency resulting from this changed preference might appear to be a memory loss.

Yet another issue is that the pattern of exploration for new stimuli might be changed. Cholinergic manipulations are known to produce effects on attention and/or habituation to novel stimuli (e.g., refs. 31,94). This issue will be addressed briefly later while noting effects of cholinergic manipulations on attention. Without the proper control experiments, such changes in preference could appear to be effects on learning or memory.

Bammer's review confirms that several noncognitive sources of change can account for alterations in passive avoidance performance. Cholinergic manipulations appear to affect them all. It should be stressed that the presence of these additional sensitivities does not disprove an effect on learning or memory. Yet the fact that cholinergic manipulations alter all these additional factors makes it difficult to simply describe their effects as "amnestic."

There is, in fact, an interesting nonreciprocity between tests of noncognitive and tests of cognitive sources of effect: the appropriate control procedure to dissociate these sources is useful only if the noncognitive factor does not have an effect. When the control procedure confirms that the noncognitive factor is effective, it does not provide evidence for or against the presence of a cognitive effect. Using an apical test of cognitive function such as passive avoidance, then, will often leave the issue of specificity of effect unresolved. Having argued persuasively that we should not do so (as discussed above), we will assume, in the discussion below, that the effect on learning or memory has been demonstrated, though such is often not the case.

Distinguishing Learning from Memory Effects

Cholinergic blockade has been shown many times to produce an anterograde amnesia for passive avoidance. For example, Rush (121) found that a moderate dose of scopolamine administered just prior to the learning trial consistently impaired retention of passive avoidance when tested 24 hr later. Though moderate doses of pretrial scopolamine prevented retention, however, only a very high dose (10–20 mg/kg) impaired this 24 hr retention if the scopolamine was administered immediately after the learning trial. That is, moderate doses of scopolamine specifically affected *acquisition* of passive avoidance and were ineffective in disrupting the consolidation of this learning (i.e., memory storage). It is worth noting that the impaired retention of passive avoidance was not merely a result of state-dependent learning in this case, since administering the same moderate dose of scopolamine just prior to the 24-hr test further impaired the performance rather than reinstating it (see below for a further description of the state-dependency issue).

However, scopolamine's effects are not restricted to acquisition. Retrieval of the learning at the time of the test could also be disrupted. While Rush (121) confirmed that scopolamine affected the acquisition more than consolidation of passive avoidance, he also found that a moderate dose of scopolamine administered just prior to the 24-hr test also impaired performance. Since this same dose had not affected initial-trial latency when given before the learning trial, this impaired retention could not be due to increased locomotor activity or dark preference during the test. Because it demonstrates an impaired ability to utilize past experience, it can be classified as a memory effect.

Though assessed with a memory test, the primary effect of scopolamine on passive avoidance performance is anterograde, affecting learning and retrieval rather than the persistence of the learning (memory). Decker et al. (40, see Table 1) demonstrated that CD-1 mice given moderate doses of scopolamine (e.g., 1.0 mg/kg) before learning acquired passive avoidance more slowly than controls. To make this assessment, they reduced shock levels to allow the rate of acquisition to be assessed more adequately. Once the mice had acquired the passive avoidance under the influence of a moderate dose of scopolamine, however, they showed that retention of the learning was as good for drugged as for control mice. There was no difference in response latency when the performance was tested after delays of 2, 16, or 28 days, indicating that there was no increased rate of forgetting. In other words, the drug affected learning but not memory.

Decker et al.'s study is included in the growing list of studies which suggest that the primary effect of scopolamine is on initial learning rather than on the ability to retain this learning. It may be that scopolamine qualitatively changes *what is learned,* an effect that is better described as a change in attention than as either learning or memory. The impact of such a change will differ depending on the task. Different tasks will undoubtedly require different sorts of memories and thus show different kinds of impairment following administration of scopolamine. Quartermain and Leo (111) have provided a striking example of the complexities involved. These investigators gave moderate doses of scopolamine immediately after training Swiss Webster mice to perform an active avoidance response (one-way, step-through shock avoidance with 30 sec intertrial intervals). These mice were given one training session, which continued until escape from a flashing warning light occurred on nine of ten trials. In their case, posttrial scopolamine produced a well-developed amnesia when performance was tested 1, 3, or 14 days after training (separate subgroups of mice were tested at these intervals). Remarkably, however, scopolamine-treated animals actually *improved* their retention when testing was given a week (7 or 10 days) after training. At this time, retention was *comparable* for scopolamine and control groups. The

TABLE 1. *Training data from mice trained to a criterion of 4 min in the safe compartment of the inhibitory avoidance apparatus using continuous-trial criterion training*

Drug	N	Training latency mean(s) ± SE	Trial 2 latency median(s) (interquartile range)	Trials to criterion mean ± SE
Control	28	18.6 ± 2.7	78.5 (59–162)	2.3 ± 0.1
Scopolamine	28	20.2 ± 2.3	8.5 (6–15)[c]	6.5 ± 0.7[c]
Diazepam	27	11.3 ± 2.2[a]	26.0 (15.5–80.8)[b]	3.0 ± 0.2

[a] $P < 0.05$, significantly different from control.
[b] $P < 0.001$.
[c] $P < 0.0001$.
From Decker et al., ref. 40, with permission.

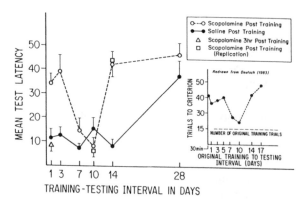

FIG. 1. Differential effects of length of retention interval on memory in scopolamine and saline-injected mice. Retention score is the mean of the five test trials. The insert illustrates the marked similarity between the scopolamine data and results obtained from rats undertrained on brightness discrimination and tested at the same intervals following learning (from Quartermain and Leo, ref. 111, with permission).

deficit had disappeared. This result represents an intriguing "release" from the amnesia induced by posttrial scopolamine. By the next test-period (28 days), however, amnesia could no longer be assessed, since neither scopolamine-treated nor control mice showed retention.

How can such a pattern of initial retention deficit followed by improved performance be interpreted? Since a comparable pattern was seen when the avoidance was only partially learned (training to only a four out of five trial criteria) or when memory for a partially trained brightness discrimination was tested, Quartermain and Leo (111) suggest that posttrial scopolamine might have prevented the storage of an intermediate-range memory that is used for the first few days after an experience. The storage of a different kind of memory that is used subsequently, however, might not have been impaired. This longer-term memory might have been "unmasked" by the passage of time. Such a complex set of assumptions must currently be made to account for their pattern of results, that considerable clarification of these effects will be required before a consensus forms. Despite the lack of such a consensus, however, the point clearly made by these data is that cholinergic blockade does NOT simply impair "memory," but appears to have complex effects on attention and learning as well (see ref. 73 for a similar assertion).

Assessing Learning with Repeated Acquisition Procedures

To make the subtle distinctions described above, it is essential that we utilize a very refined measure of learning. Since individual differences between subjects are often qualitative (e.g., strategic) rather than merely quantitative, a within-subject experimental design is best. Further, most procedures used to study learning are not very analytic since they involve the acquisition of a great many aspects of the performance at the same time (e.g., habituating to the situation, learning about the reinforcer, learning the best preresponse posture, etc.). If we use a procedure that allows us to focus on the learning of just one aspect of the test situation, our analytic power is considerably enhanced. To maximize analytic powers, several investigators have advocated the use of a repeated acquisition procedure (28,100,145,146) to permit quantification of the learning rate in individual animals from a stable baseline. A toxic effect can then be measured against this stable baseline. The cost of this increased analytic power is the often considerable training necessary for establishing the baseline. Our goal for the present section, then, is to describe some repeated acquisition procedures and to offer an evaluation of their merit.

Clearly, the ability to quantify learning repeatedly in the same subject provides a great advantage for the scientist interested in the effects of chemicals on learning. In addition to focusing on learning in a well-defined set of procedural conditions, this baseline approach to learning facilitates assessment of the time course of effects during or following prolonged exposure to a toxicant, and the acute effects of a short-acting chemical can be assessed using a given subject as its own control. The within-subject design reduces sample sizes, though at the expense of large investments in subject preparation (training). Once trained, however, a subject can serve in a series of determinations of the effects of drugs and/or toxic chemicals.[2]

By their nature, repeated acquisition procedures tend to focus on the rate at which behavior changes in response to alterations in experimental contingencies, rather than on whether or not behavior is altered by this contingency change. In this sense, repeated acquisition tests contrast markedly with passive avoidance tests, in which one-trial learning is typically measured as either having occurred or not. Incremental learning thus lends itself to analysis of the graded effects of chemicals on learning, while one-trial procedures are better suited as indices of incidence of response. The choice of method may therefore depend on whether dose-effect data (severity of a graded effect) or dose-response data (incidence of affected individuals) are needed.

Of course, the process of training acquisition baselines involves learning of another sort: learning to learn the daily problem. This kind of interproblem learning was conceptualized in the 1940s as a learning set (63), aspects of which have undergone extensive analysis since (e.g., refs. 8,11,84,86). These learning sets are fascinating in their own right, and their formation has occasionally been used to evaluate the effects of toxic chemicals (e.g., refs. 25,80,114), CNS lesions (4,153), and aging (142). Nevertheless, this review will focus on the post-learning set "steady-state" performance of animals on three exemplary repeated acqui-

---

[2] This inverse relationship between sample size and training and testing time may in fact generalize to a variety of methods (103). However, Peele and Vincent's analysis of this relationship does not take into account the fact that animals trained on a complex baseline are considered a valuable preparation by the experimenters, who then take advantage of the preparation and evaluate performance for prolonged periods, perhaps longer than would otherwise be necessary.

sition methods: serial reversal learning, repeated acquisition of response chains, and repeated acquisition of spatial locations in the radial-arm and water mazes.

Serial Reversal Learning

Reversal of a learned discrimination appears to engage different processes than does acquisition of the discrimination itself, as indicated by several facts. First, the first reversal of a discrimination usually requires more training to reach the same learning criterion than does the original discrimination. This indicates that the original response tendencies inhibit the new learning and must be extinguished, or at least redirected, before new learning occurs. Second, various CNS lesions, most notably in the limbic system, frequently disrupt reversal learning while sparing the initial discrimination. Finally, repeated (serial) reversals of the same discrimination lead eventually to a sort of steady state of acquisition in which successive reversals are learned more quickly than was the original discrimination, and often in a single trial.

This last point suggests that reversal learning at steady state (i.e., learning after the subject has acquired a set for reversal learning) may differ qualitatively from that observed at the beginning of the series, if a more efficient solution strategy develops across problems (reversals) in the series. This change in strategy would reflect the many cognitive processes that are involved in reaching this asymptote, including learning (associating stimuli with reward, learning to shift responses when contingencies change), memory (remembering which stimulus is currently positive), and attention (focusing on the relevant stimuli and ignoring irrelevant ones) (see ref. 84 for a review of these factors). Caution is therefore advised in contrasting learning of problems early in training with learning at steady state.

Once a steady state of reversals has been achieved, this "reversal baseline" then provides a means of generating a learning curve during each test session. Calhoun and Jones (28) proposed this method for the study of the effects of drugs on learning, showing that amphetamine facilitated reversal of a go/no-go discrimination between a tone and a light by reducing responses to the negative stimulus, while scopolamine impaired reversal by suppressing response to the positive stimulus (28,29). Facilitation of reversal learning by amphetamine has been attributed to its effects on attention to critical elements of discriminative stimuli, an important factor in successful reversal learning (85).

Shimizu and Hodos (135) used serial reversals of a color discrimination in pigeons to determine the functional specificity of the Wulst, a region of the avian cerebral hemisphere corresponding in some ways to the mammalian striate cortex. Birds with lesions in certain regions of the Wulst learned the original discrimination normally, but made three to four times as many errors as controls on each of a series of 20 reversals. To determine which phase of learning was affected, errors were categorized as either reflecting a position habit, or occurring in one of three phases of the learning process. Phase 1 (perseverative) errors occurred prior to two consecutive correct responses; phase 2 errors occurred prior to the first six consecutive correct responses; and phase 3 errors occurred between phase 2 and the final learning criterion of 15 consecutive correct responses. Errors in the last two phases of learning accounted for the reversal retardation, since no position preferences were observed, and the slight increase in phase 1 errors was not significant. These data were interpreted to suggest that the lesions impaired the birds' ability to maintain consistently correct responding due to interference from previously rewarded (incorrect) responses, and are consistent with other studies in which intrusion of incorrect responses impaired reversal acquisition (e.g., refs. 21,53,117,119).

In the Shimizu and Hodos study, error analysis provided a means by which different mechanisms of dysfunction could be contrasted. Other measures can also be used to assess potential sources of change in the learning process. For example, relative rates of response to two discriminanda during daily reversal acquisition can be used to characterize reversal learning (Fig. 2). A spatial dis-

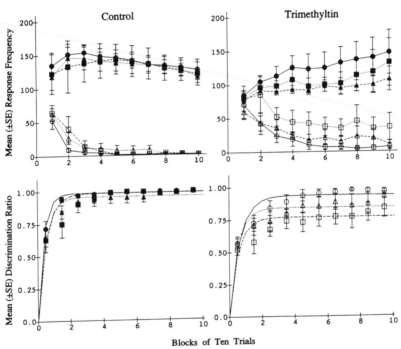

FIG. 2. Effects of trimethyltin (TMT) and delay on reversal learning in rats. Two retractable levers were extended repeatedly into an operant chamber on independent random schedules. Each retraction of one lever (S[+]) was followed immediately by delivery of food; retractions of the other lever (S[0]) were not paired with food delivery. Frequencies of lever presses had no programmed consequence. On alternate days, the lever which was paired with food was changed (reversal). In different sessions, delays of 0 (circles), 2 (triangles) or 4 (squares) seconds were imposed between retraction of S[+] and food delivery. Rats responded preferentially to S[+] (*upper panels, filled symbols*) under all conditions, demonstrating accurate discrimination of the contingency. Responses to S[0] (*upper panels, open symbols*) decreased across trial blocks as each reversal was learned. Reversal acquisition is also shown by the increase across trial blocks in the Discrimination Ratio (the frequency of responses to S[+] divided by the sum of the frequencies to S[+] and S[0] in each trial block; *lower panels*). TMT impaired reversal acquisition when delays of 2 or 4 sec were imposed between S[+] and food (*right panels*). The curves in the lower panels represent best-fitting negative exponential functions for each delay. (From Bushnell, ref. 21, by permission).

crimination was trained using an automaintenance schedule in which retractions of one lever (S^+) predicted food, while retractions of another (S^0) were unpaired with food. Pairing the S^+ with food elicited responding to the S^+, but not to the S^0. Rats learned a series of reversals of this discrimination and reached asymptotic performance levels prior to a single injection of trimethyltin (TMT) (7 mg/kg, iv). Rates of responding to the S^+ and S^0 were used to calculate a discrimination ratio, by which reversal criterion was defined. When a 2 or 4 sec delay was introduced between retraction of the S^+ and food, TMT-treated rats' reversals were impaired such that their discrimination ratios did not reach control levels during twice-weekly reversals (Fig. 2). This reduction in discrimination accuracy resembled the intrusion errors described by Shimizu and Hodos (135) in that reversal impairment resulted from an increase in the rate of response to S^0 throughout the reversal session, rather than a change in S^+ responding. A similar reversal acquisition deficit was observed following long-term treatment with styrene monomer (22), in that responding to the S^0 was related to a decline in discrimination ratio. In this case the impairment was evident with no delay imposed between the S^+ and reward.

With this procedure, analysis of response rates to the two discriminative stimuli takes advantage of the gradual nature of reversal acquisition to characterize the source of a learning deficit. This more molecular approach permits analysis of component processes involved in reversal learning, i.e., the shift from one discriminative stimulus to another, rather than a more global measurement of choice accuracy.[3]

In these studies, discriminations involved a single stimulus dimension, and no irrelevant cues were present. Use of irrelevant cues can exacerbate the effects of neurotoxic compounds, as has been shown by Rice (114). In this study, monkeys were trained to discriminate color and form patterns in an operant chamber. Monkeys with a history of exposure to lead were assessed using a series of three discrimination reversal tasks: form discrimination, color discrimination with form irrelevant, and form discrimination with color irrelevant. Deficits induced by lead were revealed only in the latter two tasks, in which some of the treated monkeys committed more errors than controls across a series of reversals. For example, while only 2 of 13 treated monkeys committed more errors than any control on the first task, 6 of those 13 monkeys did so on the second task, when form cues were used as irrelevant distractor stimuli. These results underscore the importance of attention to appropriate stimuli in reversal learning, and suggest that lead exposure may impair learning by interfering with attention, as has been suggested for childhood lead poisoning (126).

The stimulus modality used for the discrimination task can greatly affect whether a reversal learning set, and hence a baseline for repeated acquisition, can be formed. For example, Duncan and Slotnick (45) found that pigeons' learning improved across successive reversals (i.e., a reversal learning set was formed) when they were trained on a color discrimination, but not when they

---

[3] An elegant analysis of decision making in discrimination learning and reversal has been reported using response time measurements in addition to response accuracy (97).

were trained on an odor discrimination. Further, when birds were trained using compound visual-olfactory cues, transfer to new olfactory cues was negative, while transfer to new visual cues was positive. This differential transfer occurred even in birds pretrained with olfactory cues, prior to training on the compound. These data provide strong evidence that the stimulus modality used in discrimination tasks can greatly affect learning and reversal, and indeed determines whether a learning set can be acquired. They also caution that care should be exercised in comparing cues across modalities within a reversal task, as suggested by the method of Calhoun and Jones (28).

As a final example, Raffaele et al. (112) used a reversal baseline to detect covert effects of repeated exposure to an acetylcholinesterase (AChE) inhibitor, diiso-propylfluorophosphate (DFP). Rats were trained to discriminate a striped goal box from one with large black spots in a Y-maze for food reward; this discrimination was then reversed daily. At the beginning of each session, the rats' retention of the discrimination was first assessed by counting errors and trials to criterion on the previous day's discrimination. After reaching criterion, the contingencies were reversed, and the rats were trained to criterion again. After a stable reversal baseline was established, DFP (0.5 mg/kg, ip) was injected three times per week to half the rats. Challenges to the cholinergic system were conducted with scopolamine injections before, during, and after DFP treatment. DFP alone had no effect on either retention or reversal of this discrimination. As predicted,[4] however, rats treated with the AChE inhibitor were supersensitive to the disruptive effects of scopolamine at some doses and at some times during the course of DFP exposure.

Serial reversal learning may provide a suitable baseline for repeated assessment of learning, since many mammalian and avian species appear to form reversal learning sets, and performance at asymptote does not differ greatly across species. The phenomenon has been extensively studied, yielding a wealth of behavioral and parametric data to guide development of models for toxicological research. It is desirable, however, to avoid initiating reversals at the outset of each session, as subjects under those conditions tend to "sample" each stimulus at the outset of a session to determine which is correct for that session. Midsession reversals, as used by Raffaele et al. (112) and Bushnell (22), particularly if they occur at an unpredictable point in the session, reliably yield usable learning curves. Such curves may be modeled mathematically (21) and compared to performance during "nonlearning" sessions in which reversals are not administered.

Disadvantages of serial reversal learning as a repeated acquisition task include the long training time to reach stable baselines and the difficulty of training some individual subjects to reverse the initial discrimination. This latter problem of response extinction (25,142) complicates interpretation of learning reversals early in the set, but may not continue when stability is reached. Finally, it appears that reversal learning set formation, and thus the possibility of using reversals as a learning baseline, depends on the stimulus modality, which is, in turn,

---

[4] The interactions of DFP with the cholinergic nervous system are detailed in the section on memory, below.

species-dependent. For example, pigeons will form visual sets readily, but fail completely at olfactory reversal learning sets (45). In contrast, rats learn olfactory sets readily, but are relatively poor at visual, auditory, and mixed visual-auditory sets (92,137). On the other hand, assuming stable baselines can be achieved, the stimulus modality might provide a dimension of reversal difficulty which could be used to advantage in characterizing effects of toxicants on learning.

Repeated Acquisition of Response Chains

Boren (12) approached the problem of establishing a baseline of learning in the individual subject by training monkeys to learn chains of responses in a test chamber containing 12 response levers, arranged in four sets of three. Illumination of a given set of levers indicated that a response was required on that set. A response to the correct lever in that set illuminated the next set, and so forth; the fourth correct response yielded food reward. The designation of the "correct" lever in each set of levers was arbitrarily determined prior to each session, so that an animal might, for example, learn L-C-L-R in a given test session, and C-R-L-C in the next. Sequences with repeated positions were omitted to avoid generation of position preferences. Errors, recorded for each response on an incorrect lever, resulted in a brief time-out period with the lights off, but did not change the animals' position in the sequence. With repeated training, monkeys learned to acquire sequences of responses which then yielded a series of problems to be learned across test sessions. Since the number of usable sequences for these studies is large (24 four-element sequences of three levers are possible when repeated responses to a given lever are prohibited), the list can be repeated indefinitely with little possibility of memorization; in addition, performance can be averaged across sequences, avoiding the problem of idiosyncratic sequence preferences identified by Howard and Pollard (72).

Boren's (12) task, with modifications by Thompson (145), established a paradigm by which repeated acquisition tasks have since been conducted with response chains. The equipment requirements have been eased by reducing the number of response manipulanda to three: instead of groupings of levers, discriminative stimuli (e.g., the color or flash rate of cue lights or the pitch of a tone) have been used successfully to indicate the current position in the sequence. Modification of the schedule of reinforcement with the repeated acquisition response chain and use of shock as an aversive motivation have also been reported (for review, see ref. 146).

One important supplement to the study of the effects of drug and toxicants was introduced by Thompson and Moerschbaecher (146); it involved presenting an invariant response sequence in the presence of an appropriate discriminative stimulus. Once learned, performance of this sequence does not require new learning, so it may be used as a "performance control" condition to assess the sensory, motor, and motivational requirements of the task independent of learning. Using this multiple task baseline, robust, replicable, and dose-related decrements in accuracy in the learning component were observed in monkeys injected with amphetamine or cocaine at doses that did not impair accuracy in the performance component.

Learning and performance are measured primarily as changes in error frequency across trials in the two components of this task. Learning is defined by within-session reduction in error frequency on a variable sequence and performance by a low error rate on an invariant sequence (146). For reporting effects of drugs or toxicants, total errors or percent errors in a session are typically reported. A decline in the rate of trial completion or response rate indicates that the high end of the dose-effect function has been reached, above which nonspecific behavioral impairment precludes quantification of effects on learning *per se.*

While very few studies of toxicants have been conducted using these procedures, a number of studies of drug effects have been reported. In most cases in which a performance component is used, drugs impair learning at doses which do not affect performance of an invariant chain (10,90,146). Nevertheless, effects specific to learning are not always reported. For example, in studies not using an invariant chain, response rate (or pausing) is often taken as a measure of performance not specifically related to learning. In such cases, learning may be affected only at doses which reduce response rates (e.g., refs. 52,105) or increase pausing (104) in the variable chain.

To study the role of stimulus control in repeated acquisition, a faded learning (FL) component was added to the standard learning and performance components to modulate the rate at which acquisition of a variable sequence occurred (90). In the FL component, the cue light over the correct lever was illuminated at the beginning of each trial; as the animal's accuracy increased across trials, the lights above the other levers were "faded in" over eight increments of brightness until they were as bright as the light above the correct lever. As predicted, FL progressed faster than unassisted learning. In addition, it proved to be more sensitive than performance, but less susceptible to disruption by amphetamine, cocaine, or phencyclidine than was unassisted learning.

Repeated acquisition of response sequences can be learned readily by humans, monkeys, and pigeons; training rats has proved more difficult (see ref. 72). In an attempt to devise a repeated acquisition baseline for rats, Paule and McMillan (100,101) introduced an incremental repeated acquisition (IRA) task in which a series of chains of progressively longer sequences is presented during each session. Rats were first required to complete 40 sequences of a single-element chain, each of which consisted of 40 responses (FR40) to one of three levers (randomly chosen for a given session). After training to a criterion of 40 errorless one-lever sequences, the task was incremented such that FR40 was required on a different lever, followed by FR40 on the lever in the single-element sequence to obtain reward. As before, training proceeded to criterion, after which the sequence was incremented to three levers, and so on up to a maximum of five levers. At all stages, auditory and visual feedback were provided to the rat regarding the correctness of each response. Serial position indicator lights were used to indicate the current position in the sequence, and food was provided at completion of each sequence.

This IRA task yields several performance measures, including the percent task completion (reflecting the length of the sequence reached by the animal), response rate, efficiency (percent correct for each sequence length), and two types of errors. Between-sequence errors (those occurring prior to the first correct response of a sequence) reflect initial discovery of the correct lever which begins each newly incremented sequence, and within-sequence errors (those occurring after the first correct response of a sequence) reflect shifts in responding from the correct lever to an incorrect lever during completion of the required ratio in a given sequence.

Although this procedure did not include responding to an invariant sequence as an index of performance, the incremental nature of the task in principle permits a comparison of learning of longer sequences with performance of previously learned, shorter ones. As the authors propose, "drug effects on both acquisition of new responses and performance of previously acquired responses can be studied in the same behavioral session" (ref. 100, p. 437). However, three problems are evident with this proposal: first, performance in this analysis was always measured on a shorter, and thus presumably easier, sequence than was acquisition. Second, rats rarely exceeded sequences of three levers before termination of the session (60 min time limit), limiting the number of such comparisons as well as the difficulty of the sequences used. Third, due to the large ratio requirements and training to a stiff criterion of 40 errorless sequences before incrementing, the progression through the sequences frequently covered a substantial fraction of the time course of short-acting chemicals. Thus, task difficulty and chemical pharmacokinetics were confounded, which complicates interpretation of the effects of such chemicals on learning. The same problem was encountered by Penetar (104) who compared accuracy during acquisition of a sequence (learning) to accuracy after acquisition of the sequence (performance).

As suggested (100), comparison of errors between sequences versus errors within sequences should provide evidence for sensitivity of new learning versus performance of previously acquired responses, respectively. However, animals rarely completed many long sequences with the parameters selected; sequences were typically two elements in length for rats (100) and three sequences in length for monkeys (130). For longer sequences, data are described as "extremely variable" (130). One is forced to question just how well learning was demonstrated by acquisition of a two-element sequence, particularly when immediate feedback was provided for each error.

In a series of assessments with drugs, for example, Paule and McMillan (100) found that increasing doses of amphetamine, diazepam, morphine, pentobarbital, and chlorpromazine each reduced response rates and percent task completion, while efficiency (percent correct) was less sensitive to the drugs. Evidence for changes in within-sequence error reduction by these drugs was either limited to the highest doses of each compound, at which response rates were greatly suppressed, or these effects were not dose-related (as for morphine). Further, greater sensitivity of between-sequence errors (acquisition of a new response) compared to within-sequence errors (performance of the old response) was found only at the highest doses of amphetamine and diazepam. At these doses, response rates were also reduced, and amphetamine induced strong perseverative (and error-producing) response patterns. Thus, the ability of the IRA to detect changes in behavior reflecting learning is somewhat questionable.

The same concerns apply to the IRA procedure currently included in the operant test battery that has been developed for assessing the cognitive behavior of primates (99). For a series of studies evaluating the acute effects of tetrahydrocannabinol (130), marijuana smoke (131,136), diazepam (134), and amphetamine (132), the IRA component of the battery was shortened in terms of the number of errorless sequences to criterion (from 40 to 24), and expanded by adding a fourth lever and a sixth possible sequence. However, it is not clear from these reports whether the task really assessed learning, since no evidence of error reduction across trials within a sequence was presented. The task also appears to be less sensitive to the effects of amphetamine than nonincremental versions of the task described above. Thompson and Moerschbaecher (146), for example, demonstrated clear impairment of acquisition without effect on performance, with doses of 0.3 to 1.0 mg/kg d-amphetamine. In contrast, accuracy in the IRA task fell only at 1.0 mg/kg, at which dose only two of nine animals responded, and only a fifth of the task was completed (132). This difference in sensitivity is probably attributable to the length of the sequence being trained (two in the IRA vs. four in the former study).

Despite its successful use in characterizing effects of drugs on learning, repeated acquisition of response sequences has not been applied frequently to characterize learning in toxicant-exposed subjects. Dietz et al. (42) found that pigeons trained on an invariant sequence (performance group) were more disrupted after injection with lead than were pigeons trained on a series of variable sequences (acquisition group). This difference was attributed to a more variable baseline in the latter group, against which the effects of lead were obscured. Finally, Paule and McMillan (101) dosed rats trained on the IRA with TMT and observed changes in performance that were related neither to dose nor to the length of the sequence being trained (which ought to reflect task difficulty). As with the pigeon study, the stability of the baseline appeared to determine the magnitude of the effect of the toxicant: since the longest sequence had the least stable baseline, disruption of acquisition decreased in magnitude with increasing sequence length.

An advantage of this type of repeated acquisition task is its applicability across species. While efforts to train rats have not been entirely satisfactory, repeated acquisition of response sequences has been administered successfully to humans (9,10,67,68) as well as to birds and monkeys. Indeed, humans learn the task quite well, reaching stability of baseline on a 10-response sequence in about 15 sessions, on key measures of acquisition and performance errors, quarterlife of acquisition errors (a measure of the rate of within-session error reduction), and response rate (10). Further, acute benzodiazepine administration slowed acquisition of variable sequences, with lesser effects on an invariant (performance) sequence and small effects on response rate (Fig. 3). Effects faded with repeated exposure to diazepam, reaching placebo levels in 3 days (9).

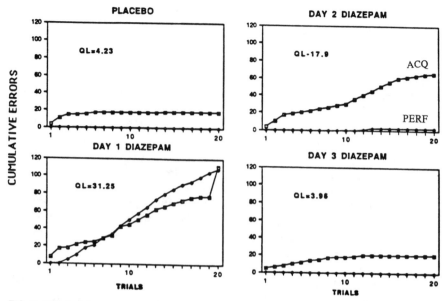

FIG. 3. Effect of diazepam on repeated acquisition in humans. Volunteers were trained to acquire a novel series of 10-member sequences of the digits 1 to 3 during acquisition (ACQ) or to produce an invariant sequence of digits (performance control, PERF). In both conditions, feedback regarding response accuracy and the current step in the sequence was provided after each response. Errors were cumulated across trials of each session; trained subjects learned a new sequence with approximately 20 errors (*upper left*) and with a quarterlife (percent of trials before which 25% of errors occurred) of about 4. The previously-learned invariant sequence was performed without error. Diazepam (80 mg) impaired acquisition and performance baselines 1 hour after administration (Day 1, *lower left*). The performance baseline was unaffected after a second dose of diazepam on Day 2, while acquisition of a new sequence was still impaired (*upper right*). After a third dose, both acquisition and performance baselines were normal (*lower left*). These data show the greater sensitivity of acquisition, as compared to performance, to disruption by diazepam in that tolerance developed more quickly to its effects on performance than to its effects on acquisition (redrawn from Bickel et al., ref. 9, by permission).

In summary, the repeated acquisition of variable response sequences provides a powerful tool for the assessment of learning in primates and in birds, but has met with less success in rats. The fact that the subjects are highly trained in the technique permits analysis to focus on learning *per se* as reduction in the frequency of errors during a test session. The ability to dissociate effects on learning (accuracy on a variable chain) from effects on performance (accuracy on an invariant chain) minimizes a crucial difficulty in interpreting the results of many learning tests: gauging the extent to which a change in performance reflects a change in learning ability, as distinct from a change in sensorimotor function or motivation. Even so, to be interpreted unambiguously, the task must be carefully constructed and administered with due consideration of appropriate parameters. Some variant of this task might well find a place in neurobehavioral test batteries for field-testing humans.

Maze Procedures for Repeated Acquisition

Toxicologists will likely identify a serious disadvantage of both the serial reversal and response chain acquisition baselines: the long training time required to bring the baseline to stability. Three maze procedures have recently been developed which alleviate this problem to some degree and broaden the scope of tasks suitable for assessing repeated acquisition as well.

For a maze-based response sequence task, Furuya et al. (59) developed a three-panel runway device which contained four sequential choice points, each of which required the rat to choose one of three doors for passage (the other two being blocked). Thus sequences of choices (e.g., L,C,R,C) could be trained; to evaluate learning, a different sequence was trained each day. Scopolamine (0.56 mg/kg and higher) and intrahippocampal AF64A (2.5 nmole/side) slowed the decline in within-session errors; latencies were increased after scopolamine only. Yatsugi et al. (155) used this device to show that cerebral ischemia of 5 or 10 min duration essentially prevented the normal within-session reduction in errors which occurred across six trials. Two min of ischemia exerted a small but non-significant effect on error reduction; affected rats recovered normal acquisition with repeated testing over 6 days (Fig. 4).

Although Yatsugi et al. did not do so, it should be possible to use an invariant sequence as a performance control in this device, given an appropriate discriminative stimulus. Training time was short, with stability in the rate of within-

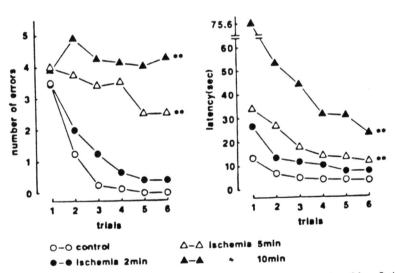

FIG. 4. Repeated acquisition in a straight alley. Rats learned a different series of four 3-choice spatial discriminations each day. In normal rats (open circles), the number of errors (*left panel*) declined quickly across 6-trial sessions and the time to finish each trial (*right panel*) was less than 15 sec. Both the number of errors and trial completion time increased with the duration of cerebral ischemia (given 24 hr prior to testing); within-session error reduction was absent after the longest duration of ischemia (from Yatsugi et al., ref. 155, by permission).

session reduction in errors and response latency achieved in five to ten sessions of six trials apiece.

A repeated acquisition procedure developed by Peele and Baron (102) for the eight-arm radial maze (8-ARM) resembled Furuya's task in that a sequence of selections was used. In the 8-ARM task, however, the rat was required to locate those four of the eight arms that were repeatedly baited with food during each test session. Rats were given 14 consecutive trials per session, during which errors (entries into nonbaited arms) declined regularly, approaching zero by the sixth trial. This within-session reduction in error frequency defined learning, and could be repeated on a daily basis by baiting a different set of four arms each day. A stable baseline was achieved after 20–30 sessions. Scopolamine interfered with the within-session reduction of errors in a dose-related manner, with evidence of disruption at a dose as low as 0.056 mg/kg, i.e., about a tenth that needed to disrupt performance in Furuya's three-panel runway. The inverse relationship between training time and difference in sensitivity to scopolamine in these two tasks is intriguing, and indicates that the radial maze task may be the more difficult of the two.

In a followup study, Bushnell and Angell (24) trained rats on Peele and Baron's procedure, requiring about 20 sessions to reach stability. After training, TMT injection seriously disrupted learning, as shown by a flattening of the within-session reductions in error frequency. This effect recovered with daily testing over an ensuing 6-week period, despite permanent damage to the hippocampus. As with the Furuya maze, a performance control condition could in principle be added to this task, assuming that an appropriate discriminative stimulus could be employed, although this has not yet been attempted.

A related approach to repeated acquisition has been described by Whishaw (152), using the Morris water maze. As in the 8-ARM, learning is based on the location of reward. In this procedure, the position of an invisible platform is fixed for a given session, and rats are trained to swim to it. Across sessions, the platform may occupy any of four locations, one of which is randomly selected for each session. Thus the task involves finding the platform each day, across a series of 8 to 16 trials. Measures of acquisition included latency to find the platform and heading errors, both of which declined regularly across trials. Normal rats acquired the "place learning-set" quickly, reaching asymptotic accuracy and latency scores within 7 days of training.

Deficits in daily acquisition of the location of the platform were observed after hippocampal lesions in rats produced either by aspiration or by infusion of kainic acid or colchicine (153). Rats with hippocampal lesions did not improve across trials (on either latency or error scores), even if trained to asymptote on the learning set prior to treatment. Some improvement across 48 postoperative sessions was observed, but within-session error reduction remained minimal. In contrast, rats with lesions in the parietal cortex showed control-like within-session error reduction, but still made more errors than controls even after 48 days of postoperative training.

In a further development of the procedure, Auer et al. (4) tested rats with ischemic lesions of the forebrain either on a standard place-learning task (platform always in one location) or on the variable-location repeated acquisition task. While slower to acquire both tasks, asymptotic performance of the ischemic rats did not differ significantly from that of controls on the fixed-location task, but was reduced on the variable-position task. The pattern of deficit resembled that seen by Whishaw (153) in rats with parietal cortex lesions, and by Tilson et al. (148) in rats treated with TRCP [tris(2-chloroethyl)phosphate]. It is intriguing that the neuronal damage induced by ischemia involves primarily the CA1 region of the hippocampus, an area particularly affected by TRIS, and that the behavioral deficits on this task induced by the two treatments were also very similar.

The three repeated acquisition procedures discussed—serial reversal learning and repeated acquisition of response chains and maze locations—all show promise as tools to quantify changes in behavior that would qualify as "learning," despite a lack of theoretical underlying process(es) called upon for their explication. Reversal learning has been studied the most thoroughly in regard to underlying processes, and appears to require several components including, for example, attention to relevant stimuli, inhibition of responding to the previously positive stimulus, and formation of new associations to the newly positive stimulus (see, e.g., ref. 84). However, the ways in which these processes change across formation of the learning set (during development of a serial reversal baseline) have not been fully articulated. Indeed, it is very likely that the properties of the associations formed and responses used in asymptotic reversal learning differ considerably from those used in a single reversal of a discrimination.

Even less is known about the processes underlying the learning of response sequences or locations in space, which may be equally as complex as reversal learning. For example, at least three distinct but related mnemonic processes appear to be required in asymptotic performance of Peele and Baron's (102) 8-ARM repeated acquisition task (24). Perhaps the complexities of changing baselines of behavior have retarded our understanding of the processes involved in these kinds of learning. In addition, some researchers have been reluctant to attempt analyses beyond the phenomenological aspects of the task at hand. For example, Thompson and Moerschbaecher (ref. 146, p. 257) contend that "learning is an observable change in behavior (e.g., within-session error reduction in a repeated acquisition procedure) rather than a hypothetical intervening construct." Nevertheless, if we are to advance beyond a descriptive neurotoxicology based on purely procedural grounds, it will be necessary to carry out such analyses and develop a predictive theory of process, as suggested by Overmeier (98).

In contrast to the situation in the area of learning, interest in the effects of aging and disease on memory has led to important advances in both procedures for studying memory and processes underlying retention of information over time. A particularly well-developed hypothetical mnemonic process, working memory, will provide a theme for the following section, as well as for the area of neurotoxicant-induced memory dysfunction.

Assessing Working Memory

Amnesia! Baby, what I knew . . . —*The Fabulous Thunderbirds*

The expression of this theme in popular music suggests that memory loss is now viewed by the general public as a likely effect of aging and exposure to drugs and environmental pollution. While appreciation of the debilitation of "oldtimer's disease" is growing, research into memory disorders continues to reveal additional complexities of character and etiology. Thus, normal cognition encompasses many forms of memory (116), and patterns of memory loss in humans vary with etiology as well (70,151). Heuristic taxonomies of memory abound in the literature, and have been discussed previously in this context (89). Global retrograde and anterograde amnesias are the most dramatic, if least common, forms of memory loss; more subtle and insidious forms of amnesia occur in progressive dementias, Korsakoff's syndrome (related to chronic alcoholism), viral encephalitis, and AIDS. Memory loss of varying severity can also follow from surgical intervention, head injury, hypoxia, and stroke (35,70). Mnemonic sequelae of exposure to neurotoxic chemicals typically take subtle forms which do not present clinically as global amnesias and thus require sophisticated neuropsychological assessments for their detection and characterization.

While patterns of memory loss differ across etiologies, disruption of memory for recent events is present in most, if not all, cases of amnesia (35,70,151). That is, amnesic patients usually have great difficulty acquiring and retaining new information, while recall of old information may or may not be affected. Memory for old physical skills and the ability to acquire new skills are seldom affected (32). Thus, declarative memory ("knowing that") appears to be much more easily disturbed than procedural memory ("knowing how"). This distinction also seems to apply to primates with temporal lobe lesions (156). Tests for memory dysfunction designed for primates appear to be sensitive to human memory disorders as well (141).

The ability to acquire new information, utilizing declarative memory, appears to require a form of recent memory which has been operationalized in the animal laboratory as *working memory* (71,96). For the present discussion, working memory will be taken to represent a process by which recurrent information is acquired and categorized according to its specific spatio-temporal context; that is, the information is derived from, and applicable to, a specific place and time (95). In this sense it resembles *episodic memory,* a term coined by Tulving (149) to describe a process by which humans categorize experiences into episodes, thereby facilitating their retrieval. However, it differs from episodic memory in its capacity and transience: while episodic memories comprise part of the permanent memory of the individual, working memory spans a restricted domain of time and quantity of information. Information in working memory may be transformed for permanent storage in *reference memory* under appropriate circumstances; however, it must be retained in its transient form long enough to survive the transfer to more permanent storage.

It is clear that working and reference memory differ with regard to the time over which each applies. Working memory involves continuous processing of information appropriate to short, discrete units of experience (e.g., within an experimental trial, or for a few moments of time). In contrast, reference memory involves more permanent storage of information over a longer, less definite time (e.g., across experimental trials or sessions, or for hours or days). Thus, information in working memory is frequently changing (reset) in response to stimulus input, while information in reference memory remains applicable for prolonged periods. One corollary of the transience of working memory is that it may not provide "storage" at all; if so, then retrieval may not play the role necessary for access to information in reference memory (see below).

Olton (95) provides a clear example of an operationalized distinction between these two mnemonic processes in a single task. A T-maze is constructed with a double-width stem that is partitioned into two parallel corridors by a solid wall. Behind a curtain, and thus invisible to the rat, one corridor ends in a cul, while the other opens into the choice area between the two arms of the T. Each trial consists of two runs from the start box at the base of the stem. To obtain food on the first run, the rat must choose between the two corridors, only one of which leads directly to the arms of the T; it is then forced to one or the other arm of the T, where it finds a food pellet. To find food on the second run, the rat may again choose either corridor, though again only the same corridor provides a direct path to the choice area where a choice must be made. If a matching (or win-stay) rule is being trained, food is found in the arm that was entered on the first run; if a nonmatching (or win-shift) rule is being trained, food is found in the opposite arm. The blocked corridor is always the same: thus, a correct corridor choice demonstrates accurate reference memory. In contrast, the arm to which the rat is forced, and thus the arm with the food, changes from trial to trial in a random fashion. A correct arm choice thereby requires intact working memory. Simultaneous determination of both processes in the same task yields the possibility of differentiating effects of a given experimental manipulation on these separate memory processes.

Using this procedure, a clean dissociation was obtained between the effects of lesions of the fimbria-fornix and the amygdala (Fig. 5). Rats were trained to asymptotic accuracy on both components of the task. Following recovery from surgery, rats with fimbria-fornix lesions were not impaired on the discrimination in the stem of the maze, indicating intact reference memory, but were completely unable to perform the working memory task required in the arms of the maze. In contrast, rats with amygdala lesions showed very little disruption in either component, indicating that neither mnemonic process was affected by this lesion. The preserved accuracy on the reference memory component did not result from mere insensitivity of this measure, since other manipulations produce reference memory loss. For example, aged rats were impaired on both task components, but were nevertheless unimpaired on a simple brightness discrimination (83). These data were interpreted to indicate that in addition to affecting working memory, aging impairs rats' ability to respond appropriately under multiple response rules in the same task.

Working memory has also been modeled in the radial-arm maze (reviewed in ref. 89), and outside the maze environment using a variety of multiple-trial procedures, including delayed response (e.g., refs. 65,77,107,143), delayed matching-to-sample (64,106,127,140,154), and delayed matching-to-position (19,20,22,46,48,49,50). Because working memory in these tasks has a short temporal duration (on the order of seconds), it is possible to generate functions representing the decay of memory over time in relatively short test sessions. These retention gradients have been used to characterize working memory and to describe the influence of drugs, toxicants, lesions, and aging on working memory, as will be discussed next.

The slope of a retention gradient reflects the rate at which information in working memory is lost over time; consequently, changes in this slope can be inferred to indicate a change in the rate of forgetting. Interpretation of the y-intercept of the gradient (i.e., accuracy at zero delay) is, however, somewhat less straightforward. It has been argued (20,66) that this value reflects the degree or accuracy of stimulus encoding, that is, the probability with which the stimulus was registered at the outset of the delay. Logically, this inference is true only if retrieval factors do not affect choice accuracy after the delay. In one view, information is not "stored" in working memory; it is either present and available, or stored elsewhere, or simply lost with the passage of time and continuing stimulus input. In this model retrieval does not play a role in working memory, which then requires just the two subprocesses of encoding and retention.

An alternate view holds that information may be stored in working memory, in which case its subsequent use by the subject requires not only accurate encoding and retention, but retrieval as well. This view would seem more applicable to the working memory described to account for behavior in the radial-arm maze, where information may be retained for hours (e.g., ref. 7). At this time, however, the means to differentiate between encoding and retrieval in procedures yielding rapid decay of working memory has not been devised. For purposes of characterizing the mnemonics of working memory in these procedures, the separate roles of encoding and retrieval will for the time being remain indiscriminable.[5]

---

[5] Differentiation of encoding and retrieval might be possible if an independent measure of encoding (e.g., an evoked potential, or increased unit activity in hippocampus) continued to reflect activity appropriate to the stimulus, while ultimate choice accuracy was disrupted. Another possible approach to this problem might involve the use of procedures which generate slower decay functions (e.g., the radial-arm maze), which would permit manipulation of the information in working memory after its encoding, but before its retrieval.

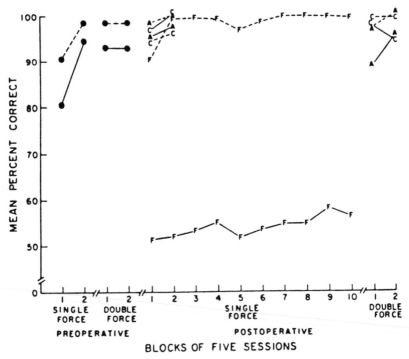

FIG. 5. A split-stem T-maze was developed (95) to differentiate between reference and working memory. The correct response in the stem of the maze was constant, thus requiring reference memory, while the correct response at the base of the T differed from trial to trial, thus requiring working memory. Prior to lesions in either the amygdala (A) or fimbria-fornix (F), accuracy of rats on both stem choices (dashed lines) and arm choices (solid lines) approached 100%. Rats with amygdala lesions performed both aspects of the task as well as controls after the lesion. Rats with fimbria-fornix lesions maintained highly accurate stem choices, but performed at chance levels on arm choices, despite extended postoperative testing. Thus lesions of the fimbria-fornix disrupted working memory while sparing reference memory (from Olton, ref. 95, by permission).

Two major types of procedures have been used to characterize working memory outside the maze environment: delayed response and delayed matching-to-sample. These procedures differ with respect to the type of information which must be retained for accurate choice (66). In delayed response tasks, the location of the correct response is determined at the beginning of the delay interval: to obtain reward, the subject must respond there when given the opportunity. In contrast, delayed matching (or nonmatching) tasks require the subject to choose a stimulus after a delay based on information supplied before the delay. Thus, delayed response tasks permit the subject to anticipate a response, while delayed matching tasks require the subject to remember a stimulus. For pigeons, at least, anticipating a response appears to be easier than remembering a stimulus, since a task permitting response anticipation yielded higher accuracy than did a closely analogous task requiring memory of a stimulus (107).

A classic difficulty with delayed response problems lies in the fact that subjects can utilize positional mediating strategies to bridge the delay between sample and choice (56). That is, the subject can reduce the memory load by first assuming a position near the location of the current response at the beginning of the delay, and then simply making the closest response when given the opportunity to choose. For this reason, spatially-cued delayed response tasks have until recently been little used.

Since its introduction (76,113), a sort of hybrid delayed response procedure has become a popular tool for analysis of the effects of drugs, toxicants, and aging on working memory in rats. In this task, called delayed matching-to-position (DMTP), the location of the choice response is determined at the outset of each trial. The task (and its win-shift analog, delayed nonmatching-to-position, DNMTP) is thus conceptually a delayed response problem (66); however, its operational similarity to delayed matching procedures has led to use of its descriptive, if formally incorrect, label. In addition, DMTP appears to enjoy some functional similarities to delayed matching tasks, as will be discussed below.

With DMTP, it is possible to circumvent the positional mediation problem and use the procedure to characterize deficits in working memory induced by neurotoxic compounds. In a study of the effects of TMT on working memory, Bushnell (19) also studied the effect of mediating behaviors on accuracy of choice responding in the DMTP task. In this study, one of two retractable response levers was extended into the chamber (sample) and retracted after being pressed by the rat. After an ensuing delay of 1 to 20 sec, during which pellets were available on a variable-interval schedule of nosepokes into a centrally located food cup, both levers were extended into the chamber (choice). Pressing the lever previously extended as the sample yielded a pellet, while pressing the other lever yielded a 5 sec period of darkness without food.

The role of mediating behaviors on choice accuracy was assessed by analyzing the distribution of responses to the retracted levers during the delay ("delay presses"). This analysis showed that (a) delay press frequency was not related to choice accuracy; (b) the lever to which a delay press was directed did not predict the choice response made after the delay (Table 2); and (c) resetting the delay for each delay press had no effect on choice accuracy. Under the conditions of this task, then, positional mediating behavior appeared not to influence choice accuracy. It is also the case that delay presses can be reduced in frequency by terminating the delay with the choice opportunity (rather than use of a food-rewarded variable-interval schedule during the delay), which increases the rate at which nosepokes are emitted into the food cup and consequently decreases the opportunity for delay presses (20,46). Further reduction in delay presses can be achieved by moving the food cup to the wall opposite the levers, thus making it more costly for the rat to engage in them (20,23). By these means, the problem

TABLE 2. *Delay presses in delayed matching-to-position*

Lever pressed during delay	Outcome of trial		Total
	Correct	Incorrect	
Correct	0.68	0.25	0.93
Incorrect	0.04	0.03	0.07
Total	0.72	0.28	1.00

Role of responding during the delay on delayed matching-to-position (DMTP) accuracy. Rats were trained to press one of two retractable levers after a delay. A press on the lever previously presented as the sample in that trial caused delivery of a food pellet, while a press to the other lever did not. To determine whether the rats mediated the delay period with anticipatory responding to the sample lever, presses to both retracted levers were recorded during the delay. The values in the table show the proportion of responses to each of the retracted levers during the delay. Rats directed 93% of their responses during the delay to the correct lever (i.e., the sample lever). However, they chose incorrectly after 25% of the correct delay presses, indicating that delay press accuracy during the delay did not ensure choice accuracy after the delay. Similarly, incorrect delay presses (i.e., those presses to the lever not presented as the sample) resulted in equal probabilities of correct and incorrect choice after the delay. These data suggest that delay presses do not greatly enhance choice accuracy after the delay (adapted from Bushnell, ref. 19, by permission).

of positional mediating strategies in this task has been reduced to the point of inconsequence.

With the DMTP procedure, it is possible to generate a retention gradient for each subject each day. Bushnell (19) showed that the detrimental effect of a single intravenous injection of 7 mg/kg TMT on choice accuracy in the DMTP task was persistent, lasting at least 4 weeks. Further, the effect of TMT was delay-dependent (the slope of the retention gradient was steeper), indicating that the stimulus information was forgotten more quickly by treated rats than it was by controls. Finally, the impairment was directly related to the degree of hippocampal damage induced by TMT, as indexed by increased concentrations of an astrocytic protein in the hippocampi of treated animals.

An important determinant of choice accuracy in working memory tasks appears to be the lingering influence of previous trials, known as proactive interference, or PI. As more than a hundred trials are typically administered in a delayed-matching test session, it is easy to see how memory for the critical stimulus on a given trial (the sample) may become confused with memory for stimuli from previous trials. Since memory for the stimulus on any trial fades with time, separating trials by longer intertrial intervals should reduce PI and thus increase choice accuracy. This improvement has been observed, both in continuous nonmatching (106) and in DMTP (19,49). However, the effect of PI in the continuous nonmatching task was independent of delay, that is, the time course of retention (slope of the gradient) was the same under conditions of both low and high PI. In the DMTP task, by contrast, slope steepened with decreasing time between

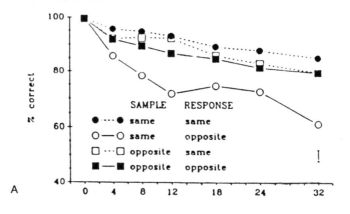

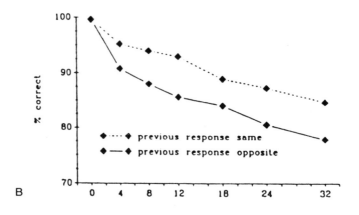

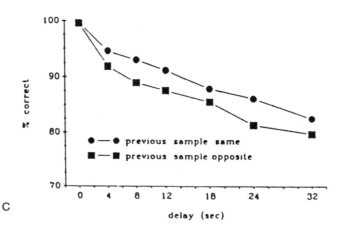

trials, i.e., with increasing PI (19,49). These results suggest that delayed matching and delayed response tasks differ with respect to the effects of PI.

A further analysis of PI effects in DMTP (49) showed that accuracy on a given trial was much lower if the response in the previous trial was to the sample (side) opposite that of the correct response on the present trial, regardless of the correctness of the previous response. In other words, the position of the previous sample did not affect accuracy on the next trial as much as did the position of the rat's previous response. It follows that the rat's previous response had more effect on its future behavior than did the stimulus which occasioned that response (Fig. 6).

The ability of the previous response to interfere with the present one in DMTP is consistent with Pontecorvo's (107) finding that the subject's choice is guided by response anticipation in a delayed response procedure. In that study, Pontecorvo compared the effects of PI on two successive discrimination tasks. One task required comparison of stimuli presented before and after a delay (delayed matching), while the other required comparison of two stimuli before a delay and responding afterward (delayed response). He found that increasing PI reduced accuracy in a delay-dependent manner in the delayed matching task (slope change only), while effects of PI in the delayed response task were delay-independent (intercept change only).

The delay-dependent effects of PI in Pontecorvo's (107) delayed matching task oppose those reported by Pontecorvo (106) as characterizing continuous nonmatching-to-sample, which is a delayed matching procedure. However, they closely resemble effects of PI on DMTP reported by both Bushnell (19) and Dunnett and Martel (49). Clearly, the role of PI does not differentiate these tasks in the way that they are classified by choice strategy (i.e., as delayed matching vs. delayed response). In summary it may be concluded that DMTP, while conceptually a delayed response procedure, also has characteristics of delayed matching, both in that mediating responses appear not to be important in choice accuracy, and in the manner with which PI affects response accuracy.

Characterizing retention gradients from working memory tests provides a useful pair of parameters to evaluate changes in retention (slope) and encoding (y-

---

FIG. 6. Proactive effects analyzed in terms of the previous sample and the previous response. **(A)** Trials subdivided into whether the previous sample (determined randomly) had been on the same or opposite side and in terms of whether the response was correct (previous sample and response both on the same or on the opposite sides) or was incorrect (previous sample same and previous response opposite, or vice versa). Note that correct trials occurred 8–10 times more frequently than incorrect trials. **(B)** Data of Panel a collapsed in terms of the side of the previous response. **(C)** Data of Panel a collapsed in terms of the side of the previous sample. Note that proactive interference from the previous trial is more clearly apparent when the data are analyzed in terms of the response rather than the sample on the previous trial. The vertical bar indicates two SEs derived from the three-way interaction of the analysis of variance.

intercept) (20). Nonlinear functions have also been applied to retention gradients (e.g., refs. 77,143,154) using a measure of sensitivity (derived from signal detection theory) to quantify accuracy. While a negatively accelerated retention gradient is probably more accurate theoretically, its derivation and fit to real data are more difficult. In consequence, some sets of scores cannot be fit to the model, leaving missing values in the matrix of data (77,143). Since untransformed scores (percent correct) tend to regress linearly with delay within a useful range of accuracy scores (100–70% correct) (20,49), the simpler linear model may be preferable for practical purposes.

Changes in slope and intercept of retention gradients are useful for characterizing changes in working memory, as different treatments may affect the two parameters independently. For example, cholinergic blockade with scopolamine appears to reduce the intercept, but not the slope, of the retention gradient in rats on a number of working memory tasks. This change, suggesting impaired encoding (or retrieval) but not rate of forgetting, has been observed in tasks based on both spatial cues (20,46) and nonspatial cues (77,140) in rats of both sexes (150) and in pigeons (127,128). Other drugs, including chlordiazepoxide (143) and amphetamine (46), also produced changes in the intercept of the retention gradient without affecting its slope. Interestingly, scopolamine steepended the slope of the gradient in monkeys, indicating an increased rate of forgetting in nonhuman primates (6,120), while slope changes in humans appear to depend on the task used (see refs. 27,79).

In contrast to the changes in encoding induced in rats by scopolamine, amphetamine and chlordiazepoxide, changes in forgetting rate, as indicated by changes in the slope of the retention gradient, have been reported to occur in rats after lesions of the fimbria-fornix (but not nucleus basalis) (46); treatment with TMT (19), nicotine, or clonidine (50); and exposure to cold (144). Thus differential effects on these parameters may be useful for characterizing the effects of neurotoxicants on working memory.

These tasks have also been used extensively to study the cognitive changes associated with aging in rats (47,48,50). In addition to valuable insights into the neurobiology of aging, these studies revealed some interesting properties of the tasks as well. In the first of this series (48), 15- and 24-month-old rats acquired both the matching and nonmatching contingencies as quickly as controls when no delay was imposed between sample and choice. However, inserting a variable delay period (0–24 sec) retarded acquisition of both tasks by both groups of older rats, and the 24-month-old rats reached a lower performance asymptote than did the other groups. Steeper retention gradients, indicating faster forgetting, were observed in both older groups on the DMTP task, while only the 24-month-old group showed a steeper slope on the DNMTP task. Little support for cholinergic involvement in these effects was observed, however, as neither arecoline, a cholinergic agonist, nor physostigmine, a cholinesterase inhibitor, was found to ameliorate the poor performance of the older rats.

While the young rats performed the two tasks equally well, variability among individuals was greater in the matching version of the task, suggesting some asymmetry between the task solutions. Using quisqualic acid lesions of the nucleus basalis magnocellularis (NBM, a source of cholinergic innervation to the neocortex), to model age-related memory dysfunction in adult rats, Dunnett et al. (47) uncovered further notable differences between the matching and nonmatching tasks. The NBM lesion reduced the activity of cortical choline acetyltransferase, a marker of cholinergic terminals, indicating significant effects on the cortical projections of the NBM. Performance of rats pretrained on either task was not impaired by the lesion; however, the lesion significantly impaired reversal of the matching rule to the nonmatching rule, but not reversal of the nonmatching rule to the matching rule. This asymmetrical sensitivity of the tasks was replicated in lesioned rats whose acquisition of DNMTP, but not DMTP, was impaired, and in a subsequent reversal of contingencies between these groups.

These observations indicate that the task contingencies in DMTP and DNMTP are not simply symmetrical. As discussed by Dunnett et al. (47), there is evidence that rats tend to adopt win-stay strategies in operant environments (54) and to adopt win-shift strategies in situations resembling foraging, such as the radial-arm maze (96). If so, then learning a nonmatching rule in the operant chamber would be more difficult than a matching rule due to its win-shift character. Dunnett et al. (47) suggest that rats with quisqualic acid lesions of the NBM evince a subtle learning deficit characterized by difficulty with acquisition of tasks requiring effortful processing (43), but no difficulty with acquisition of tasks whose solution utilizes spontaneous response tendencies.

The pattern of deficit in aged rats was further explored in terms of PI (50). As with young rats, the location of the previous response affected the probability of a correct response on the next trial, such that if the responses were to the opposite lever, accuracy was some 10% lower than if the responses were to the same lever. However, overall accuracy (regardless of the previous response) was lower in the aged rats, requiring a change in scale to make direct comparisons of PI effects across age groups. When delays for the young rats were doubled (to 0–48 sec), the effects of PI on the two age groups were identical, indicating that increased PI did not account for the impairment in accuracy due to aging. Dunnett et al. suggested rather that deterioration of septo-hippocampal pathways could be at fault since similar delay-dependent deficits in matching accuracy were observed in young rats whose hippocampal outputs were severed by knife cuts of the fimbria-fornix.

These studies illustrate the analytical power of the DMTP task in characterizing deficits in age-related cognitive decline. From these studies, it would appear that dysfunction in the hippocampus may be more directly related to working memory deficits than to loss of cholinergic innervation of the cortex from the basal forebrain, and that susceptibility to PI cannot account for the increased rate of forgetting in old rats.

These tasks can also include trials designed to assess reference memory as well as working memory. For example, discrimination trials, in which the correct response is indicated by a cue provided at the end of the delay, can be intermixed with matching trials and serve as an index of reference memory. Consistent with the lack of working memory load, accuracy on discrimination trials did not decrease with delay, as it does on matching trials in both delayed matching-to-sample in pigeons (127,128) and DMTP in rats (20). Moreover, discrimination accuracy was far less sensitive to scopolamine than was matching accuracy in both species and tasks, consistent with the low sensitivity of reference memory to cholinergic blockade in maze tasks (see ref. 82 for a review).

The combined DMTP/discrimination task was subsequently used to characterize the effects of repeated AChE inhibition by daily subcutaneous injections of DFP. After a 1-week latency period, matching accuracy deteriorated in treated animals, indicating impairment of working memory. At the same time, discrimination accuracy remained unimpaired, indicating that reference memory was not affected by the chronic inhibition of AChE, and thus that the effect was specific to working memory. The y-intercept of the retention gradient for working memory was reduced and its slope unchanged in a manner exactly comparable to the effects of scopolamine (Fig. 7). Since neurochemical analyses showed reduced density of muscarinic receptors in the hippocampus and frontal cortex, it was suggested that reduced availability of muscarinic receptors, either by down regulation resulting from chronic inhibition of AChE or by blockade from acute scopolamine, impaired the ability of rats to encode spatial stimuli for input into working memory.

These effects of repeated administration of DFP contrast with acute effects of pesticides assessed, for example, by Heise and Hudson (64,65). In these studies, no selective effects of carbamate (propoxur, carbaryl), pyrethroid (deltamethrin), or monoamine oxidase-inhibiting (chlordimeform) pesticides on working memory were observed. In contrast, both scopolamine and chlordiazepoxide impaired working memory on these tasks, showing that the tasks were sensitive to disruption of cholinergic and GABAergic systems.

Like DFP, the mode of action of the carbamates used by Heise and Hudson is to inhibit AChE. Like DMTP, the tasks used by Heise and Hudson (delayed go/no-go alternation and continuous nonmatch-to-sample) were designed to assess working memory and were proven sensitive to scopolamine, a cholinergic antagonist. Why, then, did these pesticides exert no effects on working memory? To understand this discrepancy, it is necessary to consider the nature of the interaction between cholinesterase-inhibiting compounds and the cholinergic nervous system.

Inhibition of AChE slows the degradation of the transmitter acetylcholine (ACh), increasing the levels of ACh in the cholinergic synapse and increasing its stimulation of the postsynaptic ACh receptors. Sufficient overstimulation of the postsynaptic membrane can cause an acute cholinergic crisis in the subject, including convulsions and death. Lower-level stimulation causes the receptors in the postsynaptic membrane to withdraw from activity or down-regulate (for a

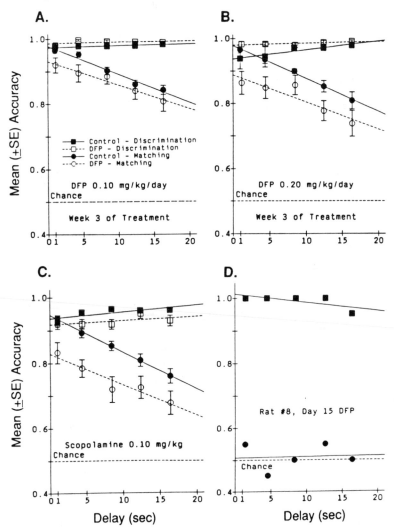

FIG. 7. Effects of acute scopolamine and repeated diisopropylfluorophosphate (DFP) on Delayed Matching-to-Position/discrimination. Discrimination trials, on which a cue light indicated the position of the correct response after the delay, were randomly presented among matching trials. Daily injections of DFP for 3 weeks, at 0.10 mg/kg (*panel A*) or 0.20 mg/kg (*panel B*) caused a dose-related decrease in matching accuracy (open circles) compared to vehicle-dosed controls (filled circles). Discrimination accuracy in treated rats (open squares) did not differ from that of controls (filled squares). Acute scopolamine caused a virtually identical effect, again specific for matching accuracy, in rats not treated with DFP (*panel C*). Both treatments caused a parallel downward shift in the matching retention gradient, suggesting that memory was impaired by an attentional deficit, rather than a direct effect on retention. The specificity of DFP for working memory is illustrated for one rat, whose ability to perform the discrimination task was preserved in the face of complete inability to perform the matching task (*panel D*) (from Bushnell et al., ref. 20, by permission).

review of this and other phenomena regarding the cholinergic nervous system, see ref. 122). Down-regulation is one of many plastic responses shown by receptors (78) in response to changes in ligand concentration or availability, and results in a reduced number of receptors available for binding to the transmitter. This compensatory change in the cholinergic nervous system appears to be a fundamental aspect of the development of tolerance, the means by which the nervous system adapts to chronic exposure to cholinesterase inhibitors, as well as perhaps to other exogenous chemicals.

This compensatory down-regulation of cholinergic receptors does not occur immediately, but requires prolonged (hours to days) inhibition of AChE. Thus acute exposure to a cholinesterase inhibitor does not cause down regulation of cholinergic receptors. It is also clear that changes in working memory do not occur immediately after exposure to an AChE inhibitor (23,64,65). In the latter study, decrements in working memory developed over the course of 2 weeks of daily DFP injection (23), which approximates the time frame for ACh receptor down-regulation with that dosing regimen of DFP. Thus it appears that the ability of AChE-inhibiting pesticides to impair working memory depends on the reduction in receptors induced in the nervous system by repeated overstimulation by their endogenous transmitter.

In addition to this change in working memory, exposure to DFP also altered the sensitivity of the treated rats to oxotremorine, a cholinergic agonist. One effect of oxotremorine is to cause hypothermia; DFP-treated rats, with fewer ACh receptors, were subsensitive to this agonist, in keeping with predictions of Russell and Overstreet's model. The model also predicts that tolerant animals should be supersensitive to cholinergic agonists, which has been amply demonstrated in other studies (reviewed in ref. 122). However, the evidence for supersensitivity of working memory to scopolamine is mixed. Raffaele et al. (112) found that DFP exposure increased the sensitivity of rats to scopolamine on repeated reversals of a nonspatial discrimination in a Y maze. Since retention of these serially reversed discriminations was evaluated daily (prior to each reversal), retention accuracy provided a measure of working memory. On the other hand, scopolamine's effects on working memory in the combined DMTP/discrimination task was additive with that of DFP, indicating a lack of supersensitivity to the antagonist (23). Moreover, treated rats' responses to scopolamine on the reference memory component of the task were greater than that of controls, suggesting supersensitivity on that measure alone.

Thus the ability of AChE-inhibiting pesticides to impair memory does not occur from a direct effect of the pesticide on the nervous system, but requires compensatory changes by the nervous system for its expression. This new aspect of tolerance, i.e., that the "tolerant individual" is not functionally normal, represents an important contribution of behavioral toxicology to the understanding of the toxicity of pesticides to mammals and to the assessment of the hazard to health and vitality posed by these important agricultural chemicals.

A striking feature of all these studies of working memory is the diversity of procedures with which similar results can be obtained. While particulars of some effects depend somewhat on procedure (e.g., PI effects differ depending on whether a delayed matching or a delayed response task is used), working memory can be measured by a number of tasks. Furthermore, effects of several manipulations—cholinergic blockade with scopolamine and lesions of the limbic system are good examples—are consistent across many procedures that share the common trait of working memory. This commonality of process now provides a sensitive indicator of an important mnemonic function, which can be quantified to assess the effects of neurotoxicants on cognitive ability.

The consistent effect of scopolamine on encoding, rather than retention, of working memory in rodents is consistent with the greater ability of anticholinergic drugs to impair acquisition of information than its retention, as discussed above for passive avoidance and repeated acquisition in the 8-ARM. That is, the memory decrement observed in working memory tasks, while clearly demonstrating reduced choice accuracy after a delay, should probably be viewed as an impairment of information input. This impairment may, in fact, relate more to attentional, than to mnemonic, dysfunction. Several lines of evidence, to be reviewed in the next section, suggest that the "amnestic" effects of scopolamine in humans can be better described as effects on acquisition, and that this acquisition deficit contains both attentional and sensorimotor components. We turn now to a discussion of the noolytic effects of anticholinergic drugs in humans, in an attempt to characterize the nature of their effects in humans and to relate them to results obtained in animals.

Does Scopolamine Affect Memory or Attention?

Several recent studies have helped to characterize the effect of cholinergic blockade on human cognitive functioning as for the animal studies reviewed above. These effects can no longer be characterized merely as "amnestic." Instead, scopolamine is seen to impair the *acquisition* of new material more than the retention of information that was previously learned. Some confusion still exists about the kind of material that suffers this acquisition deficit, but memories that are dependent on active encoding seem especially subject to impairment (see below). It is very clear, however, that cholinergic blockade affects *attention* and *sensory-motor* functioning as well as memory, and that differentiating these separate effects is complex.

Preston et al. (109), for example, found that several measures of attention were affected by intramuscular injection of scopolamine (0.4 mg) for healthy young adults (mean age 25 years). Visual sustained attention was affected (reporting a target digit against a complex, changing background: the target occurred one of nine times, with digits presented once per sec; i.e., it was not a speeded task), and choice reaction time was increased in a trial-by-trial procedure (pressing the key from a circle of four keys that corresponded to the lighted box in a visual display and then returning to a central key), though no effect of scopolamine was found for a continuous choice-reaction time version (follow the lighted box). Further, a large-movement tracking task was affected by scopolamine (7 or 14 cm movements of a joystick were required to match position of a light with a moving visual target). These sensitive tasks all required sustained attention but not memory. In this particular study, in fact, the two tasks which did require memory were not significantly impaired by scopolamine [repeated testing of memory for newly associated words (verbal) and for associated pairs of rectangles (spatial), with selective reminding for missed items].

Though retention of information is not always affected, it often is affected for healthy adults given moderate doses of scopolamine. A close analysis of the types of memory impairment has been instructive. Flicker et al. (57), for example, found that subcutaneous injections of scopolamine (0.22, 0.43, or 0.65 mg/70 kg) produced impairment for several tasks requiring working memory for verbal or spatial items. At the same time, however, tasks requiring retrieval of long-stored (semantic) information were not affected. Verbal working memory was assessed by the Guild Paragraph recall test and by a selective reminding test using shopping list items. Spatial working memory was assessed by serial recognition of repeated faces, recognition of new items in a display of household objects, and a task requiring recall of locations for objects place in a computer-drawing of a house (results were nearly significant in this last task). Semantic memory was assessed by naming objects and indicating their function and by selecting concepts (selecting from a pool of items the eight that were "most alike.").

Flicker et al. (57) also found that subjects were impaired in performing several tests that required close attention to detail or fine motor control. The tests included the WAIS Block Construction and Digit Symbol subtests, the Benton Face matching and line orientation tests, speeded finger tapping, and a video driving simulation. Performance on the WAIS digit span subtest, however, was not impaired. Subjects therefore had quite specific deficits and were not generally "confused."

Kopelman and Corn (79) offered an especially extensive survey of tasks designed to sort out effects of scopolamine on attention and memory for normal healthy adults. Effects of intravenous infusion of 0.4 mg scopolamine were assessed for

tests of verbal and visuospatial learning, measures of the forgetting rate and tests of remote, implicit, and semantic memory, [as well as] various tests of 'short-term' attention, . . . subjective arousal, and physiological degree of cholinergic blockade [visual nearpoint]" (p. 1081).

In this study, scopolamine significantly impaired the acquisition of new material but did not impair the retrieval of information that had been previously learned, either long ago [memory for news events (remote), production of words beginning with particular letter (semantic)] or learned that day prior to drug treatment (lists of words, pictures). Kopelman and Corn concluded that the learning of new material was impaired by the drug. This impairment was seen for both verbal and pictorial information. The impairment was seen immediately as well as after delayed testing. For information that was learned, the rate of forgetting was unaffected by the drug (Fig. 8). There was also no state-dependency; when tested after a 48 hr delay, the deficit was just as great when the subjects were again drugged during the test as when they were drug-free. Lastly, these investigators confirmed that scopolamine had affected attention as well as memory: choice reaction time was affected (more so when a large number of alternative response locations were being monitored), though they found no evidence for an effect on the Stroop interference or on serial addition, digit-, or block-span tasks. The lack of effects for some of these short-term attention tasks might well be due to a lack of test-sensitivity rather than to a lack of drug-effect on attention.

Subsequent evaluation of individual differences across subjects in the Kopelman and Corn (79) study provided an interesting result that calls for further research. Those subjects with greater deficits of attention were not those with greater deficits of memory. There was, in fact, no significant correlation between the two effects. On the other hand, there were significant correlations between performance on many of the learning tests and measures of both the visual near-point and subjective arousal. The authors suggest, therefore, that attention and memory influences may result from independent mechanisms.

Does the effect of cholinergic blockade on learning differ for different kinds of material (verbal, pictorial, spatial, etc.)? The data are mixed. Most of the procedures involving human subjects test verbal learning. It is clear, however, that other kinds of learning are affected as well. Kopelman and Corn (79) found effects for both verbal and pictorial material. Yet, retention of a three-block

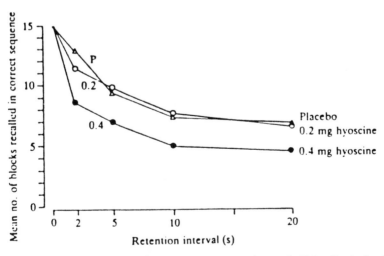

FIG. 8. Visuospatial short-term forgetting test (block retention test). Distraction is the finger-tapping task. *P*, placebo. Hyoscine = scopolamine. (From Kopelman and Corn, ref. 79, with permission.)

sequence was shown to be less sensitive to drug than was retention of a three-letter word (79). Broks et al. (17) found that the learning of spatial information was less disrupted by scopolamine in a spatial selective reminding task than was the learning of verbal information. Were these differences a result of differences in test sensitivity or differences in susceptibility of different memories to cholinergic blockade? Rusted and Warburton (124) emphasize that spatial memory can be disrupted. For example, they found that learning a pattern of locations and learning the location of eight shapes *was* impaired by scopolamine (0.6 mg injected subcutaneously), as was verbal learning. Rusted (123) also found that either verbal (digit) or spatial (mental rotation) processing was disrupted by scopolamine (1.2 mg given orally).

A useful distinction is sometimes drawn between "procedural memory" (remembering how) and "declarative memory" (remembering that). A reason for this distinction is that individuals unable to remember events can often acquire a skill (though they have no memory of practicing). Kopelman and Corn (79) found that scopolamine did not disrupt the acquisition and recall of a skill (reading backwards). Their result suggests that procedural memory, or skill learning, might be protected from the effects of cholinergic blockade. However, recent data (Fig. 9) indicate that skill learning (mirror-reversed tracking) is also impaired by administration of scopolamine (58) (0.4 mg scopolamine by intravenous infusion). These investigators make a case that skill learning can be decomposed into short-term (assembly of existing motor routines) and long-term (development of automatic skill) components and that both of these components were impaired by administration of scopolamine (while only the former was affected by noradrenergic blockade).

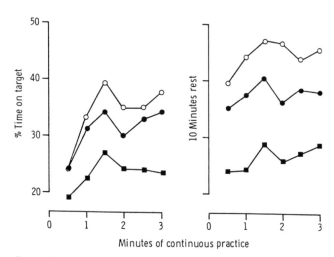

FIG. 9. The effects of treatment on the learning of the reversed horizontal component of a mirror tracking task (task 2). *Open circle:* placebo; *closed circle:* clonidine; *closed square:* scopolamine (from Frith et al., ref. 58, with permission).

It appears that the development of long-term skill does not require an active, "limited capacity" working memory (58). If so, cholinergic blockade is capable of influencing storage of memories which do not depend on active encoding or remembering. Rusted, Warburton, and colleagues (e.g., refs. 123,125), however, emphasize that most of the tests that *are* sensitive to cholinergic blockade share a dependence on an active "central executive" that encodes items and events for subsequent retrieval. Memories for items encountered during scopolamine differ by being less easily retrieved without explicit prompts. Rusted and Warburton (125) showed that for subjects given scopolamine (0.6 mg subcutaneous), words that were immediately recalled (and therefore "learned" to some degree) were remembered less consistently than for control subjects when a free-recall test was used. When a recognition test was used, however, the words were remembered equally well by both groups. The words had apparently been "stored," but were stored in a manner that was not as readily retrievable in a free-recall test. (It should be noted that additional effects of scopolamine make this result somewhat difficult to interpret: the frequency of intrusion errors was increased in the recognition test, suggesting that some of the "correct recognitions" could have been happenstantial guesses). Several investigators have provided information in agreement with the interpretation given by Rusted and Warburton (i.e., free-recall deficit without deficit in recognition or cued recall, ordered-recall deficit without deficit in unordered recall). That the picture is not simple, however, is shown by the lack of clear dependence of this "encoding" deficit on factors typically considered to modulate level of processing and encoding (e.g., "Scopolamine significantly reduced recall of items from semantically related, abstract, and unrelated control lists with a magnitude of the effect similar for all word lists," ref. 125, p. 82). It should be emphasized that this differential access to memories for items learned during cholinergic blockade does not result from a general deficit in retrieving information since items learned before drug administration are recalled well.

The effects of cholinergic blockade mimic several of the deficits observed in aging and the deficits observed in specific neurodegenerative diseases such as Alzheimer's disease and Korsakoff's syndrome. Other deficits observed in these disease states, however, are not produced by cholinergic blockade in healthy young volunteers. Kopelman and Corn (79) are in general agreement with other reviews when they note the following deficits to be absent: impairment of primary memory (e.g., digit span), remote memory (e.g., newsworthy events), impaired recall of temporal context (e.g., order-memory), and sematic knowledge (e.g., verbal fluency).

The impact of the human data that have been reviewed is to emphasize that cholinergic blockade affects attention to stimuli and encoding of experiences. These effects may be the primary source of the "amnestic" influences of these manipulations. Animal data also support this analysis of cholinergic blockade. For example, Anisman et al. (3) demonstrated that scopolamine (intraperitoneal injection of 1.9 mg/kg) prevented mice from profiting from an initial adaptation trial in a novel environment. Mice given drug during their first (adaptation) day had an increased latency to approach and lick from a water spout introduced on the second day (with or without drug during the test). Was this effect a result of impaired attention to items during the first day? Or was it due to impaired memory (habituation can be taken as evidence of memory for prior exposure)?

A clear demonstration of the insensitivity of retrieval to cholinergic blockade was provided by Sarter (129). Rats were trained to perform a conditional discrimination based on visual cues which required completion of a ratio (FR10) on the lever beneath the appropriate cue each day. Retrieval was measured as the ratio of correct to incorrect responses on the first trial, prior to any reinforcement delivery; that is, this ratio reflected the accuracy with which the response rule was retrieved at the outset of the test. In contrast, response accuracy after the completion of the first correct ratio was taken to reflect "performance." Scopolamine (0.1, 0.39, and 1.56 mg/kg) did not affect accuracy on the first trial (retrieval was spared), but impaired performance for the remainder of the session, indicating the relative robustness of retrieval to cholinergic blockade.

In an especially analytical study of this effect, Cheal (31) examined the exploration of a novel object by gerbils (10–18 week old) in an arena to which they had been partially adapted. A novel object (pen, cup) was taped to one side of the semicircular arena. Gerbils given scopolamine (0.01, 0.1, 1.0, or 10 mg/kg) in subsequent exposures to this novel object were more persistent in visiting the object than were control animals, but their individual visits were not as long. Cheal describes the data as demonstrating a lack of effect of scopolamine on "memory" since the gerbils given scopolamine continued to investigate the new stimulus only until they had accumulated the same total investigation time as shown by control animals. She claims a strong effect on sustained attention, however, since duration of individual visits was reduced. This effect of scopolamine differed from that produced by increased dopamine activity since apomorphine injection increased both the persistence of visits (frequency) and the total duration of investigation over the course of several exposures (duration of individual visits was not affected). Other interpretations of this result can also be set aside. Since "habituation" (as measured by total duration) was not affected by scopolamine, the result can not be attributed to a mere "increased activity." Further, since the effect was not observed with methylscopolamine, Cheal challenges interpreting the effect as a result of cholinergic blockade outside the central nervous system. Cheal also emphasizes that animals given scopolamine *did* effectively discriminate stimuli and were not merely "confused" (selective attention was preserved e.g., in an odor test where subjects appropriately located the presence of a novel odor).

Brazhnik and Vinogradova (13) provide physiological evidence that supports Cheal's interpretation. They found that an increase in acetylcholine produced by injection of physostigmine produced an increase in the number of medial septal-dorsal bundle neurons that remained entrained to intrinsic theta-rhythm rather than following a repeating external stimulation (electrical stimulation of

3–30 Hz explored in each case, with intensity adjusted individually for each unit studied). They suggest that this "functional deafferentiation" of the septum may be involved in aiding sustained attention. Further, cholinergic blockade by injection of scopolamine has the opposite effect—the number of cells that follow external stimulation increases—a condition they relate to the increased distractability seen in animals under the influence of scopolamine.

ATTENTION

A Review of Human Attention

In order for a stimulus to gain control over behavior, it must somehow be registered by the nervous system. The means by which the nervous system then apprehends sensory input and links this input to behavior involves processes commonly described as attention. Ultimately, attention to a stimulus is inferred from its control over behavior. Attention is also an essential prerequisite for learning and memory; in fact, the single most important determinant of whether a subject learns or remembers an event is whether she or he attends to it when it is presented. Attention is not, of course, a simple activity. Different sorts of attention can be distinguished and different impairments of attention can be described. This section reviews some recent ideas regarding models of human attention and methods that have been used to dissociate different sorts of attention (or inattention) in human and nonhuman subjects.

Knowledge of human attentional systems has grown steadily over the last several decades, particularly for processes of visual attention. Until recently, work in this area was based on a "selective filtering" model developed by Broadbent and others (e.g., refs. 14,15,16). This model posited that attention is necessary because of a fundamental constraint on processing capacity. That is, attention involved selecting relevant stimuli by filtering out extraneous information, which allowed the central nervous system to manage the otherwise overwhelming rate and volume of incoming information.

However, the filtering model does not adequately describe current knowledge of attentional processes (see ref. 1). It has not been possible, for example, to resolve the question of early versus late selection of information (e.g., is verbal information filtered at a perceptual or at a semantic level?). This and other equally disputed predictions of the filter model have lead to a paradigm shift in the field. Allport (1) argues that the central, hierarchical filter model is being replaced by a distributed, heterarchical model in which attention is viewed as a set of active "selection-for-action" systems. In visual attention, for example, it is proposed that semiautonomous systems focus on particular areas of the visual field and generate visual constructs (images) based on expectancies (that is, prior historical determinants of what the subject anticipates should be there), which are then validated or corrected by the actual images perceived.

In addition to the failure of the passive filter model to account for perceptual data, development of the active selection model has been driven by two major advances in cognitive neuroscience.[6] First, neurobehavioral observations of persons with a variety of CNS lesions have identified dramatic and specific deficits in attentional function. For example, spatial neglect refers to a syndrome of attentional dysfunction in which whole perceptual areas of the visual field may be lost to the patient's awareness (41). Spatial neglect may involve either near or distant objects and one or the other lateral half of the visual field. It may be body-centered, gravitational, or centered on external objects. As a general rule, these neglected fields show a strong directional bias, with dysfunction on the contralesional side and attention drawn toward the ipsilesional side (74). In addition to spatial neglect, other discrete attentional dysfunctions exhibit specific symptomatologies which indicate the existence of a number of discrete attentional processes. This variety of separable dysfunction suggests a distributed, heterarchical network of attentional systems with separate functions, no one of which is "central" to the overall process of visual attention.

Second, noninvasive imaging techniques have been used to study the relationships between task-specific behavior and changes in CNS blood flow or glucose utilization in normal subjects. In combination with clinical observations of patients with brain lesions, this work (summarized in ref. 108) has identified three primary processes involved in attention: (a) orienting to sensory events, (b) detecting signals for focal processing, and (c) maintaining alertness (vigilance). Orienting may involve overt behavior (eye movements) or covert shifts in attentional focus to spatially oriented visual stimuli. In experiments with humans, signal detection is typically measured by the speed with which a subject can identify a target from a list of semantically categorized verbal material, while alerting is often demonstrated by increased speed and efficiency of responding when subjects are prepared to detect high-priority signals.

Orienting to sensory events is perhaps the best studied of these three component orienting processes; moreover, discrete brain regions have been identified which appear to mediate these processes. First, a region of the posterior parietal lobe appears to be necessary to *disengage* attention from a focus in one visual field to a target in another field. Second, a collicular region has been identified as necessary for *moving or shifting* attention from one focus to another. Third, lesions in a region of the dorsolateral pulvinar nucleus of the thalamus appear to interfere with the *engagement* of focused attention, as patients with dorsal thalamic lesions are easily distracted by irrelevant stimuli. Conversely, normal subjects required to filter out irrelevancies show selectively enhanced thalamic metabolism on the side contralateral to the visual field in which the stimuli for a task are presented.

These relationships between discrete anatomical areas and specific attentional

---

[6] This information, and the discussion of human attention to follow, is derived wholly from excellent reviews by Allport (1) and Posner and Peterson (108).

processes have been derived from studies of human patients and experiments with nonhuman primates. Toxicological research into chemical-induced attentional dysfunctions would be greatly facilitated by development of rodent models of these component processes. Few such models exist, however, and none enjoy the close functional-anatomical relationships demonstrated in primates. Nevertheless, some possible avenues have been explored which may provide leads to productive areas of research.

Potential tests of attention in animals

How might we measure attention in animal subjects? Perhaps a simple demonstration of the presence or absence of stimulus control will provide an approach. When a response reliably occurs in the presence of a stimulus that predicts reinforcement (S+) and is reliably withheld in the presence of another stimulus that predicts nonreinforcement or punishment (S−), we can say that the subject is attending (or has attended) to the stimuli. When performance is well controlled, therefore, attention to the relevant stimuli as well as the ability to maintain differential performance are demonstrated. When performance is no longer well controlled, it is difficult to specify whether the disruption is due to a loss of attention to the relevant stimuli or to a loss of another type (i.e., impairment of learning, memory, motor control, or motivation). To make a differential diagnosis of impairment to attention requires that specific tests be carried out. Several approaches have been used to make this dissociation and will be reviewed below.

Latent Inhibition

Typically, after a stimulus has been repeatedly presented without consequence (i.e., an irrelevant stimulus), additional training is needed to develop an association to that stimulus when it is subsequently made relevant (e.g., predicts presentation of shock). This phenomenon is called latent inhibition, though the process is better described as a learned inattention than as an actual inhibition of responding. In this procedure, selective attention is demonstrated by the fact that the subject ignores particular irrelevant stimuli and selects other, relevant ones.

Several studies from Solomon, Crider, and co-workers have used the latent inhibition paradigm to demonstrate that alterations of the dopamine system affect an animal's ability to screen out stimuli. These data have been useful in developing a rat model of the attentional impairments observed in schizophrenia. For example, 30 presentations of a tone within 15 min is typically sufficient to delay substantially the acquisition of a conditioned avoidance response (CAR), when that same tone later predicts impending, avoidable shock. Solomon et al. (138), however, showed that if this preexposure to the tone was preceded by repeated administrations of *d*-amphetamine (five daily injections of 4.0 mg/kg but not of 1.0 mg/kg), this latent inhibition effect was prevented. Further, the effect was eliminated, however, by dopamine receptor antagonists. That is, injections of 1.0 mg/kg chlorpromazine given immediately after each of these five amphetamine injections prevented the attentional impairment and allowed the latent inhibition to develop normally. Additional tests affirmed that the effect was on attention rather than sensory capacity *per se.* That is, there was no effect of this amphetamine preexposure on auditory or shock thresholds (see ref. 139 for this latter control). Further, dopamine receptor supersensitivity produced by a repeated course of haloperidol prior to tone preexposure and CAR training also produced attentional impairment (a 21-day course of 0.5 mg/kg haloperidol). In a subsequent study (139), repeated microinjections of *d*-amphetamine (10 micrograms, given on each of 5 days) into the nucleus accumbens but not the caudate putamen produced this attentional impairment as well.

Rather than the CAR, other studies have utilized changes in the acquisition of a conditioned flavor aversion (CFA) to demonstrate latent inhibition. In the CFA procedure, exposure to a particular flavor is followed by administration of a mildly toxic substance (e.g., ip injection of 0.4 M lithium chloride). Animals so treated subsequently reduce their consumption of water with this flavor (one-bottle test) or prefer unflavored water to water with this flavor (two-bottle test). Preexposure to the flavor before the CFA training, however, reduces the degree of avoidance, thus demonstrating latent inhibition. Several studies have demonstrated an impairment of latent inhibition in CFA tests following damage to the hippocampus (e.g., refs. 60,91). Further, Nicolle et al. (91) showed that this type of latent inhibition is relatively late-developing in the rat. It was shown in 32-day-old but not 25- or 18-day-old rat pups. This late development apparently reflects the dependence of learned inattention on late-maturing neural structures. The latent inhibition procedure appears to offer a relatively specific assessment of the formation of one kind of selective attention.

Blocking of Association

Typically, after an association has been developed to one relevant stimulus, no additional associative learning is seen if a new, redundant stimulus is added without an accompanying change in the outcome that is predicted. This inattention to an added stimulus demonstrates selective attention to the previously trained stimulus, and/or learned inattention to the added stimulus.

For blocking, as for latent inhibition, impairment of dopaminergic function appears to prevent selective attention (i.e., selective inattention to redundant cues). To show this effect, Crider et al. (38) trained rats to perform a two-way shuttle conditioned shock-avoidance (CAR) using a tone as the warning stimulus on the first day of training. On the second day, a light was presented in addition to the tone. Subsequent testing showed that the light stimulus did not produce a conditioned avoidance response when presented alone (i.e., the association was blocked). Prior repeated exposure to *d*-amphetamine (4.0 mg/kg), however, prevented blocking (as it had prevented latent inhibition). Amphetamine-treated rats avoided shock cued by the light as readily as did animals that had not been given the earlier tone-alone training. This prevention of blocking cannot be attributed to a general drug-induced performance decrement since the CAR was acquired at the same rate by rats in all groups. Further, all rats showed similar avoidance of the tone during the test. The only distinction was that rats preexposed to amphetamine showed avoidance of the light as well. This task therefore contains built-in control conditions for motivational and sensorimotor deficits, and thus obviates the need for the extra control conditions required in the latent inhibition task, above.

A series of studies with blocking showed pharmacological parallels to latent inhibition, including acute administration of haloperidol preventing the amphetamine-induced attentional impairment (38), and chronic administration of haloperidol inducing the impairment, attendant upon dopamine receptor supersensitivity (37). This latter study also included an additional control condition in which a novel, salient stimulus (clicking sound) was presented during the test. Haloperidol and control rats showed a low and comparable level of avoidance of this novel stimulus, indicating that avoidance of the light stimulus by haloperidol-treated animals was a true conditioned effect (i.e., a lack of blocking) rather than merely increased reactivity.

Oades et al. (93) showed subsequently that a comparable reduced blocking effect was seen in rats given 6-OHDA-induced lesions of the frontal cortex but not of the septum (pretreatment with desipramine assured that lesions were of dopamine but not noradrenaline neurons). Sound, light, and shock sensitivities did not differ between these groups, and neither frontal nor septal lesion subjects showed a disruption of the initial acquisition of the CAR to the tone. Rats given comparable 6-OHDA lesions in the ventral tegmental area, however, did not show acquisition of the CAR. The frontal lesions, therefore, appeared to produce specific effects on the blocking of acquisition of control by the light.

Vigilance

When control by relatively simple stimuli ("readiness to respond") is maintained across extended testing or in the presence of distracting stimuli, attention may be considered sustained and the performance is said to be "vigilant." One approach to assessment for this kind of attention is offered by Grilly and coworkers. Grilly et al. (62) required a rat to press one of two levers. A signal light could be presented above either lever, and which lever was correct on a particular trial was indicated by a brief flash of this light. The order of correct response locations was approximately random. Seven to 10 sec elapsed between flashes, and the duration of the flash was adjusted for each rat until accuracy was between 75 and 87% correct within the 100 trial sessions. This baseline accuracy was increased and response latency was decreased by low doses of *d*-amphetamine (0.25, 0.75 mg/kg) or cocaine (1.25, 2.5, 5.0 mg/kg). That is, these low doses enhanced the animals' ability to sustain attention to the brief stimulus. Higher doses, however, decreased accuracy and increased response latency (1.25 mg/kg *d*-amphetamine, 15.0 mg/kg cocaine). This loss of stimulus control by the brief stimulus can be described as decreased vigilance. These two psychoactive stimulants, therefore, increased vigilance at low doses and decreased vigilance at higher doses.

While a biphasic effect was found on sustained attention to a brief all-or-none stimulus, a different pattern of effect was found when accuracy depended on close inspection of stimuli (i.e., selective attention). To assess selective attention, Grilly et al. (62) trained other rats on a different task which utilized lights above each of the two levers on each trial. These lights turned on and remained on until a response was made. One of the lights was steady while the other was flashing. Sometimes the correct lever was on the left and sometimes it was on the right, but it was always designated by the steady light. The rate of flashing for the incorrect stimulus was adjusted until baseline accuracy fell between 75 and 87% correct within the 100 trial session (flashing was only somewhat faster than three per sec, so that changes in critical flicker fusion were not a factor). Accuracy of performance in this task was not changed by low to moderate doses of either cocaine or *d*-amphetamine. High doses, however, did have an effect on accuracy. At the highest dose of *d*-amphetamine (1.25 mg/kg), accuracy was decreased, and there was also a suggestion of a decrease at the highest dose of cocaine as well (15.0 mg/kg). These stimulants never increased this selective attention (although they had increased accuracy in the sustained attention task); high doses did lead to a decline in selective attention (as it had for the sustained attention).

Robbins and co-workers provide a different measure of attention that is maintained across a series of selections. They have used a "serial reaction time" procedure to study attention (e.g., ref. 118). Because accurate performance in their task requires persistent monitoring of stimulus locations across time and utilizes relatively simple component actions, we consider it a task of sustained attention. Rats repeatedly selected one from among a series of (usually five) apertures located along one wall of the experimental chamber—the wall opposite a food tray. The correct aperture on a particular trial is indicated by a brief (0.5 sec) light flash within the aperture. The stimulus occurs only after a "time-out" or intertrial interval has elapsed with no responses (pokes into apertures) having occurred for at least 5 sec. Early, delayed, or incorrect responses extended the time-out period. Correct responses (during the brief light or within 0.5 sec of light offset) produce a food pellet followed by the time out. Sessions of 100 trials (maximum of 30 min) have typically been given.

The specific focus of the Robbins et al. (118) study was to examine the effect of excitotoxic lesions of the basal forebrain on cholinergic activity in the neocortex and its relation to attention. Lesions of the substantia innominata induced by ibotenic acid or quisqualic acid reduced accuracy markedly. Loss of accuracy was caused by an increase in errors of anticipation (responses during time-out), rather than by an increase in errors of omission.

Within this basic procedure, two experimental manipulations were made which allowed the effect to be characterized further. First, increasing the stimulus duration from 0.25 to 1.5 sec improved accuracy for the rats with basal forebrain lesions, while varying the brightness of the light did not affect accuracy. Second, for some sessions a brief (0.5 sec) burst of 85 dB white noise was presented at various points during the time-out. While control subjects actually improved their accuracy somewhat under this "distraction" condition, accuracy was reduced for lesioned subjects.

The dependence of accuracy on stimulus duration, the increase in errors of anticipation, and the increased susceptibility to distracting white noise for rats with basal forebrain lesions all support an interpretation of these effects as decreased attention *to the light,* although it may reflect an increase rather than a decrease in general reactivity (i.e., response disinhibition). Since accuracy was not improved by increasing the brightness, an interpretation in terms of decreased sensory capacity *per se* is not supported. Further, the facts that body weight remained stable and omission errors did not increase (while anticipation errors did) argue against an interpretation of these effects as results of decreased motivation (even though latencies to collect earned pellet did increase somewhat).

The effects noted in this study for basal forebrain lesions differed from those that these investigators have previously shown to follow other types of neurobiological impairments. While baseline choice accuracy was affected by these basal forebrain lesions, 6-OHDA lesions of the ventral striatum increased errors of omission and response latency, but did not affect baseline choice accuracy (34). Further, lesions of the dorsal noradrenergic bundle affected choice accuracy, but only under conditions in which the ITI was varied and when bursts of distracting white noise were presented just before the visual stimulus (30,33).

Shifting of Attention

A strong case can be made that diverting attention from one location to another is a specific kind of behavior, whether overt (noted by eye movements) or covert (e.g., ref. 108). Such an ability is probably involved in the "dual-task" evaluations that have often been found to be among the most sensitive indicators of toxic effects in human laboratory assessments (e.g., ref. 87,110). Further, a test that specifically measures a subject's ability to shift attention from one dimension of stimulation to another has been found to be among the most sensitive in the widely used Neurobehavioral Evaluation System (NES) (ref. 81).

To study this phenomenon in animals, Evenden and Robbins (55) developed an animal "tracking" task which resembles tasks that require humans to shift attention from one location to another. Their task required the rat to respond on one of two levers to complete a random ratio contingency for food reinforcement. The correct lever (i.e., the one that accumulated responses toward reinforcer-delivery) was indicated by a cue light; however, the location of this light was shifted with probability 0.2 after each response. That is, after four of every five responses the correct lever remained on the same side, but after an unpredictable one of five responses, the correct lever shifted to the other location. Animals learned to track the correct lever quite accurately (e.g., of the switches made, 99% were switches to the currently correct lever). Low to moderately high doses of d-amphetamine (0.4 to 2.3 mg/kg) increased the rats' tendency to switch between the levers; however, since the added, drug-related switches were made both toward correct and toward incorrect levers (now only 90% of switches were to the correct lever), the drug actually produced a loss of attention to the light (tracking). At the highest dose of d-amphetamine given (3.2 mg/kg), the increased tendency to switch was lost and the rats perseverated on one lever rather than switching levers. In fact, the rats continued to respond on the correct lever even after pellets were delivered; responses made after the traylight was illuminated to indicate that a pellet had been delivered increased markedly in response to the drug. These changes at the highest dose represent a continued inattention to the lights with a new response-bias: an amphetamine-induced perseveration.

Of the animal models of attention identified, two (latent inhibition and blocking) utilize learning processes to demonstrate eventual control (or lack thereof) of discriminative stimuli over behavior. For example, effects of a treatment on memory or learning may affect acquisition of stimulus control, and thus a conclusion that attention has been affected requires further demonstrations that learning of other associations has proceeded normally in treated animals.

The selective and sustained attention tasks proposed by Grilly and colleagues and Robbins and colleagues rely on performance of discrimination tasks after learning has occurred, and so avoid the potential confounds inherent in the learning process. Nevertheless, they approach the problem of attention from an apical perspective, in that "attention" is viewed as a simple, undifferentiated process. In other words, the finer differentiations among attentional processes being identified in studies of humans have not been introduced into the armamentarium of animal techniques. The Evenden and Robbins task (55), in which a rat must track the location of a shifting signal, perhaps comes closest to tapping the process for shifting attention proposed by Posner and Peterson (108) as occurring in humans. There is clearly opportunity for the development of more refined methods directed at modeling specific attentional processes in animals. In this field, as well as in the area of memory research, analysis of human behavior has provided strong leads for animal researchers to follow. There exists no lack of material for a theory of processes here; the present need is for adequate models of those processes to apply to studies of the effects of toxic chemicals on attention in experimental animals.

CONCLUSIONS

It has been our intent in the present review to update current developments in the topics of learning and memory addressed elsewhere (26,89) and to alert the reader to what we see as an important emerging subfield, the study of impairments of attention. With this third category added, the field might be well designated as Cognitive Neurotoxicology. Animal research comprises a great deal of the effort in this field, and it has been a primary focus of our review as well.

For learning and memory, we reviewed data obtained with a few of the more popular procedures. Passive avoidance and other rapidly acquired reference memory tasks continue to be the most commonly used methods for assessing learning and memory. They do provide relatively clear-cut effects of drugs and toxicants in groups of animal subjects with a minimum of time and effort. That is, they are sensitive to impairment and efficient as screens. Cautions regarding the specificity of the assessments obtained with these procedures abound, however, (e.g., ref. 5). We recounted these difficulties and emphasized that current practice rarely provides data sufficient to affirm that the effects documented with these procedures constitute impairments of learning or memory rather than alterations in sensory or motor function or motivation. While the procedure is often described as providing an assessment of memory, there are several indications that effects may better be interpreted as impairments of learning instead. In fact, considerable evidence points toward changes in attentional processes as providing a primary contribution to these effects on learning, and we would encourage future investigations to test for such impairments specifically.

Repeated acquisition procedures may ultimately provide a more specific evaluation of learning than will procedures that assess initial acquisition of reference memories. Learning in repeated acquisition baselines can provide a relatively "pure" measure of acquisition since confounding influences of, for example, habituation to the first environment and failure of long-term memory can be minimized. In addition, they allow this assessment to be carried out for individuals rather than for groups and offer the greater reliability of a steady baseline against which changes can be evaluated. Several exemplary studies were noted to indicate the strength of these procedures. The cost of the specificity and analytical power of repeated acquisition procedures has been an extended training effort that has detracted from their practicality. Some shortcuts have been explored, and some maze procedures have been proposed which provide stable repeated acquisition baselines fairly quickly. We strongly advocate the continued exploration of their use.

Considerable effort has been given to the evaluation of impairments of working memory—the kind of remembering that allows us to bridge gaps in time but does not necessarily lead to long-term storage of information. We reviewed results obtained with the delayed matching-to-position procedure that is now widely and effectively used. As was emphasized for the passive avoidance procedures, however, many of the effects that have been obtained using working memory procedures may actually be better described as effects on acquisition of information rather than on its retention; these effects may also result from reduced attention. For example, cholinergic blockade produces effects that are often described as "amnestic." Human as well as animal data were reviewed to indicate that these so-called memory effects are often intermixed with (and/or confused with) effects on attention.

We turned, therefore, to a brief introduction to the literature on attention. Considerable effort is now being given to distinguishing different kinds of human attention and the biological mechanisms that underly these separable processes. This work was explored to provide counsel regarding the types of animal assessments needed to provide an adequate screening for attentional impairments. Some candidate animal procedures were explored to indicate the kinds of distinctions that can now be drawn from the animal literature. We conclude that this effort has really just begun. While several alternative procedures have been utilized, the studies of attention are rarely sufficiently analytical to assess specific kinds of attentional impairment. Again, we urge that future investigations be designed to test specifically for the different subtypes of attentional impairments.

REFERENCES

1. Allport A. Visual attention. In: Posner MI, ed. *Foundations of cognitive science* Cambridge, MA: MIT Press, 1989;631–682.
2. Anger WK. Worksite behavioral research: results, sensitive methods, test batteries and the transition from laboratory data to human health. *Neurotoxicology* 1990;11:629–720.
3. Anisman H, Kokkindis L, Glazer S, Remington D. Differentiation of response biases elicited by scopolamine and *d*-amphetamine. Effects on habituation. *Behav Biol* 1976;18:401–417.
4. Auer RN, Jensen ML, Whishaw IQ. Neurobehavioral deficit due to ischemic brain damage limited to half of the CA1 sector of the hippocampus. *Neurosci* 1989;9:1641–1647.
5. Bammer G. Pharmacological investigations of neurotransmitter involvement in passive avoidance responding: a review and some new results. *Neurosci Biobehav Rev* 1982;6:247–296.
6. Bartus RT, Johnson HR. Short-term memory in the rhesus monkey: disruption from the anticholinergic agent scopolamine. *Pharmacol Biochem Behav* 1976;5:39–46.
7. Beatty WW, Shavalia DA. Spatial memory in rats: time course of working memory and effect of anesthetics. *Behav Neural Biol* 1980;28:454–462.
8. Bessemer DW, Stollnitz F. Retention of discriminations and an analysis of learning set. In: Schrier AM, Stollnitz F, eds. *Behavior of nonhuman primates,* vol. 4. New York: Academic Press, 1971;2–59.
9. Bickel WK, Higgins ST, Griffiths RR. Repeated diazepam administration: effects on the acquisition and performance of response chains in humans. *J Exp Anal Behav* 1989;52:47–56.
10. Bickel WK, Hughes JR, Higgins ST. Human behavioral pharmacology of benzodiazepines: effects on repeated acquisition and performance of response chains. *Drug Dev Res* 1990;20:53–65.
11. Bitterman ME. Phyletic differences in learning. *Am Psychol* 1965;20:396–410.
12. Boren JJ. Repeated acquisition of new behavioral chains. *Am Psychol* 1963;18:421.
13. Brazhnik L-S, Vinogradova DS. Modulation of the afferent input to the septal neurons by cholinergic drugs. *Brain Res* 1988;451:1–12.
14. Broadbent DE. *Decision and stress.* London: Pergamon, 1971.
15. Broadbent DE. *Perception and communication.* London: Pergamon, 1958.
16. Broadbent DE. Task combination and selective intake of information. *Acta Psychol* 1982;50:253–290.
17. Broks P, Preston GC, Traub M, Poppleton P, Ward C, Stahl SM. Modelling dementia: effects of scopolamine on memory and attention. *Neuropsychologia* 1988;26:685–700.
18. Bruner JS, Goodnow JJ, Austin GA. *A study of thinking.* New York: Wiley, 1956.
19. Bushnell PJ. Effects of delay, intertrial interval, delay behavior and trimethyltin on spatial delayed response in rats. *Neurotoxicol Teratol* 1988;10:237–244.

20. Bushnell PJ. Modelling working and reference memory in rats: effects of scopolamine on delayed matching-to-position. *Behav Pharmacol* 1990;1:419–427.
21. Bushnell PJ. Delay-dependent impairment of reversal learning in rats treated with trimethyltin. *Behav Neural Biol* 1990;54:75–89.
22. Bushnell PJ. Learning deficits in rats after oral exposure to styrene monomer. *Toxicologist* 1991;592.
23. Bushnell PJ, Padilla SS, Ward T, Pope CN, Olszyk VB. Behavioral and neurochemical changes in rats dosed repeatedly with diisopropylfluorophosphate. *J Pharmacol Exp Therap* 1991;256: 741–750.
24. Bushnell PJ, Angell KE. Effects of trimethyltin on repeated acquisition (learning) in the radial-arm maze. (*submitted*).
25. Bushnell PJ, Bowman RE. Reversal learning deficits in young monkeys exposed to lead. *Pharmacol Biochem Behav* 1979;10:733–742.
26. Cabe PA, Eckerman DA. Assessment of learning and memory dysfunction in agent-exposed animals. In: Mitchell CL, ed. *Nervous system toxicology.* New York: Raven Press, 1982;133–198.
27. Caine ED, Weingartner H, Ludlow CL, Cudahy EA, Wehry S, Caine ED et al. Qualitative analysis of scopolamine-induced amnesia. *Psychopharmacology* 1981;74:74–80.
28. Calhoun WH, Jones EA. Serial discrimination reversal learning as a repeated-acquisition method to test drug effects. *Bull Psychonom Soc* 1979;13:375–377.
29. Calhoun WH, Jones EA. Methamphetamine's effect on repeated acquisition with serial discrimination reversals. *Psychopharmacologia* 1974;39:303–308.
30. Carli M, Robbins TW, Evenden JL, Everitt BJ. Effects of lesions to ascending noradrenergic neurons on performance on a 5-choice serial reaction time task in rats: implications for theories of dorsal noradrenergic bundle function based on selective attention and arousal. *Behav Brain Res* 1994;9:361–380.
31. Cheal ML. Scopolamine disrupts maintenance of attention rather than memory processes. *Behav Neural Biol* 1981;33:163–187.
32. Cohen NJ, Squire LR. Preserved learning and retention of pattern-analyzing skill in amnesia: dissociation of knowing how and knowing that. *Science* 1980;210:207–210.
33. Cole BJ, Robbins TW. Amphetamine impairs the discriminative performance of rats with dorsal bundle lesions on a 5-choice serial reaction time task: new evidence for central dopaminergic-noradrenergic interactions. *Psychopharmacology* 1987;91:458–466.
34. Cole BJ, Robbins TW. Effects of 6-hydroxydopamine lesions of the nucleus accumbens on performance of a 5-choice serial reaction time task in rats: implications for theories of selective attention and arousal. *Behav Brain Res* 1989;33:165–179.
35. Corkin S, Cohen NJ, Sullivan EV, Clegg RA, Rosen TJ, Ackerman RH. Analyses of global memory impairments of different etiologies. *Ann NY Acad Sci* 1985;444:10–39.
36. Cory-Slechta DA. Behavioral measures of neurotoxicity. *Neurotoxicology* 1989;10:271–296.
37. Crider A, Blockel L, Solomon PR. A selective attention deficit in the rat following induced dopamine receptor supersensitivity. *Behav Neurosci* 1986;100:315–319.
38. Crider A, Solomon PR, McMahon MA. Disruption of selective attention in the rat following chronic d-amphetamine administration: relationship to schizophrenic attention disorder. *Biol Psychiatry* 1982;17:351–361.
39. Decker MW, McGaugh JL. Effects of concurrent manipulations of cholinergic and noradrenergic function on learning and retention in mice. *Brain Res* 1989;477:29–37.
40. Decker MW, Tran T, McGaugh JL. A comparison of the effects of scopolamine and diazepam on acquisition and retention of inhibitory avoidance. *Psychopharmacology* 1990;100:515–521.
41. DeRenzi E, Gentilini M, Pattacini F. Auditory extinction following hemisphere damage. *Neuropsychologia* 1984;22:733–744.
42. Dietz DD, McMillan DE, Mushak P. Effects of chronic lead administration on acquisition and performance of serial position sequences by pigeons. *Toxicol Appl Pharmacol* 1979;47:377–384.
43. DiMattia BV, Kesner RP. Serial position curves in rats: automatic versus effortful information processing. *J Exp Psychol Anim Beh Proc* 1984;10:557–563.
44. Domjan M, Burkhard B. *The principles of learning & behavior.* Monterey, CA: Brooks/Cole Publishing Co., 1986.
45. Duncan HJ, Slotnick BM. Comparison of visual and olfactory reversal learning in pigeons. *Chem Senses* 1990;15:59–73.

46. Dunnett SB. Comparative effects of cholinergic drugs and lesions of nucleus basalis or fimbria-fornix on delayed matching in rats. *Psychopharmacology* 1985;87:357–363.

47. Dunnett SB. Short-term memory in rodent models of ageing: effects of cortical cholinergic grafts. In: Gage F, Privat A, Christen Y, eds. *Neuronal grafting and Alzheimer's disease.* Berlin: Springer-Verlag, 1989;85–102.

48. Dunnett SB, Evenden JL, Iversen SD. Delay-dependent short-term memory deficits in aged rats. *Psychopharmacology* 1988;96:174–180.

49. Dunnett SB, Martel FL. Proactive interference effects on short-term memory in rats: I. basic parameters and drug effects. *Behav Neurosci* 1990;104:655–665.

50. Dunnett SB, Martel FS, Iversen SD. Proactive interference effects on short-term memory in rats: II. effects in young and aged rats. *Behav Neurosci* 1990;104:666–670.

51. Dunnett SB, Rogers DC, Jones GH. Effects of nucleus basalis magnocellularis lesions in rats on delayed matching and non-matching to position tasks. *Eur J Neurosci* 1989;1:395–406.

52. Evans EB, Wenger GR. The acute effects of caffeine, cocaine and d-amphetamine on the repeated acquisition responding of pigeons. *Pharmacol Biochem Behav* 1990;35:631–636.

53. Evenden JL, Marston HM, Jones GH, Giardini V, Lenard L, Everitt BJ, Robbins TW. Effects of excitotoxic lesions of the substantia innominata, ventral and dorsal globus pallidus on visual discrimination acquisition, performance and reversal in the rat. *Behav Brain Res* 1989;32:129–149.

54. Evenden JL, Robbins TW. Win-stay behaviour in the rat. *Q J Exp Psychol* 1982;36B:1–26.

55. Evenden JL, Robbins TW. The effects of d-amphetamine, chlordiazepoxide and alpha-flupenthixol on food-reinforced tracking of a visual stimulus by rats. *Psychopharmacology* 1985;85:361–366.

56. Fletcher HJ. The delayed response problem. In: Schrier AM, Harlow HF, Stollnitz F, eds. *The behavior of nonhuman/primates,* vol. 1. New York: Academic Press, 1965;129–165.

57. Flicker C, Serby M, Ferris SH. Scopolamine effects on memory, language, visuospatial praxis and psychomotor speed. *Psychopharmacology* 1990;100:243–250.

58. Frith CD, McGinty MA, Gergel I, Crow TJ. The effects of scopolamine and clonidine upon the performance and learning of a motor skill. *Psychopharmacology* 1989;98:120–125.

59. Furuya YT, Yamamoto S, Yatsugi S, Uecki S. A new method for studying working memory by using the three-panel runway apparatus in rats. *Japn J Pharmacol* 1988;46:183–188.

60. Gallagher M, Meagher MW, Bostock E. Effects of opiate manipulations on latent inhibition in rabbits: sensitivity of the medial sepatal region to intracranial treatments. *Behav Neurosci* 1987;101:315–324.

61. Gamzu ER, Bairkhimer LJ, Hoover T, Gracon ST. Keynote Lecture 3: early human trials in the assessment of cognitive activators. *Pharmacopsychiat.* 1990;32 (Suppl):44–48.

62. Grilly DM, Gowans GC, McCann DS, Grogan TW. Effects of cocaine and d-amphetamine on sustained and selective attention in rats. *Pharmacol Biochem Behav* 1989;33:733–739.

63. Harlow HF. The formation of learning sets. *Psychol Rev* 1949;56:51–65.

64. Heise GA, Hudson JD. Effects of pesticides and drugs on working memory in rats: continuous delayed response. *Pharmacol Bioch Behav* 1985;23:591–598.

65. Heise GA, Hudson JD. Effects of pesticides and drugs on working memory in rats: continuous non-match. *Pharmacol Biochem Behav* 1985;23:599–605.

66. Heise GA, Milar KS. *Handbook of psychopharmacology.* New York: Plenum, 1984.

67. Higgins ST, Bickel WK, O'Leary DK, Yingling J. Acute effects of ethanol and diazepam on the acquisition and performance of response sequences in humans. *J Pharmacol Exp Therap* 1987;243:1–8.

68. Higgins ST, Woodward BM, Henningfield JE. Effects of atropine on the repeated acquisition and performance of response sequences in humans. *J Exp Anal Behav* 1989;51:5–16.

69. Hilgard ER. The trilogy of mind: cognition, affection, and conation. *J History Behav Sci* 1980;16:107–117.

70. Hirst W. The amnesic syndrome: descriptions and explanations. *Psychol Bull* 1982;91:435–460.

71. Honig WK. Studies of working memory in the pigeon. In: Hulse SH, Fowler H, Honig WK, eds. *Cognitive processes in animal behavior.* Hillsdale, NJ: Erlbaum, 1978.

72. Howard JL, Pollard GT. Effects of d-amphetamine, Org 2766, scopolamine, and physostigmine on repeated acquisition of four-response chains in rat. *Drug Dev Res* 1983;3:37–48.

73. Izquierdo I. Mechanism of action of scopolamine as an amnestic. *Trends Pharmacol Sci* 1989;10: 175–177.
74. Jeannerod M, Biguer B. The directional encoding of reaching movements: a visuomotor conception of spatial neglect. In: Jeannerod M, ed. *Neurophysiological and neuropsychological aspects of spatial neglect.* Amsterdam: North Holland, 1987.
75. Johnson BL. *Advances in neurobehavioral toxicology: applications in environmental and occupational health.* Chelsea, MI: Lewis Press, 1990.
76. Kesner RP, Bierley RA, Pebbles P. Short-term memory: the role of d-amphetamine. *Pharmacol Biochem Behav* 1981;15:673–676.
77. Kirk RC, White KG, McNaughton N. Low dose scopolamine affects discriminability but not rate of forgetting in delayed conditional discrimination. *Psychopharmacology* 1988;96:541–546.
78. Klein WJ, Sullivan J, Skorupa A, Aguilar JS. Plasticity of neuronal receptors. *FASEB J* 1989;3: 2132–2140.
79. Kopelman MD, Corn TH. Cholinergic "blockade" as a model for cholinergic depletion: a comparison of the memory deficits with those of Alzheimer-type dementia and the alcoholic Korsakoff Syndrome. *Brain* 1988;111:1079–1110.
80. Kreitzer JF. Effects of toxaphene and endrin at very low dietary concentrations on discrimination acquisition and reversal in bobwhite quail. *Environ Poll (Series A)* 1980;23:217–230.
81. Letz R. Occupational screening for neurotoxicity: computerized techniques. *Toxicology* 1988;49: 417–424.
82. Levin ED. Psychopharmacological effects in the radial-arm maze. *Neurosci Biobehav Rev* 1988;12: 169–175.
83. Lowy AM, Ingram DK, Olton DS, Waller SB, Reynolds MA, London ED. Discrimination learning requiring different memory components in rats: age and neurochemical comparisons. *Behav Neurosci* 1985;99:638–651.
84. Mackintosh NJ. *The psychology of animal learning.* London: Academic Press, 1974.
85. Mackintosh NJ. *Conditioning and associative learning.* New York: Oxford, 1983.
86. Medin DL. Information processing and discrimination learning set. In: Schrier AM, ed. *Behavioral primatology, vol. 1: Advances in research and theory.* Hillsdale, NJ: Lawrence Erlbaum, 1977;33–69.
87. Mihevic PM, Gliner JA, Horvath SM. Carbon monoxide exposure and information processing during perceptual-motor performance. *Int Arch Occup Environ Health* 1983;51:355–363.
88. Miller GA, Gazzaniga MS. The cognitive sciences. In: Gazzaniga MS, ed. *Handbook of cognitive science.* New York: Plenum, 1984.
89. Miller DB, Eckerman DA. Learning and memory measures. In: Annau Z, ed. *Neurobehavioral toxicology.* Baltimore, MD: Johns Hopkins University Press, 1986;94–149.
90. Moerschbaecher JM, Thompson DM. Effects of d-amphetamine, cocaine, and phencyclidine on the acquisition of response sequences with and without stimulus fading. *J Exp Anal Behav* 1980;33:369–381.
91. Nicolle MM, Barry CC, Veronesi B, Stanton ME. Fornix transections disrupt the ontogeny of latent inhibition in the rat. *Psychobiology* 1989;17:349–357.
92. Nigrosh BJ, Slotnick BM, Nevin JA. Olfactory discrimination, reversal learning, and stimulus control in rats. *J Comp Physiol Psychol* 1975;89:285–294.
93. Oades RD, Rivet JM, Taghzouti K, Kharouby M, Simon H, Le Moal M. Catecholamines and conditioned blocking: effects of ventral tegmental, septal and frontal 6-hydroxydopamine lesions in rats. *Brain Res* 1987;406:136–146.
94. Oliverio A. Effects of scopolamine on avoidance conditioning and habituation of mice. *Psychopharmacologia* 1968;12:214–226.
95. Olton DS. Hippocampal function and memory for temporal context. In: Isaacson RL, Pribram KH, eds. *The hippocampus,* vol. 4. New York: Plenum, 1986;281–299.
96. Olton DS, Becker JT, Handelman, GE. Hippocampus, space, and memory. *Behav Brain Sci* 1979;2:313–365.
97. Olton DS, Samuelson R. Decision making in the rat: response-choice and response-time measures of discrimination reversal learning. *J Comp Physiol Psychol* 1974;87:1134–1147.
98. Overmeier JB. But where is the path? A comment on the state of the art and suggestions for

future research strategies. In: Tilson HA, Sparber SB, eds. *Neurotoxicants and neurobiological function: effects of organoheavy metals.* New York: Wiley, 1987;303–313.

99. Paule MG. Use of the NCTR operant test battery in nonhuman primates. *Neurotoxicol Teratol* 1990;12:413–418.
100. Paule MG, McMillan DE. Incremental repeated acquisition in the rat: acute effects of drugs. *Pharmacol Biochem Behav* 1984;21:431–439.
101. Paule MG, McMillan DE. Effects of trimethyltin on incremental repeated acquisition (learning) in the rat. *Neurobehav Toxicol Teratol* 1986;8:245–253.
102. Peele DB, Baron SP. Effects of scopolamine on repeated acquisition of radial-arm maze performance by rats. *J Exp Anal Behav* 1988;49:275–290.
103. Peele DB, Vincent A. Strategies for assessing learning and memory, 1978–1987: a comparison of behavioral toxicology, psychopharmacology, and neurobiology. *Neurosci Biobehav Rev* 1989;13:317–322.
104. Penetar DM. The effects of atropine, benactyzine, and physostigmine on a repeated acquisition baseline in monkeys. *Psychopharmacology* 1985;87:69–76.
105. Poling A, Cleary J, Berens K, Thompson T. Neuroleptics and learning: effects of haloperidol, molindone, mesoridazine and thioridazine on the behavior of pigeons under a repeated acquisition procedure. *J Pharmacol Exp Ther* 1990;255:1240–1245.
106. Pontecorvo MJ. Effects of proactive interference on rats' continuous nonmatching-to-sample performance. *Ani Learn Behav* 1983;11:356–366.
107. Pontecorvo MJ. Memory for a stimulus versus anticipation of a response: contrasting effects of proactive interference in two delayed comparison tasks. *Ani Learn Behav* 1985;13:355–364.
108. Posner MI, Peterson SE. The attention system of the human brain. *Ann Rev Neurosci* 1990;13: 25–42.
109. Preston GC, Ward C, Lines CR, Poppleton P, Haigh JRM, Traub M. Scopolamine and benzodiazepine models of dementia: cross-reversals by Ro 15-1788 and physostigmine. *Psychopharmacology* 1989;98:487–494.
110. Putz V, Johnson BL, Setzer JV. A comparative study of the effects of carbon monoxide and methylene chloride on human performance. *J Environ Pathol Toxicol* 1979;2:97–112.
111. Quartermain D, Leo P. Strength of scopolamine-induced amnesia as a function of time between training and testing. *Behav Neural Biol* 1988;50:300–310.
112. Raffaele K, Olton D, Annau Z. Repeated exposure to diisopropylfluorophosphate (DFP) produces increased sensitivity to cholinergic antagonists in discrimination retention and reversal. *Psychopharmacology* 1990;100:267–274.
113. Rawlins JNP, Tsaltas E. The hippocampus, time and working memory. *Behav Brain Res* 1983;10: 233–262.
114. Rice DC. Chronic low-lead exposure from birth produces deficits in discrimination reversal in monkeys. *Toxicol Appl Pharmacol* 1985;77:201–210.
115. Rice DC. Principles and procedures in behavioral toxicity testing. In: Arnold DL, Grice HC, Krewski DR, eds. *Handbook of in vivo toxicity testing.* San Diego, CA: Academic Press, 1990.
116. Richardson-Klavehn A, Bjork RA. Measures of memory. *Ann Rev Psychol* 1988;39:475–543.
117. Ridley RM, Baker HF, Drewett B, Johnson JA. Effects of ibotenic acid lesions of the basal forebrain on serial reversal learning in marmosets. *Psychopharmacology* 1985;86:438–443.
118. Robbins TW, Everitt BJ, Marston HM, Wilkinson J, Jones GH, Page KJ. Comparative effects of ibotenic acid- and quisqualic acid-induced lesions of the substantia innominata on attentional function in the rat: further implications for the role of the cholinergic neurons of the nucleus basalis in cognitive processes. *Behav Brain Res* 1989;35:221–240.
119. Roberts AC, Robbins TW, Everitt BJ, Jones GH, Sirkia TE, Wilkinson J, Page K. The effects of excitotoxic lesions of the basal forebrain on the acquisition, retention and serial reversal of visual discriminations in marmosets. *Neurosci* 1990;34:311–329.
120. Rupniak NMJ, Steventon HJ, Field MJ, Jennings CA, Iversen SD. Comparison of the effects of four cholinomimetic agents on cognition in primates following disruption by scopolamine or by lists of objects. *Psychopharmacology* 1989;99:189–195.
121. Rush DK. Scopolamine amnesia of passive avoidance: a deficit of information acquisition. *Behav Neural Bio* 1988;50:255–274.
122. Russell RW, Overstreet DH. Mechanisms underlying sensitivity to organophosphorus anticholinesterase compounds. *Prog Neurobiol* 1987;28:97–131.

123. Rusted JM. Dissociative effects of scopolamine on working memory in healthy young volunteers. *Psychopharmacology* 1988;96:487–492.
124. Rusted JM, Warburton DM. The effects of scopolamine on working memory in healthy young volunteers. *Psychopharmacology* 1988;96:145–152.
125. Rusted JM, Warburton DM. Effects of scopolamine on verbal memory; a retrieval or acquisition deficit? *Neuropsychobiology* 1989;21:76–83.
126. Rutter M. Raised lead levels and impaired cognitive/behavioral functioning: a review of the evidence. *Dev Med Child Neurol Suppl* 1980;42.
127. Santi A, Bogles J, Petelka S. The effect of scopolamine and physostigmine on working and reference memory in pigeons. *Behav Neural Biol* 1988;49:61–73.
128. Santi A, Hanemaayer C, Reason W. The effect of scopolamine on reference and working memory in pigeons. *An Learn Behav* 1987;15:395–402.
129. Sarter M. Retrieval of well-learned propositional rules: insensitive to changes in activity of individual neurotransmitter systems? *Psychobiology* 1990;18:451–459.
130. Schulze GE, McMillan DE, Bailey JR, Scallet A, Ali SF, Slikker W, Paule MG. Acute effects of D-9-tetrahydrocannabinol in Rhesus monkeys as measured by performance in a battery of complex operant tests. *J Pharmacol Exp Ther* 1988;245:178–186.
131. Schulze GE, McMillan DE, Bailey JR, Scallet AC, Ali SF, Slikker W, Paule MG. Acute effects of marijuana smoke on complex operant behavior in Rhesus monkeys. *Life Sci* 1989;45:465–475.
132. Schulze GE, Paule MG. Acute effects of d-amphetamine in a monkey operant behavioral test battery. *Pharmacol Biochem Behav* 1990;35:759–765.
133. Schulze GE, Slikker W, Paule MG. Multiple behavioral effects of diazepam in Rhesus monkeys. *Pharmacol Biochem Behav* 1989;34:29–35.
134. Shimizu T, Hodos W. Reversal learning in pigeons: effects of selective lesions of the Wulst. *Behav Neurosci* 1989;103:262–272.
135. Slikker W, Paule MG. Acute effects of marijuana smoke on complex operant behavior in rhesus monkeys. *Life Sci* 1989;45:465–475.
136. Slotnick BM. Olfactory stimulus control in the rat. *Chem Senses* 1984;9:157–165.
137. Solomon PR, Crider A, Winkelman JW, Turi A, Kamer RM, Kaplan LJ. Disrupted latent inhibition in the rat with chronic amphetamine or haloperidol-induced supersensitivity: relationship to schizophrenic attention disorder. *Biol Psychiatry* 1981;16:519–537.
138. Solomon PR, Staton DM. Differential effects of microinjections of d-amphetamine into the nucleus accumbens or the caudate putamen on the rat's ability to ignore an irrelevant stimulus. *Bio Psychiatry* 1982;17:743–756.
139. Spencer DG, Pontecorvo MJ, Heise GA. Central cholinergic involvement in working memory: effects of scopolamine on continuous nonmatching and discrimination performance in the rat. *Behav Neurosci* 1985;99:1049–1065.
140. Squire LR, Zola-Morgan S, Chen KS. Human amnesia and animal models of amnesia: performance of amnesic patients on tests designed for the monkey. *Behav Neurosci* 1988;102:210–221.
141. Stephens DN, Weidmann R, Quartermain D, Sarter M. Reversal learning in senescent rats. *Behav Brain Res* 1985;17:193–202.
142. Tan S, Kirk RC, Abraham WC, McNaughton N. Chlordiazepoxide reduces discriminability but not rate of forgetting in delayed conditional discrimination. *Psychopharmacology* 1990;101:550–554.
143. Thomas JR, Ahlers ST, Schrot J. Cold-induced impairment of delayed matching in rats. *Behav Neural Biol* 1991;55:19–30.
144. Thompson DM. Repeated acquisition as a behavioral base line for studying drug effects. *J Pharmacol Exp Ther* 1973;184:506–514.
145. Thompson DM, Moerschbaecher JM. Drug effects on repeated acquisition. *Adv Behav Pharmacol* 1979;2:229–259.
146. Tilson HA, Sparber SB. *Neurotoxicants and neurobiological function: effects of organoheavy metals.* New York: Wiley, 1987.
147. Tilson HA, Veronesi B, McLamb RL, Matthews HB. Acute exposure to tris(2-chloroethyl)phosphate produces hippocampal neuronal loss and impairs learning in rats. *Toxicol Appl Pharmacol* 1990;106:254–269.

148. Tulving E. Episodic and semantic memory. In: Tulving E, Donaldson WD, eds. *Organization of memory.* New York: Academic Press, 1972.

149. van Hest A, Stroet J, van Haaren F, Feenstra M. Scopolamine differentially disrupts the behavior of male and female Wistar rats in a delayed nonmatching to position procedure. *Pharmacol Biochem Behav* 1990;35:903–909.

150. Weingartner H, Grafman J, Boutelle W, Kaye W, Martin PR. Forms of memory failure. *Science* 1983;221:380–382.

151. Whishaw IQ. Formation of a place learning-set by the rat: a new paradigm for neurobehavioral studies. *Physiol Behav* 1985;35:139–143.

152. Whishaw IQ. Hippocampal, granule cell and CA3-4 lesions impair formation of a place learning-set in the rat and induced reflex epilepsy. *Behav Brain Res* 1987;24:59–72.

153. White KG. Characteristics of forgetting functions in delayed matching to sample. *J Exp Anal Behav* 1985;44:15–34.

154. Yatsugi S, Yamamoto T, Ohno M, Ueki S. Effect of S-aderosyl-L-methienine on impairment of working memory induced in rats by cerebral ischemia and scopolamine *Eur J Pharmacol* 1989;166:231–239.

155. Zola-Morgan SM, Squire LR. The primate hippocampal formation: evidence for a time-limited role in memory storage. *Science* 1990;250:288–290.

Neurotoxicology, edited by Hugh Tilson
and Clifford Mitchell. Raven Press, Ltd.,
New York © 1992.

11

Schedule-Controlled Behavior in Neurotoxicology

Deborah A. Cory-Slechta

*Environmental Health Sciences Center, University of Rochester School
of Medicine and Dentistry, Rochester, New York 14642*

REINFORCEMENT SCHEDULES IN BEHAVIOR

Our learned operant behavior is always followed by some consequent event(s). The arrangement between a response and the consequent event that follows may be intentionally arranged or it may occur quite by accident. In either case, the nature of the consequent event then determines the future fate of that particular response. If the response is strengthened, the consequence is deemed a reinforcer; if the response is weakened, or even eliminated, the consequence is defined as a punisher. Seldom, however, does the behavioral environment provide that a specific consequence uniformly or consistently follows a particular response. For example, it would not be unusual, after dialing a telephone number, to get a busy signal rather than an answer, in which case we must repeat the response, or perhaps, give up. Nor would it necessarily be expected that each trip to the mail box will result in our receiving a letter.

It is also the case that these consequent events can occur under a wide variety of payoff schemes in that their delivery following the response may be based on time, number of responses, or both. In interval reinforcement schedules, such as the fixed-interval (FI) and the variable-interval (VI) reinforcement schedules, reinforcement availability is based on the time elapsed since the previous reinforcer delivery. In simple ratio schedules, such as the fixed-ratio (FR) and variable-ratio (VR), it is based instead on the number of responses that have occurred since the previous reinforcement delivery. The complexity of reinforcement schedules can increase when combinations of these basic schedules are utilized. Collectively, these various conditions have been designated as schedules of reinforcement and were elaborated in detail by Ferster and Skinner in 1957 (11). As might be expected on the basis of the differing conditions of reinforcement availability, the various schedules of reinforcement generate wide differences in both the rate of response, and the pattern of responding over time.

The realm of human behavior undoubtedly presents the potential for some of the most complex scheduling conditions to occur. This is because the human behavioral environment is likely to include simultaneous access to any of several different reinforcers and punishers. Moreover, these different consequences may occur under both a wide variety and even frequently varying schedules of reinforcement. On a day off from work, for example, one may have multiple options to pursue: deciding whether to go to the beach, swim, lay in the sun, and have a barbecue, or whether to stay home, finish painting the house and repair the leaking faucet. Our choices are influenced by several factors, including the strength of competing responses, past behavioral experience, the schedules under which they occur, and antecedent and motivational conditions.

In addition to the concurrently operative reinforcement schedules in the human environment, we may also frequently be engaged in the proverbial "doing two things at once," i.e., working under two or more different reinforcement schedules simultaneously. Many people watch television or listen to the radio while cooking, reading, doing housework, etc., as just one simple example. Thus, the human behavioral environment offers concurrent access to multiple consequent events arranged under a diversity of reinforcement schedules. Moreover, these schedule conditions may not be static, but may fluctuate over time as other environmental conditions change. For example, our parents may suddenly stop checking to see whether we have actually cleaned our room after each such request, if the task has been consistently accomplished in the past; they may resort to random or intermittent assessments. Taken together, these factors, i.e., the existence of multiple concurrently available schedules of reinforcement and the dynamic nature of these schedules, provide the basis for tremendously complex human behavior.

Despite this substantial role in the realm of human behavior, the significance of schedules of reinforcement as behavioral endpoints in neurotoxicology has frequently been both misunderstood and overlooked. This seems particularly surprising when one considers that schedule-controlled behavior itself, i.e., the rate and pattern of responding generated by a schedule of reinforcement, is a significant constituent of human operant behavior. Schedule-controlled behavior may also play an important role in other operant behavioral processes, including those which seem to have achieved more of an elevated level of significance in neurotoxicology because of their perceived face validity, such as learning and memory.

To use learning as an example, it should be pointed out that the rate at which a particular response is learned may very well be influenced by, among other factors, the reinforcement schedule according to which the reinforcer for that response is presented. In addition, the strength of this response, i.e., how well it is learned, may also be impacted by reinforcement schedule factors such as the frequency of reinforcer presentations or the rate of reinforcement. The response will certainly be more slowly acquired if initially reinforced only infrequently, than if it is initially consistently or routinely reinforced. Furthermore, whether the response will be learned at all may depend on the strength of competing responses, reinforcer strength and magnitude, and the concurrent availability of competing schedules of reinforcement. Thus, an understanding of schedule-controlled behavior is vital to a full understanding of behavior, since reinforcement schedules govern both the rate and pattern of behavioral responding involved in different behavioral processes.

To extend this argument a step further, changes in schedule-controlled behavior *per se* may, in fact, *underlie* behavioral changes that are attributed to impairments of other behavioral processes. For example, consider a case in which the environment is arranged such that a food reward follows responding when a yellow light is on, but not when there is no light on. Animals exposed to a toxicant may respond less frequently during the presence of the yellow light than do nontreated animals, leading to the interpretation that the toxicant has impaired learning. Such an effect could instead be attributable to a generalized decreased rate of responding produced by administration of the toxicant, resulting in a lower frequency of the rate of the response to be learned. Alternatively, treated animals may appear to have learned the discrimination more quickly by emitting more responses in the presence of the yellow light. Again, such an effect could derive from a change in schedule-controlled behavior if the compound increased rates of responding. Thus, what are actually changes in rates of responding may be mistakenly identified as learning deficits in the absence of information on the effects of a toxicant on schedule-controlled behavior. A change in rate of responding may be quite distinct from the alternative explanation of a learning deficit. To truly understand the behavioral effects of a toxicant, then, requires an understanding of the impact of the compound on schedule-controlled behavior, since behavior occurs at given rates and in particular patterns over time.

Like many other complex behavioral processes, schedule-controlled behavior can best be studied under laboratory conditions where the various reinforcement schedules can be studied independently of each other. In this way, the effects of toxicants on simple schedules of reinforcement can be understood before attempting to delineate the effects of drugs or chemicals under concurrent and complex schedule conditions. Although the scientific literature is by no means profuse to date, the effects of several different toxicants on various schedules of reinforcement, primarily simple schedules, have been examined. The purpose of this chapter, though, is not to review the sum results of those studies, but rather to emphasize the importance of schedule-controlled responding as an underlying component of all behavioral processes, as was described above, and subsequently, to illustrate how reinforcement schedules have already enhanced our understanding of the manner in which toxicants impact behavior, and finally to address several issues which have arisen with respect to their usage as behavioral baselines and interpretation of resulting effects.

SCHEDULES OF REINFORCEMENT AND BEHAVIORAL TOXICITY

Although the scientific literature to date is relatively sparse, studies of toxicant effects on schedule-controlled performance have already yielded an important base of information on aspects of toxicity, such as the manner in which various environmental factors modulate the behavioral effects of a toxicant. The studies are proving to be useful baselines for comparative sensitivity and time course studies, for a more advanced assessment of behavioral and neurobiological mechanisms of effect, and for evaluating behavioral or physiological antidotes to behavioral effects. Examples illustrating such findings are described below.

Modulation of Toxicity by Behavioral and Environmental Factors

Several studies have demonstrated that various behavioral and environmental factors can dramatically modulate the behavioral effects of a toxicant. In fact, one such modulator may be the schedule of reinforcement itself. For certain toxicants, the sensitivity of behavior to disruption may depend on the schedule according to which the behavior is being reinforced. Levine (18), for example, found that both single and repeated administrations of carbon disulfide to pigeons produced a greater disruption of fixed-interval schedule-controlled responding than of fixed-ratio performance. In her study, carbon disulfide decreased response rates on both reinforcement schedules, but the magnitude of the effect was greater and occurred at lower exposure levels on the FI schedule, as can be seen in Fig. 1. Such differential effects as a function of the baseline schedule of reinforcement are by no means limited to carbon disulfide, but are noted with other toxicants as well. Similar differential effects on fixed-interval versus fixed-ratio performance were reported for the pesticide chlordimeform by Leander and MacPhail (17), again with the fixed-interval schedule exhibiting greater sensitivity to disruption than did the fixed-ratio. Oral exposure of rats to low concentrations of lead similarly alters fixed-interval schedule-controlled responding at far lower concentrations than those at which fixed-ratio responding shows modifications (5,8,9).

Nor are the differential effects as a function of the baseline schedule of reinforcement limited to fixed-interval, fixed-ratio comparisons; they are observed with other reinforcement schedules as well. Peele and Crofton (25) examined the effects of type I (permethrin) and type II (cypermethrin) pyrethroid insecticides using a multiple schedule consisting of four different variable-interval schedules, including VI 10 sec, VI 30 sec, VI 90 sec, and VI 270 sec. In a multiple schedule, the various component schedules (here the four different VIs) alternate during the course of a single experimental session, with transitions between components signaled by some external stimulus cue and occurring usually after a fixed number of reinforcer deliveries or after a fixed period of time. Such an arrangement allows the assessment of different types of schedule performances during the same experimental session. Peele and Crofton, comparing the effects of these two compounds on VI 10 sec versus VI 270 sec performance, found that the effects of cypermethrin differed on the two VI schedules, decreasing responding to a greater extent in the VI 270 sec component than in the VI 10 sec component; in contrast, permethrin produced equivalent decrements in responding on the two VI schedules.

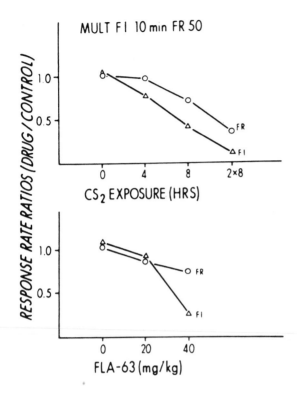

FIG. 1. Top: Ratio of response rates after carbon disulfide exposures to preexposure rates for fixed-interval and fixed-ratio components of a multiple schedule. Each point represents the mean of three birds. **Bottom:** Ratio of response rates after administration of FLA-63 to predrug rates for fixed-interval and fixed-ratio components of a multiple schedule. Each point represents the mean of three birds. (From Levine, ref. 18, with permission.)

Sometimes the modulating effects of various reinforcement schedules on behavioral toxicity may actually be manifest in the form of contrasting changes in performance as opposed to differences in sensitivity. For example, in a study by Colotla et al. (3), toluene exposure increased response rate in the DRL (differential reinforcement of low rate; a schedule in which reinforcement is available only after a period of time elapses during which the response does not occur) schedule component of a multiple schedule, while at the same time decreasing fixed-ratio response rates. From these examples it can be seen that not only the magnitude of a behaviorally toxic effect, but also the direction of that effect may depend on the schedule according to which responding is reinforced.

The contrasting effects of toxicants on various reinforcement schedules are not necessarily limited to situations in which positive reinforcers are used, that is, where the *presentation* of some stimulus serves as a reinforcer. They can also be shown to occur when responding is instead followed by a negative reinforcer, i.e., where the *termination* of the stimulus serves as a reinforcer. In that regard, Mele et al. (20), comparing across studies, suggested that fixed-ratio shock escape performance, in which responding terminated an ongoing shock, was disrupted at lower levels of ionizing radiation than those that were necessary to disrupt shock avoidance responding, in which responding postponed the onset of the shock delivery.

In addition to the importance of the reinforcement schedule *per se* as a determinant of toxic effect, the baseline rate of responding controlled by certain schedules of reinforcement may likewise modify behavioral toxicity, a phenomenon known as rate dependency. On fixed-interval schedules of reinforcement, response rates are typically lowest early in the interval when no reinforcement is available, and are highest at the time that reinforcement becomes available at the end of the interval. It is well known that a number of drugs increase low rates of responding early in the fixed-interval, while decreasing or leaving intact the higher rates of responding later in the interval. A similar effect is found with some, but not all toxicants. A study by Moser and MacPhail (23) found that the fungicide triadimefon increased the low rates of responding early in the interval on a fixed-interval schedule of reinforcement, while tending to decrease or have little effect on higher response rates occurring later in the interval, as shown in Fig. 2.

Likewise, in the study of Peele and Crofton (25) mentioned above, which contrasted the effects of permethrin and cypermethrin on a multiple schedule with four VI components, the authors reported that the response rate reductions produced by cypermethrin showed a dependence on the baseline levels of responding, with low response rates showing more pronounced sensitivity to disruption; permethrin's effects were not rate-dependent. Neither do the effects of lead (Pb) on fixed-interval performance appear to be rate-dependent; Pb exposure does not increase low rates of responding. In fact, Pb exposure even further increases the highest baseline rates of responding at the end of the interval on a FI schedule (4). As these examples collectively demonstrate, not only can the

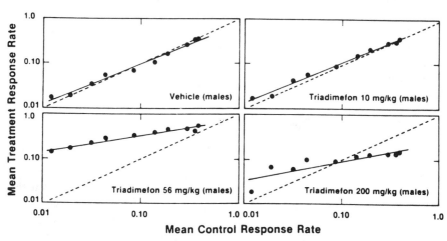

FIG. 2. Mean rates of responding (responses per sec) in successive tenths of the interval following various doses of triadimefon plotted as a function of corresponding mean local response rates under control conditions. *Dashed diagonal line:* the line of no effect (slope = 1). (From Moser and MacPhail, ref. 23, with permission.)

differences in the contingencies of reinforcement between schedules themselves modulate toxicant effects, but so can the differences in baseline rates of responding generated by a particular reinforcement schedule.

Under certain conditions, two other modulators of toxicity may be the type of reinforcement, i.e., negative versus positive, and the type of response. With respect to the former, Mele et al. (21) reported that the dose-effect function for ionizing radiation impairment of fixed-ratio escape performance in which responding was maintained by the termination of electric shock, was shifted to the right of the dose-effect function for fixed-ratio performance maintained by milk delivery, indicating the significance of the type of event maintaining responding. These data suggested that under some conditions, behavior maintained by positive reinforcement may be more easily disrupted by a toxicant than is behavior maintained under the same schedule but by negative reinforcement. In a study by Glowa (13) comparing response topographies, the effects of toluene exposure on a fixed-interval 60 sec schedule in which a nose-poke was the designated response were compared to those of a fixed-interval 60 sec schedule in which running was the response. Figure 3 shows that low toluene exposure con-

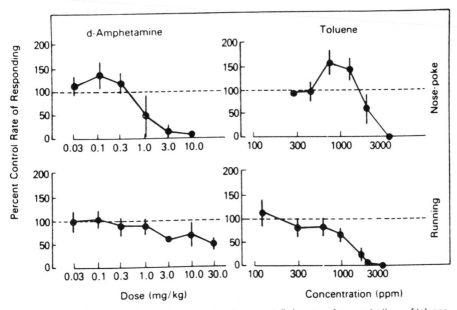

FIG. 3. Dose-effect curves for acute exposure to incrementally increased concentrations of toluene or d-amphetamine. The effect is expressed as the average percentage of the control rate of responding for two separate determinations in each of four mice, i.e., the average effect of eight determinations for each concentration or dose. Average control response rates (with the SEM expressed as a percent of control) were 0.748 (21%) and 0.894 (18%) for nose-poke, and 0.332 (14%) and 0.412 (5%) for running, for d-amphetamine and toluene, respectively. The abscissa shows concentration (toluene) during the 30 min time out (TO) preceding, or the dose (d-amphetamine given 5 min before responding was assessed.) (From Glowa, ref. 13, with permission.)

centrations increased nose-poking rates, while higher concentrations decreased those rates. In contrast, when the required response was running, only a decrease in response rate was produced by these same toluene exposure concentrations. A similar difference between response classes was seen following amphetamine administration in that study. Thus, reinforcement schedule contingencies, reinforcement schedule parameters, response rates maintained by a reinforcement schedule, types of reinforcers, and the nature of the response may all modulate the behavioral manifestations of toxicity. Such qualifying information certainly needs to be kept in mind when interpreting the outcome of any single study.

Assessment of Behavior in Transition Using Schedules of Reinforcement

As previously pointed out, the human behavioral environment is by no means static, and as such, frequent transitions of behavior may be mandated. One good example of such a requirement is the early years of school, as reading and other skills are developed entailing repeated acquisitions of new behavior or modifications of existing response repertoires. Schedules of reinforcement are particularly useful behavioral procedures for assessing the effects of toxicants on behavior in transition. One simple way to accomplish this is through repeated manipulations of the parameter values of a schedule of reinforcement.

A study carried out by Cory-Slechta (6) illustrates this point and its results revealed that effects resulting from lead exposure, which had seemingly vanished, were resurrected during behavioral transitions. In that study, rats exposed to low levels of lead in drinking water initially showed decreased rates of responding on a fixed-interval 60 sec schedule of reinforcement which disappeared over the course of 40 experimental sessions, at which time response rates of Pb-exposed and control rats were identical. Subsequently, the value of the FI schedule was changed from FI 60 sec to FI 5 min. This change induced increases in response rates during the first five sessions of the FI 5 min schedule in control rats, but marked decrements in response rates in Pb-exposed rats, as shown in Fig. 4A, even though any Pb effects on FI response rates had previously disappeared.

After response rates of controls and Pb-exposed rats equalized on the FI 5 min schedule, the reinforcement schedule was again changed, this time from the FI 5 min to an FI 5 min clock. Like the first transition, the transition to the FI 5 min clock evoked differential response rate changes in control and Pb-exposed rats over the initial five sessions, with response rates of control rats increasing, while response rates of Pb-treated animals were relatively unimpacted (Fig. 4B). These differential response rates of control and Pb-treated rats on the FI 5 min clock eventually disappeared, and the schedule was then returned to the original FI 1 min schedule with which all rats had had previous experience (Fig. 4C). During this transition, there were no differential effects between control and Pb-exposed rats on response rates. Taken together, these manipulations showed that the primary effects of Pb in these experiments appeared to be during the acqui-

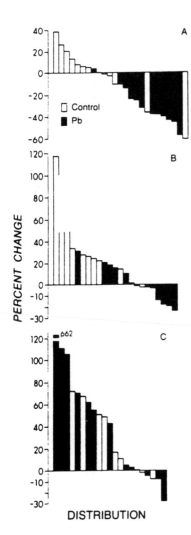

FIG. 4. Changes in overall response rate during the first five sessions following the transition from fixed-interval (FI)-1 min to FI-5 min (**panel A**), FI-5 min to FI-5 min clock (**panel B**), and FI-5 min clock to FI-1 min (**panel C**) plotted as a percent of the median response rate over the final five sessions in the preceding condition. Each bar shows the percent change for an individual animal (*open bars*, control rats; *solid bars*, Pb-exposed rats). Bars are arranged in order of decreasing percentage. (From Cory-Slechta, ref. 6, with permission.)

sition of new behavior, i.e., during behavioral transitions, and furthermore, that effects of Pb which had seemingly disappeared were only dormant, reemerging when the environment imposed transitions, i.e., alterations in the reinforcement contingencies.

A similar phenomenon shown to result from neonatal cadmium exposure was reported by Newland et al. (24). In that study, a series of increasing fixed-ratio values was imposed across sessions, i.e., a repeated transition. Effects of neonatal cadmium exposure emerged only during the transition from an FR 25 to an FR 75 schedule of reinforcement when rats that had been exposed to 3 mg/kg cad-

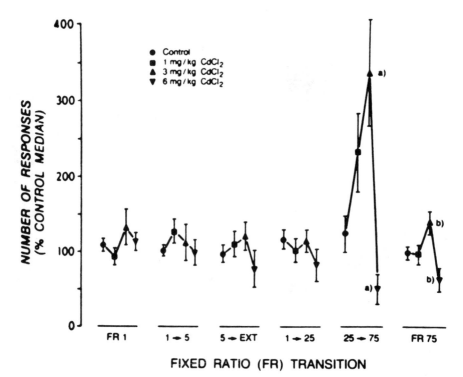

FIG. 5. Mean number of responses (±SEM) during a session as a percentage of the median obtained from a concurrent control group for the last fixed-ratio (FR) 1 session, four transitions, and the 19th FR 75 session. The rats exposed neonatally to 1 and 3 mg/kg CdCL$_2$ were compared with a concurrent control group, and those exposed to 6 mg/kg CdCl$_2$ were compared with a second concurrent control group. The error bars for the control groups were obtained by computing the rate for each control rat as a percentage of the control median, pooling both control groups, and then obtaining a standard error of the mean in the conventional manner. Points labeled with *a* are significantly different from control values based on Newman-Keuls comparisons of the raw response rates. In addition, the 3 and 6 mg/kg points for the last FR 75 session (labeled with *b*) differ from each other. The median control values for the control group concurrent with the 1 and 3 mg/kg rats were 127, 391, 262, 1747, 898, and 3,802 responses per session for the FR1, the four transitions, and the FR 75 session. Those values for the control group concurrent with the 6 mg/kg rats were 162, 768, 360, 2,197, 1,950, and 4,013 responses per session. (From Newland et al., ref. 24, with permission.)

mium chloride exhibited pronounced increases in response rates, while those exposed to 6 mg/kg showed a marked suppression of response rates relative to controls (Fig. 5). As with the lead study described above, these differences were attenuated over subsequent experimental sessions. Thus, parametric manipulations of schedules of reinforcement values provide a useful means of studying behavior in transition and have revealed, in the two cases cited here, behavioral effects of a toxicant which had either previously seemed to disappear or to be nonexistent.

Delineating Neurobiological Mechanisms of Toxicity

Two methods by which to begin to understand the neurobiological bases of a toxicant's effect on behavior are, first, to directly compare the effects of that toxicant to drugs with well-known mechanisms of action and, second, to study the interactive effects of the toxicant with drugs selectively interacting with particular neurotransmitter systems. The data resulting from such studies can be expected to provide guidance as to the neurochemical mediation of the toxicant's behavioral activity. For example, Moser and MacPhail (23) found that triadimefon and amphetamine similarly affected fixed-interval schedule-controlled behavior. Both resulted in inverse U-shaped dose-effect functions for overall response rate. In addition, both amphetamine and triadimefon showed rate-dependent effects, increasing low rates of responding early in the interval, while decreasing the higher response rates later in the interval. These facts, along with the similarity of the two compounds on other behavioral indices, led the authors to speculate that triadimefon might exert its effects through the same systems mediating the effects of CNS stimulants such as amphetamine, perhaps dopaminergic or noradrenergic neurotransmitter systems. Rastogi et al. (26) investigated the interactions of d-amphetamine and chlorpromazine with triethyltin in rats responding under a multiple FR 30 FI 5 min schedule of food reinforcement. They found that prior treatment with triethyltin produced an attenuation of the rate-increasing properties of d-amphetamine on the FI schedule, leading them to speculate that triethyltin may have some subtle long-term effects on adrenergic and/or dopaminergic systems. Information on the potential involvement of particular neurotransmitter systems in toxicant effects on behavior provides the rationale for the development of further hypotheses aimed at understanding both the neurobiological and behavioral mechanisms of toxicity.

Delineating Behavioral Mechanisms of Toxicity

Schedules of reinforcement also serve as useful behavioral baselines for evaluating behavioral mechanisms of toxicant action. This statement frequently provokes the question, "What exactly are behavioral mechanisms of toxicity?" In addressing that point, Thompson and Boren (29) stated that the rate, pattern, and form of current operant behavior are determined by certain antecedent factors, by the current stimulus conditions, and by the consequences maintaining behavior. Drugs or toxicants, as independent variables, may interact with any or all of these classes of factors in modifying behavior. An understanding of the behavioral mechanisms of drug or toxicant effects is sought, then, by asking the questions, "With which of these factors that regulate behavior has the drug (or toxicant) interacted? Has the drug (toxicant) altered the deprivation state (an antecedent variable), stimulus control (a current stimulus variable), response topography (a property of the response), or the reinforcer or schedule control (a consequence variable)?"

With respect to schedule-controlled behavior, one might frequently ask how a toxicant has altered either the rate or patterning of responding under the reinforcement schedule. In examining the rate suppressing effects of carbon monoxide exposure on differential reinforcement of low rate (DRL) performance, Ator et al. (1) examined the possibility that such effects were due to the loss of temporal stimulus control over performance, i.e., a loss of temporal discrimination, since elapsed time is thought to serve as the discriminative stimulus for responding on this schedule. However, whereas carbon monoxide indeed suppressed response rates on the DRL 21 sec schedule, the measures of temporal discrimination, that is, the distribution of times between successive responses [interresponse times (IRTs)], were not disrupted by the compound. This can be seen in Fig. 6 which shows the conditional probabilities of IRTs of various lengths for individual rats under control and carbon monoxide exposure conditions. Thus, Ator et al. (1) stated that their results are not consistent with interpreting a disruption of temporal discrimination as a behavioral mechanism of CO-produced response rate suppression.

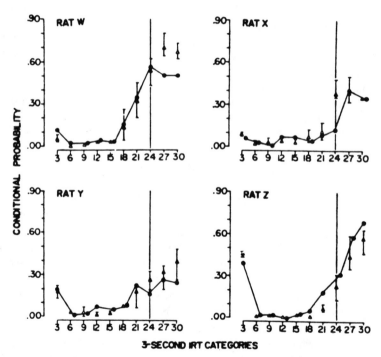

FIG. 6. Distribution of interresponse times (IRTs)/Opportunity (Op.) on differential reinforcement of low rate (DRL) 21 sec schedule of reinforcement with 750 ppm CO present (*circles*) and in control sessions. *Triangles* and *vertical lines* represent the means and ranges, respectively, of the IRTs/Op. in control sessions. All IRTs longer than 21 sec were reinforced as indicated by the vertical lines at the 21- to 24-sec category. The conditional probability of IRTs greater than 30 sec was always 1.0 and was not plotted. (From Ator et al., ref. 1, with permission.)

In contrast, the behavioral mechanisms underlying the well-documented changes in FI schedule-controlled responding produced by lead have been far more elusive. Such effects do not appear to be rate-dependent, as described above (4). Nor do the changes in response rate appear to reflect a loss of temporal discriminability as measures of the temporal patterning of responding in the interval, i.e., the postreinforcement pause time and the index of curvature do not reliably change with lead exposure (8,9). Further studies will obviously be necessary to achieve an understanding of the behavioral mechanisms by which Pb alters FI schedule-controlled performance.

Understanding behavioral mechanisms of toxicity is important not only to understand the behavioral bases of the effects of toxic compounds on schedule-controlled responding, but to understand toxicant effects on other behavioral processes as well. For example, in the case of lead, similar behavioral mechanisms could conceivably be involved in the alterations of FI schedule-controlled behavior and the deficits reported in learning processes. For example, a general disruption of the control of stimuli over behavior could disrupt both FI performance and learning. In the former case, this could reflect a loss of the control of response feedback on responding during the interval, while in the latter, it could reflect a loss of the control of the learning task stimuli upon behavior. Clearly, efforts designed to elucidate behavioral mechanism(s) of toxicant effects, including those for changes in schedule-controlled behavior, will provide the most rapid advancement to an understanding of and capabilities for dealing with behavioral toxicity.

Devising Behavioral or Physiological Antidotes to Behavioral Toxicity

As behavioral baselines, schedules of reinforcement also offer the opportunity to determine the efficacy of antidotes to toxicant effects of both a behavioral and physiological nature, which have obvious clinical relevance. Cory-Slechta and Weiss (7) examined the ability of CaEDTA, a chelating agent which binds metals and accelerates their excretion, to reverse lead-induced increases in response rates on FI reinforcement schedules. Since the most consistent effect of Pb on the FI schedule is to increase the frequency of short IRTs, the change in the proportion of short IRTs before and after a 5-day administration of CaEDTA served as the primary dependent variable in that study. As shown in Fig. 7, contrary to what might be predicted, CaEDTA administration did not reverse the Pb-induced increase in the frequency of short IRTs. In fact, CaEDTA administration even increased further the proportion of short IRTS, as if it were somehow enhancing Pb toxicity. These findings were consistent with the fact that the 5-day course of CaEDTA chelation produced no net loss of Pb from brain, and a single injection of CaEDTA actually doubled the brain Pb concentration. Such results certainly suggest that further studies of CaEDTA chelation for reversing the behavioral impact of Pb are warranted before prophylactic CaEDTA chelation becomes commonplace in human Pb-exposed populations.

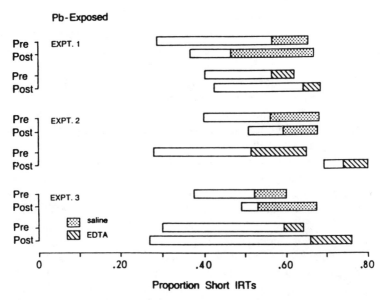

FIG. 7. Modified stem and leaf plots (minus the 10th and 90th percentile bars) of the proportion of short interresponse times (IRTs) of Pb-exposed animals for the five sessions prior to (pre) and the final five session following (post) injections of saline (*stippled*) or CaEDTA (*diagonals*). In each stem and leaf plot, the shaded area encompasses the 50–75th percentiles of short IRTs values, and the unshaded area represents the 25–50th percentile; the group median value is shown by the point where the shaded and unshaded areas meet. (From Cory-Slechta and Weiss, ref. 7, with permission.)

In another study, Chambers and Chambers (2) found that high levels of the centrally acting muscarinic receptor antagonist atropine could partially attenuate the rate-decreasing properties of paraoxon on fixed-ratio schedule-controlled responding.

Although to date no such studies appear to have been done, the same types of experiments could be used to evaluate "behavioral therapeutic" approaches to behavioral toxicity. For example, if one suspected that a toxicant was producing changes in rates of responding on a schedule of reinforcement by minimizing the strength of the conditioned reinforcer, it might be possible to remedy such a situation by arranging for the conditioned reinforcer to be paired more frequently with the unconditioned or primary reinforcer. By subsequently determining whether such a manipulation then eliminates or attenuates the rate-altering properties of the toxicant, a better understanding of the behavioral mechanism of toxicant action will have been achieved. In addition, if successful, such strategies have obvious clinical implications. For example, if this behavioral manipulation had proven successful in reversing Pb-induced changes in response rates in experimental animals, such a strategy might be attempted in a classroom situation for children who had been identified as having elevated Pb burden and associated classroom performance problems. Obviously, the better defined the behavioral and neurobiological mechanisms of behavioral toxicity are, the more precise and efficacious the behavioral and physiological therapeutics can be. Conversely, the use of such behavioral and physiological interventions can provide guidance and assistance in further delineating behavioral and neurobiological mechanisms of toxicity. An understanding of behavioral mechanisms of toxicant action and the concept of behavioral therapeutics thus go hand-in-hand and offer alternatives to consider, particularly where further reduction of toxicant exposure may not be realistically achieved.

Evaluating Time Course of Toxicant Effect and Comparisons Between and Within Classes of Compounds

Many different studies have utilized schedule-controlled performance as a baseline from which to examine the time course of toxicant action upon behavior. Normally, after behavior stabilizes on a particular reinforcement schedule, it can remain relatively invariant for prolonged periods of time in the absence of any imposed parametric changes. This relative stability of performance over time permits the determination of delayed effects of toxicants (time to onset of effect), the duration of effect, and the reversibility of that effect following exposure cessation. Moreover, if a toxicant induces behavioral changes only with cumulative dosing, one can investigate the cumulative dose required to produce functional changes by intermittently dosing across experimental sessions until changes in schedule-controlled behavior become evident. By performing such experiments using multiple rather than simple reinforcement schedules, effects on different types of schedule-controlled performances can be evaluated concurrently.

Such an approach was utilized by Wenger et al. (31) to examine the effects of trimethyltin (TMT). In those experiments, a single ip injection of trimethyltin chloride was administered to mice after performance on a multiple fixed-ratio 30, fixed-interval 600 sec schedule of milk presentation had stabilized. Behavior was evaluated 3 hr after TMT administration and every 24 hr thereafter. The authors found that a dose of 3 mg/kg TMT disrupted responding for periods up to 6 or 7 weeks, as can be seen in Fig. 8. Moreover, by utilizing these same reinforcement schedules in another mouse strain, the authors were able to provide information on comparative strain sensitivity to TMT, with resultant data suggesting a greater sensitivity of the BALB/c mouse to TMT effects as compared to the C57BL/6N strain.

Similar procedures have also been used to compare the time course and differential potency of toxicants both within and between classes of compounds. For example, Wenger (30) also examined the time course of triethyltin (TET) on the multiple fixed-ratio, fixed-interval schedule of reinforcement described above and found differences in both potency and the period of behavioral activity, with TET being far less potent than TMT and exhibiting a differential time

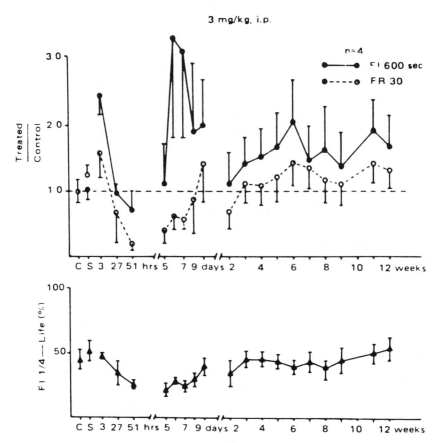

FIG. 8. The effect of 3 mg/kg trimethyltin (TMT) on the rate of responding and fixed-interval (FI) quarter-life in C57BL/6N mice responding under the multiple FR30 FI600 sec schedule. *Abscissa:* time after TMT administration; *ordinate* (**top**): the rate of responding after TMT administration divided by the mean rate of responding prior to TMT administration; *ordinate* (**bottom**): FI quarter-life expressed as a percentage of the total 600-sec interval. Points above 3 to 51 hr and above 5 to 9 days represent the mean of each daily test session for the four mice. Points above 2 to 12 weeks represent the weekly means of the four mice. Points above C and S represent noninjection control and saline injection (3 hr), respectively. Mean control rates of responding were 0.20 and 1.33 responses per sec for the FI and fixed-ratio (FR), respectively. Vertical bars indicate ±SE. (From Wenger et al., ref. 31, with permission.)

course of effect. Glowa (12) compared the effects of a variety of solvents on performance under a fixed-interval 60 sec schedule of reinforcement and found a wide range of concentrations over which the rate-decreasing effects of solvents occur, with their comparative sensitivity exhibited in Fig. 9. Moser and Balster (22) compared exposures to various volatile agents on a fixed-interval 60 sec schedule in the mouse and found evidence supporting the assertion that two industrial solvents, namely toluene and 1,1,1-TCE (trichlorethane) were qualitatively similar to those of two volative CNS depressants, halothane and ethanol.

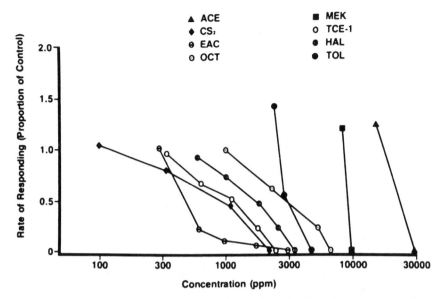

FIG. 9. Average rates of responding (proportion of control) as a function of exposure concentration (ppm) for the rate-decreasing effects of cumulative exposure to acetone (ACE), carbon disulfide (CS$_2$), ethyl acetate (EC), n-octane (OCT), methyl ethyl ketone (MEK), trichloroethane (TCE-1), halothane (HAL), and toluene (TOL), for fixed-interval (FI) 60 sec responding in mice. (From Glowa, ref. 12, with permission.)

Issues Regarding Schedules of Reinforcement in Neurotoxicology

Several issues have arisen repeatedly in the neurotoxicology community concerning the use of schedule-controlled performance in behavioral toxicology and the interpretation of data from such studies. One such issue seems to derive from the lack of any obvious face validity of these procedures, resulting in the frequently posed question: "What does schedule-controlled behavior mean?" In essence, this problem seems to represent a lack of understanding of the fact that schedule-controlled performance is a major constituent of learned behavior, and also plays a major role in other behavioral processes. In contrast, behavioral processes such as learning and remembering have a more obvious meaning to a broader range of the scientific community. As a result, the scientific community also attributes more importance to, and has a better understanding of, the consequences of impaired learning and remembering processes. A grasp of the consequences of altering schedule-controlled behavior is more elusive.

Part of this failure may be due to the tendency to consider schedule-controlled behavior as an independent entity. If, instead, schedule-controlled behavior is considered in the context of human behavior, and the importance of response rate and pattern changes to other behavioral processes is pointed out, then the significance of schedule-controlled performance becomes evident even to those without extensive behavioral training. A study by Mele et al. (20) helps to illustrate this point. In that study, monkeys were exposed to PCBs early in life and tested at 60 months of age on fixed-interval schedules, with an increasing interval value across experimental sessions. There were no performance differences on the FI schedules between control and PCB-exposed monkeys during that time. Subsequently, an FI 60 sec schedule was imposed with a reinforcement omission procedure wherein 25% of the scheduled reinforcers were omitted randomly. While there were no differences in response rates during the FIs following reinforcement delivery, the high dosed PCB group had significantly higher response rates and greater interanimal variability of response rate than did controls during the FIs following those FIs in which reinforcement had been omitted, as shown in Fig. 10.

Reinforcement omission procedures analogous to those described above might be fairly typical occurrences in a classroom situation, i.e., in a human context where certainly every instance of a given behavior will not be followed by reinforcement, but instead reinforcer dispensing may depend on how busy the teacher is at any particular time. If one can extrapolate directly from monkeys to children, such a situation might then be expected to foster higher response rates and greater variability in children with elevated PCBs in the classroom in periods following reinforcement omission. This would be disruptive for the entire class and could require the teacher to waste instructional time on disciplinary activity. If that particular response happened to be time out of seat and was being inadvertently maintained by any sort of teacher attention which followed it, the teacher might then decide to try to ignore such behavior. As a reinforcement omission procedure, this could then further increase the rate of a behavior which was clearly incompatible with the learning demands of the classroom.

Studies examining the behavioral toxicity of compounds with protectant effects against known neurotoxicants provide another set of examples. Liu and Shi (19) examined the effect of the pyridinium aldoxime cholinesterase reactivator HI-6, which, of the organophosphate antidotes, is among the best tolerated by humans. However, the compound suppressed fixed-ratio response rates in a dose-dependent manner for periods up to 2 hr postadministration. Moreover, these effects occurred in the absence of any overt signs of toxicity. If HI-6 was administered to soldiers to protect against organophosphate poisoning during military battles, it might prevent such poisoning but render him/her unable to perform effectively those functions necessary to protect themselves in battle.

In addition to putting schedule-controlled behavior into a human context, it is also important to remember the significance of schedule-controlled behavior to other behavioral processes. In certain situations, for example, changes in rates of responding could in fact underlie what are erroneously interpreted to be changes in learning and/or memory. If a toxicant increased response rates, these could have detrimental effects on learning or memory paradigms. For example, in a learning situation in which various stimuli were presented and a response to the correct stimulus was required, an increase in response rate might also

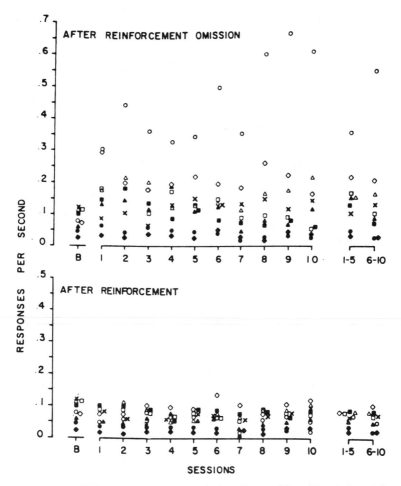

FIG. 10. Individual FI 60 sec response rates under baseline conditions (*B*) and after reinforcement and reinforcement-omission. Data are for the five control and four 2.5 ppm postexposure PCB monkeys for experiment II. (From Mele et al, ref. 20, with permission.)

decrease latency to respond to the stimuli, with a response occurring before the subject had had adequate time to assess the relevant stimulus conditions. In a delay task, such an effect on response rate might result in too short a latency following the delay interval, with the net result being a decline in overall accuracy of performance which could be mistakenly interpreted as a memory deficit. Decreases in response rate might likewise have detrimental consequences under such conditions. In either a learning or memory paradigm, increased latencies to respond after a delay interval ends may effectively increase the delay interval across which the correct response is to be remembered and thus impair accuracy of performance.

A related issue often raised in regard to studying schedule-controlled performance as a behavioral baseline is the definition of adverse effects: what constitutes an adverse effect? Like the previous issue, the definition of adverse effects is best addressed by considering such behavioral outcomes in a human context. For example, it is well known that low levels of lead exposure can produce response rate increases on fixed-interval schedules of reinforcement. The frequently raised question regarding such an outcome is whether this should be defined as behavioral toxicity. After all, what difference does it make since the rat still obtains all the available reinforcers at the appropriate time? While some behavioral toxicologists have adopted the position that any change in behavior is an adverse effect, this argument is not particularly instructive, and is not convincing when one is faced with what appear to be temporary changes in response rates.

Even from a theoretical point of view, either increases or decreases in response rate on fixed-interval schedules can be considered impaired performance. Normally, the fixed-interval reinforcement schedule generates what is described as a scalloped pattern of responding, with low rates of responding early in the interval when no reinforcers are available, and a gradually increasing rate of responding as the time of reinforcement availability approaches at the end of the interval. This increasing response rate across the interval guarantees that reinforcement delivery will occur with minimal delay upon its availability. Further increases in rates of responding under these conditions reflect extremely inefficient performance in that the normal response pattern already guarantees the minimum delay in reinforcement delivery; further increases in response rate do not further minimize that delay. Decreased response rates are of course detrimental since they can further increase the delay to reinforcement and decrease the rate of reinforcement. Since the rate and pattern of performance engendered by reinforcement schedules are typically that which minimizes delay of reinforcement, similar arguments apply to response rate changes on other schedules as well.

It is necessary to remember that the responses studied in operant chambers under schedules of reinforcement are arbitrarily chosen responses. Putting response rate changes in a human behavioral context is another way to understand the importance of changes in response rate under reinforcement schedules. It is not difficult to think of situations in the human behavioral environment where alterations in rates of responding *per se* would have deleterious consequences. On a manufacturing assembly line, it is evident that increases in rates of responding may actually cause physical harm to the subject and increase the probability of becoming entangled in the machinery. Decreases in response rates may have a different type of consequence. In a situation such as an assembly line where productivity is closely monitored, decreased rates of responding may be grounds for termination of employment. In fact, there are probably few types of employment conditions in which productivity is not somehow monitored. Certainly increases in rates of inappropriate classroom behaviors, or conversely, decreases in rates of learning-compatible responses, are undesirable.

An additional point with respect to defining what constitutes an adverse change in schedule-controlled behavior arises when the performance modification appears to be temporary. This often leads to the contention that such changes are insignificant. However, the temporary nature of an effect may be quite misleading, as was illustrated on the basis of studies described above. Further modifications, such as schedule transitions with lead (6) and reinforcement omission procedures with PCBs (20), were able to resurrect what had appeared to be only temporary effects of Pb in the first case, and to uncover treatment-related differences resulting from PCBs in the second case. Thus, effects that seem temporary may instead only be dormant.

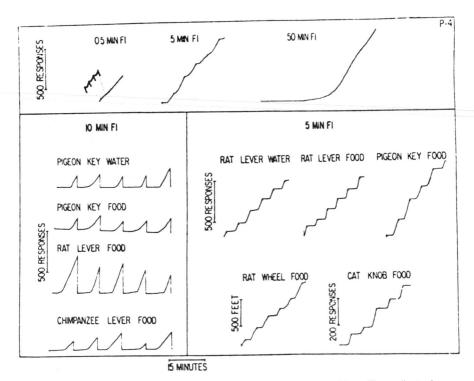

FIG. 11. Generality of fixed-interval (FI) performance under various conditions. The *ordinate* shows the cumulative number of responses; time is represented on the *abscissa*. The typical performance is characterized by little or no responding after reinforcement, followed by a gradually accelerating rate of responding. In all examples, a FI schedule of food or water presentation was in effect. Shown is a performance under a 10 min FI schedule (*lower left*). Each reinforcement delivery reset the pen to the baseline. The comparability of performance of the pigeon, rat, and chimpanzee, either pecking a key or pressing a lever, is evident. Likewise, during a 5 min FI schedule (*lower right*), the comparability of performance of rats, pigeons, and cats, pecking a key, pressing a lever, wheel-running, or pulling a knob is illustrated. (From Kelleher and Morse, ref. 15, with permission.)

A third related issue is cross-species extrapolation of changes in schedule-controlled responding. What is the generality of schedule-controlled behavior? Does schedule-controlled performance engendered in experimental animals such as rats and mice have any correspondence to human schedule-controlled behavior? Such questions were first addressed many years ago. Innumerable studies have documented the comparability of various types of schedule-controlled performances across a wide variety of species, including humans (10,14,15,16,27,28). The point was particularly well made in a monograph by Kelleher and Morse (15), in an illustration shown here as Fig. 11. It shows the comparability of fixed-interval performance across different species, across different responses, and across different reinforcers. In fact, without the descriptive labels accompanying each individual cumulative record shown, one would be hard pressed to identify any of these experimental parameters on the basis of the performance itself. Thus, it is clear that a basis for extrapolating such results across species already exists.

Summary

All learned operant human behavior occurs in time, and both the rate and the pattern of that behavior are governed by the prevailing schedules of reinforcement. As a result, schedule-controlled behavior *per se* is a major constituent of human behavior. The nature of responding, the rate and pattern generated by reinforcement schedules, can also play a major role in other behavioral processes such as learning or remembering, behavioral processes which also occur in time. For these reasons, a true understanding of the behavioral toxicity of a compound requires an understanding of its impact on schedule-controlled behavior. Further support for this assertion is provided by studies which have demonstrated that the profile of behavioral effects produced by a toxicant may, in fact, be determined or modulated by several different reinforcement schedule considerations, including the nature of the governing reinforcement schedule itself, the baseline rates of responding generated by that schedule, and the transition of schedule parameter values over time. To understand the relevance of changes in schedule-controlled behavior, it is only necessary to consider the results of laboratory studies in a human behavioral context, remembering that the lever press response is an arbitrarily chosen one, and that schedule-controlled behavior shows remarkable similarity across a wide range of species, including humans.

ACKNOWLEDGMENTS

Partial support for this chapter was provided by grant #ES05017 from the National Institute of Environmental Health Sciences.

REFERENCES

1. Ator NA, Merigan WH, McIntire RW. The effects of brief exposures to carbon monoxide on temporally differentiated responding. *Environ Res* 1976;12:81–91.
2. Chambers JE, Chambers HW. Short-term effects of paraoxon and atropine on schedule-controlled behavior in rats. *Neurotoxicol Teratol* 1989;11:427–432.
3. Colotla VA, Bautista S, Lorenzana-Jimenez M, Rodriguez R. Effects of solvents on schedule-controlled behavior. *Neurobehav Toxicol* 1979;1(Suppl. 1):113–118.
4. Cory-Slechta DA. Behavioral toxicity of lead: problems and perspectives. In: Thompson T, Dews PB, Barrett JE, eds. *Advances in behavioral pharmacology,* vol. 4. New York: Academic Press, 1984;211–255.
5. Cory-Slechta DA. Effects of lead exposure beyond weaning on fixed-ratio performance. *Neurobehav Toxicol Teratol* 1986;8:237–244.
6. Cory-Slechta DA. Exposure duration modifies the effects of low level lead on fixed-interval performance. *Neurotoxicology* 1990;11:427–442.
7. Cory-Slechta DA, Weiss B. Efficacy of the chelating agent CaEDTA in reversing lead-induced changes in behavior. *Neurotoxicology* 1989;10:685–698.
8. Cory-Slechta DA, Weiss B, Cox C. Delayed behavioral toxicity of lead with increasing exposure concentration. *Toxicol Appl Pharmacol* 1983;71:342–352.
9. Cory-Slechta DA, Weiss B, Cox C. Performance and exposure indices of rats exposed to low concentrations of lead. *Toxicol Appl Pharmacol* 1985;78:291–299.
10. Dews PB, Wenger G. Rate-dependency of the behavioral effects of amphetamine. In: Thompson T, Dews PB, eds. *Advances in behavioral pharmacology,* vol. 1. New York: Academic Press, 1974;167–227.
11. Ferster CB, Skinner BF. *Schedules of reinforcement,* New Jersey: Prentice Hall, 1957.
12. Glowa JR. Behavioral effects of volatile organic solvents. In: Seiden LS, Balster RL, eds. *Behavioral pharmacology: the current status.* New York: Alan R. Liss, 1985;537–552.
13. Glowa JR. Comparisons of some behavioral effects of d-amphetamine and toluene. *Neurotoxicology* 1987;8:237–248.
14. Holland JG. Human vigilance. *Science* 1958;128:61–67.
15. Kelleher RT, Morse WH. Determinants of the behavioral effects of drugs. In: Tedeschi DJ, Tedeschi RE, eds. *Importance of fundamental principles in drug evaluation.* New York: Raven Press, 1969;383–405.
16. Laties VG, Weiss B. Effects of a concurrent task on fixed-interval responding in humans. *J Exp Anal Behav* 1963;6:431–436.
17. Leander JD, MacPhail RC. Effect of chlordimeform (a formamidine pesticide) on schedule-controlled responding. *Neurobehav Toxicol* 1980;2:315–321.
18. Levine T. Effects of carbon disulfide and FLA-63 on operant behavior in pigeons. *J Pharmacol Exp Therap* 1976;199:669–678.
19. Liu W-F, Shih J-H. Neurobehavioral effects of the pyridinium aldoxime cholinesterase reactivator HI-6. *Neurotoxicol Teratol* 1990;12:73–78.
20. Mele PC, Bowman RE, Levin ED. Behavioral evaluation of perinatal PCB exposure in rhesus monkeys: fixed-interval performance and reinforcement omission. *Neurobehav Toxicol Teratol* 1986;8:131–138.
21. Mele PC, Franz CG, Harrison JR. Effects of ionizing radiation on fixed-ratio escape performance in rats. *Neurotoxicol Teratol* 1990;12:367–373.
22. Moser VC, Balster RL. The effects of inhaled toluene, halothane, 1,1,1-trichlorethane, and ethanol on fixed-interval responding in mice. *Neurobehav Toxicol Teratol* 1986;8:525–531.
23. Moser VC, MacPhail RC. Neurobehavioral effects of triadimefon, a triazole fungicide, in male and female rats. *Neurotoxicol Teratol* 1989;11:285–293.
24. Newland MC, Ng WW, Baggs RB, Gentry GD, Weiss B, Miller RK. Operant behavior in transition reflects neonatal exposure to cadmium. *Teratology* 1986;34:231–241.
25. Peele DB, Crofton KM. Pyrethroid effects on schedule-controlled behavior: time and dosage relationships. *Neurotoxicol Teratol* 1987;9:387–394.
26. Rastogi SK, McMillan DE, Wenger GR, Chang LW. Effects of triethyltin and its interaction with d-amphetamine and chlorpromazine on responding under a multiple schedule of food presentation in rats. *Neurobehav Toxicol Teratol* 1985;7:239–249.

27. Richelle M. Combined action of diazepam and d-amphetamine on fixed-interval performance in cats. *J Exp Anal Behav* 1969;12:989–998.
28. Tews PA, Fischman MW. Effects of d-amphetamine and diazepam on fixed-interval, fixed-ratio responding in humans. *J Pharmacol Exp Therap* 1982;221:373–383.
29. Thompson T, Boren, JJ. Operant behavioral pharmacology. In: Honig WK, Staddon JER, eds. *Handbook of operant behavior.* New Jersey: Prentice-Hall, 1977;540–569.
30. Wenger GR. The effects of trialkyl tin compounds on schedule-controlled behavior. In: Seiden LS, Balster RL, eds. *Behavioral pharmacology: the current status.* New York: Alan R. Liss, 1985;503–518.
31. Wenger GR, McMillan DE, Chang LW. Behavioral effects of trimethyltin in two strains of mice. *Toxicol Appl Pharmacol* 1984;73:89–96.

Neurotoxicology, edited by Hugh Tilson
and Clifford Mitchell. Raven Press, Ltd.,
New York © 1992.

12

Developmental Neurotoxicology

Charles V. Vorhees

*Institute for Developmental Research, Children's Hospital Research Foundation and
Departments of Pediatrics and Environmental Health, University of Cincinnati,
Cincinnati, Ohio 45229*

Developmental neurotoxicity refers to adverse health effects of exogenous agents acting on neurodevelopment. An agent inducing such an effect is, therefore, a developmental neurotoxin.[1]

While development may be broadly considered as encompassing all stages of ontogeny from conception to senescence, a more restricted usage will be employed here. The present discussion emphasizes stages of development from conception through infancy although occasionally events extending as far as puberty will be included.

The term developmental neurotoxicity has shared meaning with research categorized as behavioral teratogenicity or neuroteratogenicity. As a term, developmental neurotoxicity has the advantage of being more inclusive in its implications than these other terms, and perhaps less controversial (10). However, in reality there are only minute, if any, differences in meaning between them (60,66).

SPECIAL CONSIDERATIONS IN DEVELOPING ORGANISMS

A number of neurotoxins have proven to be more toxic to the developing nervous system than to that of the adult. This has been shown for methylmercury, which can induce congenital mental retardation and cerebral palsy while inducing no or few and reversible neurological symptoms in these children's mothers (8). Similar effects have been shown for lead, which induces psychological deficits and interference with academic development in children of mothers who are asymptomatic (14). In the fetal alcohol syndrome, ethanol induces developmental delay, mild to moderate mental retardation, attention deficit–hyperactivity dis-

---

[1] The present terminology follows the precedent of Koob and Bloom (34), of D'Amato, Lipman, and Snyder (13), of *Dorland's Medical Dictionary* (24th edition), and of the evolution of the meaning of toxin and neurotoxin according to science lexographers at Merriam-Webster (R. W. Pease, personal communication).

order, and reaction time and coordination impairments in children whose mothers are not sufficiently alcoholic to exhibit ethanol-induced disease (16). Finally, in an important and dramatic recent example, the toxic effects of maternal exposure to the prescription medication isotretinoin (Accutane) have been studied. At maternally therapeutic doses, isotretinoin not only induces a high prevalence of congenital malformations (35,36), but new data reveal that it also induces a high prevalence of CNS dysfunction (2). The effects range from profound mental retardation to neuropsychological impairments that are not revealed with psychometric instruments of global intelligence. Other examples include the special vulnerability of the developing brain to polychlorinated biphenyls (PCBs) (54) and to ionizing radiation (50).

While the mechanism of action of these agents has yet to be determined, the basis for the developing nervous system's unique vulnerability may be understood based on the ontological processes which regulate growth and development. One of the first stages of nervous system vulnerability occurs during early organogenesis when neuronal proliferation is rapid. This stage, neurogenesis, is followed by cell migration in which nascent neurons travel to their destinations through intervening cell layers, some guided by radial glia. This is followed by differentiation into specific neuronal phenotypes. Differentiation includes neurite growth (elongation of axonal and dendritic process at the growth cone through extension, substrate adhesion, and actin-driven traction), and arborization of dendritic spines. Neurotransmitter expression also occurs during this phase. These processes culminate in synaptogenesis, and integration of synaptic function (45). These processes are regulated by developmental genes and are also influenced by surrounding tissues (epigenetic factors). Whether the regulatory processes are gene-mediated or the result of the action of growth factors and various morphogens, such as retinoic acid, all of these signals are targets for potential interference by drugs and environmental agents. Therefore, vulnerability to perturbation is a function of the complexity, timing, and sequential ordering of these developmental processes.

Perhaps one of the most crucial characteristics of the developing CNS which causes it to be especially vulnerable is the immaturity of the blood-brain barrier (48). The consequence of the inability of the embryonic and fetal brain to exclude exogenous (blood-borne) molecules is that drugs and environmental agents have ready access to the embryonic CNS once they cross the placenta. Since the placenta is permissive to the transport of many molecules, agents in maternal circulation will reach the developing brain in concentrations at least equal to those attained in the embryo or fetus as a whole. Once in the brain, such compounds have access to the host of critical events in neuronal ontogeny mentioned above. Thus, the developing CNS possesses greater vulnerability to disruption than is the case for the mature organism because exclusionary processes are not yet developed at the time when cells are rapidly dividing, migrating, differentiating, and connecting. Deflection of these processes can occur despite the fact that the tissues involved possess excess capacity. Neuronal reserves exist because cellular proliferation generates greater numbers of cells than are ultimately needed. Later, these excess cells are pruned to sustainable numbers through apoptosis, a process apparently controlled by their target tissues (45).

Aside from the processes of normal ontogeny, cells possess an array of repair mechanisms which can be triggered to help them maintain their developmental course in the face of adversity. Yet despite these corrective forces, some agents have proven to be potent developmental neurotoxins, such that some types of injury are not able to be remediated and/or produce cell losses that exceed a critical threshold necessary to sustain the normal sequence of morphogenetic events.

An important ramification of the heightened sensitivity of the nervous system is that investigations of developmental neurotoxicity may be valuable for predicting general neurotoxicity. In this respect, developmental neurotoxicity may represent a useful biomarker for many types of CNS toxicity (40). The implications of this perspective should be considered carefully. If developmental neurotoxicity presages adult neurotoxicity, then it follows that investigations of developmental neurotoxicity should precede investigations of adult neurotoxicity in research on preclinical/premarketing safety assessment of new products. This stands in contrast to current regulatory practice, which generally accords primacy to adult neurotoxicity and invokes developmental neurotoxicity assessments only when evidence of adult neurotoxicity is obtained first.

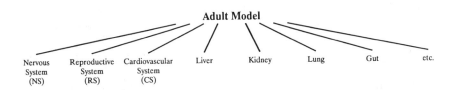

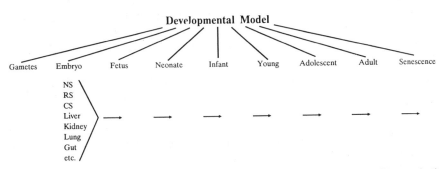

FIG. 1. In the adult model, the reproductive system becomes the target organ for effects on both reproduction and development, including teratogenesis and developmental neurotoxicity. By contrast, in the developmental model all organs are viewed in the context of their stage of maturation. In this view treatment effects are seen as influences on developmental processes rather than as pathological changes in fully developed target organs.

The contrast of this approach to the traditional view is depicted in Fig. 1. Current regulations emphasize adult toxicity because it has been around longer and is more deeply entrenched, and because of the view that development is a specialty, rather than a mainstream area. Using developmental neurotoxicity tests as bellwethers for adult toxicity would, therefore, require a substantial reorganization of current concepts. Reorganization of such a complex process might prove unsettling and would have to overcome the inertia of current safety assessment guidelines. However, if it leads to improved risk assessment, it may be justified.

RATIONALE FOR FUNCTIONAL EMPHASIS WITHIN DEVELOPMENTAL NEUROTOXICITY

The principle reason that developmental neurotoxicology has focused on functional manifestations of CNS injury is straightforward: functional assessment reflects the final output of the CNS and involves processes too complex to be comprehensively assessed individually. Furthermore, function can be used as a metric of the magnitude and meaning of a dysfunction. No other form of analysis can capture the nature and impact of a disorder as well, since this is the level at which health and well-being are experienced. While measurement of function can be complex in its own right and subject to intervening influences, it remains central to the characterization of all forms of abnormality; in short, it is the yardstick of brain integrity. In addition, since functional assessment is noninvasive, a second strength is that it can be performed repeatedly. In humans, functional evaluation may be one of the few types of assessment possible. Finally, functional assessment often detects adverse effects well beyond the reach of current cellular or biochemical approaches. Despite substantial progress in cellular and molecular neuroscience, functional assessment still serves as the vanguard of neurotoxicity detection. Functional neurotoxicity may ultimately be buttressed by the availability of improved neuroimaging techniques, but these will not replace the need to measure function. Together with behavioral indices, neuroimaging should further advance the entire process of neurotoxicological detection and characterization. Regarding humans, Hartman (26) argues that the combination of neuropsychological, psychometric, neuroimaging, and neurophysiological testing provides the key ingredients to understanding clinical neurotoxicity. This comprehensive approach is an important one and should be followed more often.

The next logical step from this, especially pertaining to use of neuropsychological approaches to assessing neurotoxicity, is to extrapolate this approach to preclinical assessment in animals. It is the thesis of this author that exactly this approach is required in animal neurotoxicologic evaluation, i.e., neurobehavioral tests are needed which are organized around CNS structure-function relationships analogous to those constructed for human neuropsychological tests. These should be amplified by techniques for neuroimaging as the resolution of instruments improves to permit their application to small laboratory species. Tests are already available for neurophysiological assessment in several species (19). The idea of organizing animal neurotoxicity assessment on a different basis instead of the current "off-the-shelf" approaches represents a new way of viewing this issue. This idea will be developed later in the chapter.

This chapter focuses on (a) a description of the organizing concepts currently extant in developmental neurotoxicity; (b) presentation of a series of examples of developmental neurotoxins which illustrate the key organizing concepts; (c) a brief discussion of government regulations relevant to developmental neurotoxicity assessment; and (d) a discussion of the neuropsychological approach to CNS assessment and its implications for preclinical evaluation in nonhuman mammalian species.

TABLE 1. *Organizing concepts for developmental neurotoxicity*

1. *Genetic Determination:* The type and magnitude of neurotoxic effect depends on the genetic milieu of the organism.
2. *Critical Periods:* The type and magnitude of neurotoxic effect depends on the stage of development of the organism when affected.
3. *Specific Mechanisms:* Agents that are neurotoxic act on the developing nervous system by specific molecular mechanisms to alter development.
4. *Type of Response:* Developmental neurotoxicity is ultimately expressed as impaired cognitive, affective, social, arousal, reproductive, and sensorimotor behavior, as delayed behavioral maturation of these capacities, or related indices of compromised behavioral competence.
5. *Target Organ Access:* The type and magnitude of the neurotoxic effect depends on the agent or a metabolite reaching specific sites in the developing nervous system.
6. *Dose-Response:* The type and magnitude of the neurotoxic effect depends on the dose of the agent reaching the site in the nervous system where it alters critical developmental events.
7. *Environmental Determination:* The type and magnitude of a neurotoxic effect depends on the nature of the environmental influences it receives, including both prenatal and postnatal influences.
8. *Types of Developmental Neurotoxins:* Only those agents that are central nervous system teratogens or are neuroactive are capable of producing developmental neurotoxicity.
9. *Response Relationships:* For prenatal exposure, developmental neurotoxic effects are demonstrable at doses below those causing gross malformations if the agent is capable of neurotoxicity. For postnatal exposure, no similar systematic relationship is known.
10. *Maximum Susceptibility:* The period of susceptibility to developmental neurotoxic effects is isomorphic with CNS development, and the period of maximum susceptibility corresponds to the period of maximum susceptibility of the CNS to structural injury, i.e., during neurogenesis. This relationship is modified for those instances where the CNS has a later window of vulnerability produced by specific late occurring developmental processes which are only vulnerable when present.
11. *Limits of Susceptibility:* Not all agents producing developmental toxicity are capable of producing developmental neurotoxicity.
12. *Preconception and Transgenerational Effects:* Some agents can induce damage through the germ line that is transmissible to their offspring following exposure to the male or female parent prior to conception and that are manifest in the progeny. Such effects may occur across one or more generations.
13. *Patterns of Exposure:* Some agents produce developmental neurotoxicity as a result of total exposure (area under the dose-time curve), while others produce their effects based on peak concentration irrespective of cumulative exposure.

PRINCIPLES OF DEVELOPMENTAL NEUROTOXICITY

Through a combination of extrapolation from organizing concepts derived from research in experimental teratology and studies directly within the field of developmental neurotoxicology, a set of conceptual principles has been derived which serves to define the basic attributes of mammalian CNS toxic response. These are presented in Table 1. In the sections that follow several of these principles will be illustrated with examples of established developmental neurotoxins. It should be noted that the first seven of the principles listed in Table 1 are widely accepted, while the remainder represent areas which possess varying degrees of certainty. For these, the primary reason for their inclusion is to stimulate research testing their validity. These are in the realm of hypotheses: confirmation or disconfirmation of each one is needed. However, this should not be taken to imply that they are purely speculative, because they are not. Each one has evidence to support it, some of which will be cited.

METHYLMERCURY

The developmental neurotoxicity of methylmercury has been reviewed recently by Burbacher, Rodier, and Weiss (8). Methylmercury represents a prototypic compound for developmental neurotoxicity. Methylmercury induces neuropathological and neurobehavioral effects which exhibit striking comparability across a number of species, including humans. In terms of the organizing concepts presented in Table 1, methylmercury is perhaps the best example currently available, therefore it is appropriate to review briefly each concept as it applies to methylmercury.

While neither the effect of genetic influences (generality 1) nor the relationship to critical periods (generality 2) have been specifically tested for methylmercury, the evidence suggests that prenatal exposure produces more severe neurobehavioral symptoms and more severe and widespread neuropathology than do infantile or adult exposures (8).

Mechanisms of action of methylmercury (generality 3) remain poorly understood and no attempt to summarize current evidence will be undertaken here. The reader is referred to the recent review of Choi (11), which provides a detailed discussion of the neuropathological effects and current theories of methylmercury's developmental neurotoxicity.

A broad range of effects are associated with developmental exposure to methylmercury (generality 4) (8). Burbacher et al. provide both descriptive and tabular summaries of the findings (see especially their Table 2) comparing effects across species. Their review will serve as the basis for the description that follows.

Humans exposed developmentally to methylmercury exhibit sensory deficits (primarily blindness), mental deficiency, cerebral palsy, seizures, and spasticity at high levels. At these same levels, monkeys and rodents exposed developmentally similarly exhibit blindness, cerebral palsy, spasticity, and seizures. Concentrations of methylmercury in brain causing these effects are 12 to 20 ppm, irrespective of the dose administered. As with many other compounds, methylmercury must be administered in higher doses in rodents to achieve the same brain concentrations found in primates given lower administered doses. At slightly lower doses of exposure (3–11 ppm), humans exhibit mental deficiency, reflex and muscle tone abnormalities, and delayed motor development. Monkeys at the same brain exposure levels exhibit impaired object permanence and visual recognition memory, deficits in social behavior, and visual disturbances. Rodents at the same brain exposure levels exhibit increased escape failures in a complex water maze, swimming abnormalities, heightened acoustic startle reflex amplitude, altered visual evoked potentials, reduced active escape and/or avoidance performance, impaired operant conditioning, and abnormal locomotor activity. At still lower exposure levels (approximately 0.1–3 ppm), humans exhibit evidence of delayed psychomotor development, while rodents exhibit the same types of effects as seen at higher exposure levels, but the magnitude of the effects is smaller. These effects include altered locomotor activity to d-amphetamine challenge, and reduced active avoidance and operant conditioning. No data from monkeys are available for low-level exposure. In sum, a broad range of neurotoxic effects are exhibited following methylmercury exposure, illustrating most of the categories of CNS dysfunction that have been listed for generality 4.

The CNS is a target organ (generality 5) of methylmercury and its selective sequestration in the brain has been described in detail (11). Methylmercury also exhibits renal toxicity.

Dose-dependent relationships (generality 6) are well-documented for methylmercury, as has already been illustrated for neurobehavioral effects in the discussion above on types of response as a function of brain concentrations (generality 4). A similar dose-dependent comparison may been seen for neuropathological changes (8), although available histopathological data are less complete than are those for neurobehavioral effects (see also discussion below of generality 9).

Critical environmental factors influencing methylmercury's degree of effects (generality 7) include the sources of the contaminant, such as food, whether the children were breastfed, and the type of environment in which the children were raised. However, these factors, although presumably important, have never been systematically investigated in developmental methylmercury poisoning cases.

Methylmercury is not a CNS structural teratogen inasmuch as it does not induce neural tube defects. However, it is neuroactive in the sense that it causes CNS symptoms in adults at high doses. Therefore, it adheres to the second part of generality 8, that developmental neurotoxins are either CNS teratogens, causing overt brain malformations, or are psychoactive, causing changes in thought or mood.

Methylmercury is a prime example of a developmental neurotoxin that causes dysfunction at doses below those causing other types of effects (generality 9).

Methylmercury is not a teratogen, as noted above, i.e., it does not induce major malformations. While methylmercury does induce neuropathological changes, these are effects not classified as terata. Methylmercury induces reduced brain size, ventricular dilation, disorganized cortical layering, ectopic cells, necrosis, and improperly oriented neurons in humans and similar changes in monkeys and rodents (8). Of these, only reduced brain size approaches the definition of a teratogenic or dysmorphic effect. Based on this view, methylmercury should be classified as a primary developmental neurotoxin.

The period of susceptibility to methylmercury-induced developmental neurotoxicity (generality 10) has not been rigorously tested, but some trends are apparent. For example, it appears that proliferation, migration, and differentiation are the stages at which neurons are the most vulnerable to methylmercury's effects and these are stages characteristic of early brain development, primarily those occurring during embryogenesis. Methylmercury causes neuronal necrosis, arrested or displaced migration, and morphologic changes.

Generality 11, which states that not all developmental toxins are developmental neurotoxins, does not apply to methylmercury, since it is clearly a developmental neurotoxin. Generality 12, which refers to preconception and transgenerational effects, has not been investigated for methylmercury. Generality 13, which refers to the pattern of exposure, is one for which no direct experimental information is available, although methylmercury-induced neurotoxicity appears to be cumulative. To the extent that its effects are cumulative, the area under the concentration time curve would appear best suited to describing its toxicity.

In conclusion, methylmercury not only demonstrates one of the clearest cases of human developmental neurotoxicity, but it also serves to illustrate several of the key organizing concepts in the field. Methylmercury also epitomizes the situation in which the developing organism exhibits greatly increased vulnerability to injury compared to that of the mature organism. This increased vulnerability of the developing nervous system is a hallmark of methylmercury-induced toxicity. It is also a hallmark of most, if not all, developmental neurotoxins.

LEAD

Evidence of the neurotoxicity of lead has recently been reviewed by Davis et al. (14). Lead studies illustrate several important aspects of what is currently known about developmental neurotoxicity. Lead induces encephalopathy in children at high exposures and may even be life threatening. While lead exposure is pervasive in modern culture, the developmental neurotoxicity of moderate lead levels has become the central focus of most investigations. As Davis et al. (14) note, there have been many exposure paradigms, species, and dependent variables examined following lead exposure, but taken collectively some patterns emerge. Most striking are those in the area of cognitive development, where multiple studies in humans, monkeys, and rats have shown performance deficits on tests of intelligence and learning. In humans, deficits of 2 to 8 points on the Bayley Scales of Mental Development and on the McCarthy scales have been reported, at blood lead concentrations of 10 to 15 μg/dL. In monkeys, deficits in discrimination reversal learning, and in rats, changes in operant rates or variability of responding on conditioning schedules have been found (14). The effects in monkeys were obtained following exposures of 15 to 25 μg/dL early in life and tested after the blood lead levels had returned to untreated levels. The comparability of the exposure-effect relationship between humans and monkeys leaves little doubt about lead's cognitive neurotoxicity.

The magnitude of the cognitive effects in humans has been small in most studies and some well-designed and well-controlled prospective studies have failed to find any significant cognitive effects (15). The reasons for this are not clear, but it is prudent to bear in mind that the exposure levels being investigated in recent years are quite low, especially in studies conducted in the United States since the US government eliminated lead from fuel, paint, and other manufactured products. Therefore, US prospective studies are operating near the detection limits of the psychological instruments being employed. The question arises from these studies: how meaningful are 2 to 8 point IQ point deficits? One perspective on this comes from the long-term followup study of Needleman et al. (42). They found that 10 years after initial evaluation, lead-exposed children continued to show a significant inverse relationship between previous dentine lead concentrations and school performance (dropping out, having a reading disability, lower class standing, increased absenteeism, and lower vocabulary and grammatical test scores). The followup data also revealed several neuropsychological deficits in lead-exposed children (poorer eye-hand coordination, longer reaction times, and slower finger tapping rates). One implication is that tests of mental abilities, such as the Bayley, administered early in life do not capture the full extent of the cognitive deficits actually present in these children. Such tests may, however, serve as predictors of later, more extensive functional difficulties. It could also be that the lesion underlying the cognitive dysfunction has momentum such that it continues to take a toll on intellectual development beyond the point where the early tests scores are obtained. Alternatively, the later effects may be the product of continued exposure. The factors which dispose a child to early exposure probably do not end abruptly, so that those with higher body burdens presumably continue to be the most heavily exposed for a period of many years. This may result in a cumulative effect of lead. Regardless of which situation prevails, the data of Needleman et al. (42) demonstrate that with some neurotoxins, small early deficits on cognitive measures can ultimately translate into significant later dysfunctions having a substantial impact on the affected individual's life.

In terms of the organizing concepts which are the focus of this review, lead illustrates several points. Direct evidence for genetic differences in susceptibility to lead (generality 1) has never been systematically studied in relation to cognitive development. However, lead undoubtedly acts by binding to some receptor.

Receptors and enzymes exhibit genetic polymorphisms (41). Therefore, lead will undoubtedly be found to exhibit varying effects as a function of underlying differences in genetically determined substrate binding.

There are very little data regarding critical periods of lead neurotoxicity (generality 2). While human and monkey studies suggest that neonatal, infantile, and early childhood are probably the periods of greatest vulnerability, followed by prenatal and later childhood, no clear-cut critical period for functional effects has been demonstrated. In monkeys, where such a susceptibility could be rigorously evaluated, studies have suggested that early postnatal exposure is a more sensitive period for some effects (46) but not for others (47). However, the design of these experiments and many others in the field is not optimal for testing critical period hypotheses. For example, the usual design of these studies is to investigate varying overlapping exposure periods, while the optimal design for determining critical periods would be to test three or four nonoverlapping periods using a few well-established, lead-sensitive dependent measures.

The mechanisms of action of lead on neural tissue may involve effects on calcium regulation at the synapse (39) but whether such work can eventually be related to cognitive effects is unclear.

Lead-induced effects on cognitive ability and attention illustrate two of the major types of functional effects (generality 4) associated with lead exposure. Lead's entry into the developing CNS has been well-documented and this exemplifies the target organ access principle (generality 5). Dose-dependent effects of developmental lead exposure (generality 6) warrant some discussion. There is little doubt that there is a coarse dose-response relationship between neurotoxicity and lead exposure levels, since high exposures produce encephalopathy and lower exposures produce cognitive deficits. Furthermore, several studies have used the higher and lower ends of the moderate lead exposure spectrum, as in Needleman's studies, and have generally found the higher exposed groups to be more affected on cognitive measures than the lower exposed groups. Distributions of lead exposure levels have also been examined without resort to dichotomizing exposure groups, and have also reported an inverse relationship between blood lead levels and cognitive performance. Despite this, the dose-dependency of lead's effects has not been striking in such studies. The reason for the absence of a strong relationship is not apparent. One possibility is that the relationship may be nonlinear, but at present there is insufficient information to support this idea. Rodent and monkey studies have not reliably attempted to determine dose-response patterns, and those that have made the attempt have obtained relationships that shed little light on the matter. For example, some have found effects at one dose level and no effect at the other; or one type of effect at one dose level and a different effect at the second; or even one temporal pattern of effect at one dose and a different temporal pattern at the other. Such outcomes produce more confusion than clarity. This problem stems from the fact that the effects of lead are inadequately described. If the behavioral effects of lead were well characterized, problems such as these would not be as prevalent, but typically studies have attempted to both identify and characterize lead-associated functional effects within a single experiment. To further complicate matters, most studies have used inadequate numbers of dose groups. The combined effect of doing too much and using too few groups is that the experimental design is stretched beyond what it can reasonably accomplish, and the results are proportionately disappointing. Such overly ambitious designs produce findings with little generality, and the net effect of a number of these studies is that there is little agreement concerning lead's effects.

There is little doubt that environmental factors (generality 7) influence lead's neurotoxicity, but experiments which demonstrate this systematically are not available. Lead is a neuroactive substance (generality 8) in that it induces neurological effects in adults at high exposure levels. Lead nicely demonstrates generality 9 inasmuch as lead does not produce malformations in humans even at high exposure levels, and in animals where lead can induce malformations, the effects on CNS function occur on the dose-response continuum well to the left of the effects on morphology. In sum, lead is a developmental neurotoxin but there is no evidence to support the idea that lead is a teratogen in the usual sense of the word.

The stage of development at which the CNS is most susceptible to lead (generality 10) has not been determined. Lead is a developmental neurotoxin, therefore it is not covered by generality 11 which states that not all developmental toxins are developmental neurotoxins. No preconceptional or transgenerational effects of lead (generality 12) have yet been shown.

That pattern of exposure to lead is a determinant of lead's effects (generality 13) is supported by evidence that peak lead concentrations may be critical to lead's adverse CNS effects. This evidence comes from studies of monkeys in which animals were exposed to either chronic moderate levels of lead or chronic moderate levels of lead with superimposed episodic pulses of higher lead exposure. Laughlin (37) points out that the chronic plus pulse lead exposure regimen produced more severe long-term adverse effects on behavior than chronic exposure alone. This suggests that peak lead levels play a crucial role in the ultimate effects induced by lead, thereby supporting a peak effect concept for lead neurotoxicity. We will return to this concept below in the discussion of ethanol's effects.

ETHANOL

Ethanol-induced teratogenicity has been characterized clinically as the fetal alcohol syndrome (FAS). There are numerous reviews of this literature, two of the more recent being those by Driscoll et al. (16) and Vorhees and Mollnow (70). FAS is described as consisting of three clusters of effects: those related to the CNS, those related to growth, and those related to craniofacial development. The CNS effects include mild to moderate mental deficiency, attention deficit–hyperactivity disorder, learning deficiencies, altered reaction times, altered reflexes, and delayed development. The growth effects include small for gestational age and failure to grow at normal rates postnatally. The facial effects include midfacial hypoplasia, epicanthus, flattened philtrum, thin vermilion, reduced head circumference, and low nasal bridge, collectively resulting in characteristic dysmorphic facies (52).

No attempt will be made to review ethanol's neuroembryopathy in relation to all of the generalities presented in Table 1 as has been done for methylmercury and lead. Rather, for ethanol and the remaining agents (anticonvulsants and retinoids), only those concepts that are especially noteworthy for each case will be presented.

From a comparison of the neurobehavioral effects of ethanol in humans and animals based on the recent review of Driscoll et al. (16), the following similarities were noted. FAS children have been identified as being hyperactive and to have attention deficits. Animals, primarily rats, exposed prenatally to ethanol are also found to be more active and to display impaired attention. Some FAS children exhibit mental deficiency, and *in utero* ethanol-exposed rats exhibit impaired learning on tests such as active avoidance acquisition. Both humans and rats exhibit inhibitory or impulse control impairments, and both species exhibit perseveration and diminished habituation to novelty. Some FAS children exhibit gait abnormalities, and some rat studies have found changes in stride dimensions. FAS children frequently have fine motor movement difficulties and reduced gross motor coordination has been reported in intrauterinely ethanol-exposed rats. FAS children are frequently developmentally delayed in attaining maturational milestones, and some studies in rats have reported delays in reflex development, although the results in this area from animal experiments are mixed. Finally, some FAS children have hearing impairments and in a few studies in rats, changes in acoustic startle reactivity have been reported following prenatal ethanol exposure.

Ethanol-induced developmental neurotoxicity illustrates several of the organizing concepts that are the focal point of this discussion. In addition, consideration of several other of the organizing concepts for which evidence is lacking will serve to illustrate remaining gaps in our knowledge about this drug.

No critical period (generality 2) for FAS neurobehavioral effects or equivalent effects in animals has been identified. Indirectly the evidence suggests that some exposure periods may be more sensitive than others, but the data are far from definitive. For example, the dysmorphic facies that are considered a key diagnostic feature of FAS are controlled by morphogenetic events occurring during embryogenesis. From an embryological point of view, this points to first trimester exposure. Growth retardation, by contrast, is associated most strongly with second and third trimesters of development. The observation that interventions which effectively curtail ethanol consumption by mid-pregnancy result in newborns whose weight falls within normal limits suggests that second and third trimesters may be important to ethanol-associated growth deficiency. Sensitive stages for the induction of neurobehavioral effects are not as readily estimated using embryological markers because CNS damage may be induced during all trimesters due to the protracted development of the brain. Therefore, CNS dysfunction does not lead to a predicted stage of vulnerability as do the other symptom clusters. An important experiment conducted in monkeys sheds some light on the subject. Clarren, Astley, and Bowden (12) found that neurobehavioral effects were more prevalent in pigtailed macaque progeny exposed to ethanol during early gestation than in those with exposure delayed to later stages of gestation. These data provide evidence that generality 2 is valid, *viz.,* ethanol-induced neurobehavioral injury is most consistently associated with early (i.e., during organogenesis) rather than later exposure. There is considerable support for this in rodent experiments with a variety of agents other than ethanol. These data have been reviewed elsewhere (60) and will not be restated here, but the examples include salicylates, hydroxyurea, phenytoin, trimethadione, methylazoxymethanol (MAM), and retinyl esters.

Animal studies on prenatal ethanol exposure have largely ignored the issue of critical periods, and those that have dealt with it have done so with an experimental design in which the exposure levels were too broad to provide much insight. Early postnatal studies have been done in rats using artificial rearing and ethanol gastric infusions as models of human third trimester exposure, and these experiments have resulted in evidence of late-occurring neuroanatomic and behavioral effects (21). This demonstrates that late exposures can be neurotoxic, but provides no glimpse as to which periods of development are most sensitive or which types of effects are associated with different exposure periods.

Dose-dependency, generality 6, is an area of considerable importance in ethanol embryotoxicity, but thus far, investigations addressing this matter have yielded results subject to conflicting interpretations. One view is that humans exposed to ethanol exhibit a continuum of effects. These effects are characterized as full or complete FAS at the most severely affected end of the spectrum, through partial FAS in the middle zone (those exhibiting two, but not three, of the symptom clusters), to fetal alcohol effects (FAE) at the lower end of the spectrum (those exhibiting primarily neurobehavioral effects). FAE represents those types of ethanol-associated effects that are most frequently found in prospective studies of women who are not alcoholic. When evaluated longitudinally for a variety of neurobehavioral functions, one large study, the Seattle prospective study, found graded effects and a progressively increasing number of affected individuals as self-reported ethanol dose increased (53). These data imply that a dose-response relationship exists for ethanol. However, the data establish no threshold for these effects, and hence no apparent NOEL (no observable effect level). Such information is critical for an ultimate understanding of this drug's developmental neurotoxicity.

In contrast to the findings from the Seattle prospective study on ethanol, a series of recent reports from a prospective study in Cleveland provides a different view. It should be noted at the outset that the socioeconomic status (SES) of the Cleveland cohort was lower than that of the Seattle cohort, and this may influence the interpretation of the findings from the two studies.

What is most intriguing about the Cleveland findings is that despite careful neurobehavioral testing, attention to ethanol intake self-report methods, and the use of multiple covariates in the data analyses, no significant relationship between maternal ethanol and offspring neurobehavioral outcome up to the age of 4 years and 10 months emerged (6,22,23). A trend was reported for body weight and craniofacial anomalies in relation to ethanol. The authors found one child in whom the symptoms of FAS were fully manifest and this child was severely mentally retarded. The conclusion reached by the authors was that ethanol, when it induces FAS, produces clear effects on cognition, but in the absence of the full syndrome, no demonstrable neurobehavioral effects exist. These authors speculate that the relationship between maternal ethanol use and neurobehavioral deficits in their children are determined by the presence of a few cases in each of these large studies of severely affected children. These children produce an anchoring effect on the data, in effect pulling the lower end of the test results downward for the children of women with the heaviest ethanol consumption. This produces a significant inverse slope. This implies that ethanol-induced embryopathy has a sharp and relatively high threshold.

Given the differences in the human prospective studies thus far reported, what light can animal data shed on ethanol dose-response patterns? In the review of Driscoll et al. (16), a case is made that dose-response relationships have been identified but the authors acknowledge that data are few. This conclusion appears well-justified. This is not to say that some attempts have not been made. Dose-effect studies have been reported from the laboratories of Abel and of Riley [see review by Meyer and Riley (38)]. However, the dose-dependency information that has emerged from these investigations has been difficult to interpret because often a step function is obtained. The role of nutritional confounders in ethanol-induced neurobehavioral effects; the role of postnatal maternal factors in modifying such effects; the optimal dose, route, and frequency of ethanol administration for producing effects; internal dose; characterization of the types of neurobehavioral functions affected; and the permanence of effects, i.e., whether the effects are reversible or irreversible, have received greater emphasis than dose-response investigations. However, the determination of ethanol's threshold is a cornerstone in the process of estimating the risks associated with developmental ethanol exposure, and threshold, in turn, can be determined only by dose-effect information. It is essential that this area receive greater attention in the future in order to resolve these issues.

Ethanol is unequivocally a neuroactive agent, and therefore fits the second part of generality 8 on types of developmental neurotoxins.

When it comes to generality 9 (response relationships) and the proposition that ethanol is predicted to produce effects at doses below those inducing malformations, no definitive conclusions are possible. The human data on this are split as noted above, and the rat data are difficult to apply to the concept because rats are refractory to ethanol-induced teratogenesis. The best data on this comes from the primate work of Clarren et al. (12). In that experiment, pigtailed macaques were found to exhibit neurobehavioral deficits at doses below those causing physical abnormalities. On the basis of this finding, it is concluded that ethanol does not violate the proposition that neurobehavioral effects are the most sensitive indicators of ethanol's developmental toxicity. However, Greene et al. (23) came to the opposite conclusion based on their human data from the Cleveland prospective study. They maintain that body weight and anomalies are more sensitive indices of ethanol-induced embryotoxicity than neurobehavioral effects. The data available do not allow this point to be resolved. However, it is worth noting that Greene et al.'s interpretation would be more persuasive if it was based on clear evidence of neurobehavioral effects. One could then actually compare the sensitivity of the various endpoints. Instead, we are left to compare a modest increase in physical effects to no increase in neurobehavioral effects, rather than comparing relative degrees of effects.

As noted previously, the data of Clarren et al. (12) support the view that early prenatal exposure to ethanol is more decisive in inducing CNS functional damage than is later prenatal exposure. This may be regarded as support for generality 10, that the period of maximum susceptibility is early, during neurogenesis, rather than later.

Finally, Streissguth et al. (51) have examined the long-term outcomes of some of the first children ever diagnosed with FAS. They found that the characteristic facies that were important in the initial identification of this syndrome had changed substantially by the time these individuals reached adolescence and adulthood. Often the features changed so dramatically that they no longer had a recognizable pattern of facial anomalies. By contrast, the CNS dysfunctions present in these individuals did not dissipate with age. The FAS individuals continued to have intellectual, emotional, and behavioral problems. In fact, one of the authors' major conclusions was that these individuals' dysfunctions were sufficiently disruptive to their lives that none were found to be functioning fully independently. These data reveal that recovery of function did not prevail in individuals with ethanol-induced developmental neurotoxicity. The lack of recovery after early CNS injury has also been found in more recent studies in animals with induced focal lesions (30). Such data raise serious questions about the long-held concept that recovery is better after early brain injuries compared to those after adult brain injuries. We will return to this important point later in this chapter.

In conclusion, ethanol represents one of the most thoroughly investigated agents for its effects on development. The data, although individually containing deficiencies, collectively provide strong support that ethanol is a developmental toxin. Furthermore, the data show that among the effects induced by ethanol, neurotoxicity is prominent. In fact, ethanol's neurotoxicity may ultimately be the most serious of its long-term consequences. Despite this, ethanol remains an incompletely understood agent, and one whose effects are proving difficult to fully characterize. This is due in part to ethanol itself, because it is a drug for which the basic biological effects remain incompletely understood. It is also due to ethanol's use in society, where exact usage is difficult to document because of denial, underreporting, and the absence of biomarkers of ethanol consumption. Finally, as noted above, there has been a lack of attention to some of the important issues needing clarification if the area is to progress, such as critical period effects and dose-response relationships.

ANTICONVULSANTS

Prenatal exposure to anticonvulsants has been associated with a collection of developmental syndromes, each one named after the drug presumed to have caused it. The principal disorders have been the fetal hydantoin, trimethadione, barbiturate, valproate, and carbamazepine syndromes. Each syndrome is characterized by the presence of dysmorphic facies, CNS effects, and sometimes growth retardation. All the syndromes share common features, and these have been reviewed recently by Adams, Vorhees, and Middaugh (5), with emphasis on the fetal hydantoin syndrome (FHS). FHS will also be the focus of the present discussion.

The primary hydantoin anticonvulsant drug for the control of epilepsy is phenytoin. Phenytoin is one of the oldest drugs in this therapeutic category, one of the most efficacious, and remains the most widely prescribed worldwide. The human data on FHS, however, are not nearly as ample as for methylmercury, lead, or ethanol. The fetal hydantoin syndrome was originally described and named by Hanson and Smith (24). Subsequently in a small, prospectively ascertained series, Hanson et al. (25) reported that 11% of infants exposed to phenytoin manifested the syndrome, followed by another 31% with partial expression of the major symptoms (the number of exposed cases in this study was 35). In a recent report, VanOverloop et al. (55) found in a case-control study that children exposed *in utero* to therapeutic doses of phenytoin and tested cognitively at 4 to almost 8 years of age had significantly reduced scores on the Wechsler intelligence test subscales for block design and mazes. These effects persisted using different types of control groups and despite a lack of significant effects on full-scale IQ scores. There were also significant effects on several other cognitive and behavioral measures, including lower performance on a visual motor integration test, and a measure of gross motor activity (time spent in various room quadrants in a controlled environment). The authors also noted that on the two affected subscales of the Wechsler, children exposed prenatally to phenytoin combined with phenobarbital were more affected and affected on more subscales than those exposed to phenytoin alone. This raises the possibility that the syndrome is really the result of an interaction between two anticonvulsants and, if correct, should more correctly be called the fetal hydantoin-phenobarbital syndrome (FHPS). Notwithstanding the possible contribution from phenobarbital, these data appear to support the idea that phenytoin is associated with cognitive deficits. However, the data also suggest that the effects are rather specific, unlike those found with methylmercury and ethanol which affect global intelligence.

The human data on FHS are suggestive but not conclusive because they are based on small sample sizes and only a few studies. For these reasons it is not possible to evaluate the syndrome in relation to the organizing concepts in any comprehensive way beyond noting that phenytoin appears to fulfill the basic requirements necessary to be considered a human developmental neurotoxin.

The animal data on phenytoin help clarify this picture, however, and unlike the situation for some of the other agents reviewed thus far, the animal data on phenytoin can be used to address a number of the organizing concepts.

The animal data on phenytoin directly address generality 2 on critical periods. In an experiment in rats at a dose of phenytoin previously established to be developmentally neurotoxic, it was shown that midpregnancy (rat embryonic days 11–14) was critical for the induction of neurobehavioral abnormalities, whereas earlier and later exposures showed only small effects by comparison (57).

The animal data on phenytoin for generality 4 (types of response) reveal that in rats the effects of exposure to this drug include increased early pivoting locomotion, delayed swimming ontogeny, delayed development of the dynamic air-righting reflex, hyperactivity, inhibited startle reactivity, abnormal circling behavior (evident in only a minority of the offspring), severely impaired maze learning, some evidence of impaired memory on several types of retention tests, and coordination impairments (17,57,63).

The animal data on phenytoin also address generality 6 on dose-response relationships. Experiments in rats have shown that the functional neurotoxic effects induced by phenytoin are linearly dose-dependent. This dose-dependency is revealed in two ways: (a) the phenytoin offspring exhibit graded severity of effects among those showing abnormalities, and (b) numbers of affected individuals vary as a function of administered phenytoin dose (5,63,71,72).

The animal data on phenytoin illustrate generality 8 in that this drug is not a CNS teratogen in rats. On the other hand, phenytoin is neuroactive. Therefore, phenytoin fulfills one of the two primary criteria expressed in this generality for being a developmental neurotoxin.

The animal data on phenytoin also illustrate that the developmental neurotoxic effects of the agent are manifest at doses along the dose-response continuum to the left of those for malformations (generality 9). It was shown in rats that doses of phenytoin not causing malformations or growth retardation cause severe behavioral abnormalities (ref. 57, *cf.* experiments 1 and 2).

The animal data on phenytoin are consistent with generality 10, that the period of maximum susceptibility to insult is during organogenesis. This is suggested by the finding noted above that the critical period for phenytoin-induced neurotoxicity was rat embryonic days 10 to 14, a period when neuronal proliferation is at its peak in many brain regions (49).

Evidence for the remaining generalities is not yet available. However, the phenytoin findings available thus far illustrate how the concepts in Table 1 can be tested experimentally and how this leads to a more complete characterization of an agent's neurotoxicity.

RETINOIDS

The last example in this series is vitamin A and its congeners. Vitamin A (retinol) and its esters have long been known to be teratogens in experimental animals [reviewed by Geelen (20)]. Furthermore, retinoids are CNS teratogens (28). In fact, the retinyl esters were the compounds first used to describe instances of developmental neurotoxicity in animals and led directly to the formulation of generalities 8 and 9 [see Butcher (9)].

Despite the clear evidence in animals of retinoid-induced developmental neurotoxicity, these data were largely of theoretical interest until the 1980s when a drug that was a congener of retinoic acid, 13-*cis*-retinoic acid (isotretinoin), was marketed. Sold by Hoffmann-LaRoche under the trade name Accutane, this compound was efficacious for the treatment of cystic acne. The drug was known to be an animal teratogen prior to marketing and was designated by the manufacturer and the Food and Drug Administration (FDA) as a category X drug, i.e., contraindicated for use during pregnancy. However, once on the market, the drug's inappropriate prescription, drug sharing, and inadequate contraceptive monitoring led to the birth of a number of children in the United States with isotretinoin-induced embryopathy (35). The relevance of this to the present discussion is that the animal data on retinoids, together with the concepts described here as generalities 8 and 9, led directly to the prediction that isotretinoin would be found to be a human developmental neurotoxin (62). This prediction has since been confirmed (2). The isotretinoin data reveal more than the simple fact that this agent is developmentally neurotoxic in humans. However, before discussing some of these ramifications, a brief summary of the basic observations is needed.

Using the isotretinoin-exposed cases ascertained retrospectively and prospectively by Lammer et al. (35,36), Adams (2) has evaluated 31 of the children and a similar number of matched controls using the Stanford-Binet (4th edition) test of intelligence and a series of neuropsychological tests. Considering the IQ data first, one would expect 2.5% of the cases to be more than two standard deviations below the mean, whereas among the isotretinoin children, 19.4% had scores more than two standard deviations below the mean. Another 13.5% would be expected to have scores more than 1 and less than 2 standard deviations below the mean, whereas the isotretinoin-exposed children had 32.2% with scores in this range. In the middle range, with scores within ±1 standard deviation from the mean, there should have been 68% of the cases, but for the isotretinoin-exposed children, 41.9% actually fell within this zone. Thus, the lower ranges were significantly overrepresented with low scores, while the middle (average) range was underrepresented. When the upper ranges were inspected, the opposite effect was seen. The isotretinoin-exposed children were underrepresented compared to what would normally be expected. Specifically, there were only 6.4% of the cases in the range of more than 1 but less than 2 standard deviations above the mean, and no cases more than 2 standard deviations above the mean, compared to expected values of 13.5% and 2.5%, respectively.

When these children were examined on a series of neuropsychological tests, again marked performance reductions were evident. Without detailing all the findings, the areas of deficient functioning in the isotretinoin-exposed children were language processing (vocabulary, comprehension, picture vocabulary, sentence memory, story memory, and alphabet), visual/perceptual processing (bead memory, pattern analysis, quantitative, visual-motor integration, draw-a-person), executive control (animal house, continuous performance), auditory discrimination (word discrimination), articulation (words), and motor functions (finger localization, pegboard). In all these areas the isotretinoin-exposed children were heavily overrepresented in the subnormal functioning range.

Adams (2) also compared the IQ data with the morphological findings on these same isotretinoin-exposed children. She found that there was a rough relationship between major malformations and IQ in that just as there was an overrepresentation of low IQs in the distribution of isotretinoin-exposed children, so too was there an overrepresentation of low IQ children with major malformations. There was also a weak trend toward more low IQ children having CNS malformations. Thus, in a very gross way, malformations and cognitive dysfunction tended to covary. However, within this coarse relationship, another observation emerges. Adams' (2) data show that there is a considerable degree of dissociation between IQ and major malformations. For example, even though all the cases with the lowest IQs (i.e., those below 70) had major malformations, less than half (40%) of those with IQs between 70 and 85 had major malformations. In other words, the correspondence between these endpoints was so poor that if one had based a prediction of cognitive dysfunction on the presence of major malformations, one would have had an 83.3% correct hit rate, because of the 12 cases with major malformations, 10 had significantly reduced IQs. However, one would also have had a 16.7% false positive rate, since 2 out of 10 of these children had normal or above normal IQs but had major malformations. An even larger problem arises when one looks at the data in terms of misses, i.e., cases with reduced IQ without major malformations. Here the predictive value of malformations is low, since fully 37.5% of the children with low IQs had no malformations, i.e., 6 out of 16, with the remaining 62.5% showing both effects. In reality the situation for the predictive value of malformations is even worse than this implies since only 1 of 13, or 7.7%, of the children with IQs between 85 and 115 had major malformations, yet many of these children had lesser cognitive deficits. In a regression analysis many of these children would undoubtedly fall along the regression line for drug-related IQ decline yet have no malformations, thereby reducing the correlation between malformations and IQ even further. Obviously, as Adams adds more cases to her cohort, the full scope of the relationship between malformations and psychometric deficits and between malformations and neuropsychological deficits will become better delineated. Nevertheless, the Adams (2) data show that children exposed to a developmental neurotoxin may be cognitively abnormal and have no major malformations, especially when not severely cognitively affected (60% with IQ of 70–85 had no major malformations).

The isotretinoin human data do not illustrate the critical periods concept because virtually all of the cases Lammer has collected were first trimester exposures. However, they hope to eventually gather a cohort of second and third trimester exposed children. Until that occurs, the animal experiments available have shown the most sensitive period for neurotoxic effects to be during neurogenesis (43,67).

Neither the human nor animal data are very satisfactory for the demarcation of dose-response relationships. For humans, the problem is that the therapeutic dose range for isotretinoin is very narrow; hence, all the identified cases have been prescribed doses of 0.5 to 1.5 mg/kg. Within such a narrow range and with such small numbers of affected cases, establishing a dose-dependent response pattern has thus far been impossible. In the case of animal experiments where a wider range of dose levels can be investigated, the data are limited because the range in which one obtains dysfunction without obtaining malformations is narrow. Thus, Nolen (43) found that even small increases in the dose of retinoic acid during organogenesis resulted in high mortality in the treated offspring. Vorhees (56) found that within the narrow range of doses required to obtain survivors, retinyl ester exposure produced no clear dose-dependency on tests of avoidance learning or open-field activity. These studies should not be the last word in experimental investigations on the matter, however, and more assiduous efforts may ultimately overcome the difficulties just described. What the animal data suggest is that the dose-response curves for malformation and dysfunction lie close together, so that the upper end of one curve overlaps the lower end of the other. Thus, at most doses which induce embryotoxicity, a mixed set of outcomes is obtained that includes fetal death, malformation, and dysfunction. If the dose is lowered sufficiently to prevent fetal death and malformation, then one is at a point along the curve at which only a minority of the offspring exhibits dysfunction. When a high percentage of the offspring are unaffected, finding the few abnormal individuals among the many unaffected becomes more difficult. However, because of recent advances in behavioral methodology, this kind of differentiation may now be within reach.

In conclusion, isotretinoin is clearly a CNS teratogen in both humans and rodents. Furthermore, isotretinoin's functional effects often occur at doses not causing malformations. The animal data on this are clear, since in these studies the dose is decreased to the point where few or no malformations are present, but dysfunction is evident. The human data also support this interpretation, although not definitively. This is probably due to the fact that the dose-response curves for cognitive dysfunction and malformations lie close together. An alternate view is that there is really only one dose-response curve with varying degrees of dysmorphic effects, some of which induce dysfunction. For this to be true, there would have to be a tight correspondence between CNS malformation and dysfunction, which there is not. Some of Adams' isotretinoin children had low IQs and no CNS malformations. In sum, isotretinoin is a teratogen and developmental neurotoxin that conforms to the concepts presented in Table 1.

REGULATORY ASPECTS OF
DEVELOPMENTAL NEUROTOXICITY

Pharmaceuticals

Preclinical developmental neurotoxicity safety assessment guidelines for the evaluation of new therapeutic drugs were first introduced in 1974–75 (61). In those years the reproductive and developmental safety assessment guidelines of Japan and Great Britain were amended to include behavioral effects on the offspring following prenatal and prenatal/neonatal exposure to test agents. The history and rationale for these additions has been reviewed elsewhere (61) and will not be presented here. Instead, the focus of this discussion will be on a new proposal to develop a set of harmonized international reproductive test guidelines that would simplify the variety of national guidelines that now prevail. This new guideline would not replace current national regulations, but rather would serve as an alternative approach that each participating government would agree to accept in place of its own guidelines in order to facilitate the review of drugs intended for international marketing. Those portions of the international draft guideline that bear on developmental neurotoxicity assessment will be reviewed. However, first a brief discussion of existing reproductive and developmental test guidelines in general, and of neurotoxicity assessment in particular, is in order.

Reproductive and developmental toxicity assessment has traditionally been divided into three segments. Each segment represents a different aspect of reproduction. The segmental approach divides ontogeny into the following phases: segment 1 includes reproduction, preimplantation, and implantation; segment 2 includes organogenesis; and segment 3 includes fetogenesis and neonatal development.

Assessments of behavior were added to these existing segments but not uniformly. Different governments included functional assessments in different segments. For example, the Japanese included tests for behavioral effects in segments 2 and 3, while the British included them in segments 1 and 3. Why the difference? The reason has never been completely clear. As for what should be evaluated, the Japanese originally listed functional areas for assessment of motor, learning sensibility, and emotion, but in 1986 this language was amended to state simply that tests of behavior were required. The British government included evaluation for auditory, visual, and behavioral impairment. This language was widely adopted throughout Europe during the 1970s. In 1985 the British model became part of guidelines adopted by the entire EEC (European Economic Community).

In the United States, the parallel regulatory agency with responsibility for pharmaceuticals and foods is the Food and Drug Administration (FDA). Unfortunately, their record on neurotoxicity test guidelines has been abysmal. From 1974 to about 1986, different entities within the FDA took separate positions. The Center for Food Safety and Applied Nutrition supported developmental neurotoxicological research and argued that the field was potentially important but not ready for regulatory action, while the Center for Drugs ignored the field almost entirely. Since about 1986, the FDA has taken the position that the need for neurotoxicity testing is under internal review. In the meantime they assert there is no reason for concern because existing guidelines are adequate to detect any serious neurotoxicity that might occur. The notion that current testing is adequate is entirely unsubstantiated and is not credible. As of the spring of 1991, the FDA has been contemplating the entire area in one form or another for about 20 years, 17 years of which have elapsed since the Japanese government put the matter into regulatory practice. After such an interval it becomes difficult to imagine that action is imminent, but it is equally difficult to believe that inaction can endure forever.

There seem to be three possible ways in which change may come to FDA. (a) There could be a severe adverse drug reaction, i.e., a drug may be approved that is said to be safe but that induces developmental neurotoxicity. Those injured will be justifiably incensed when they discover that the agency could have, but did not, require tests that would have forecast such effects. Isotretinoin is such an agent in some ways but since it also produces major malformations the issue is not an uncomplicated case. (b) A combination of FDA internal review, new leadership, congressional surveillance, and further accretion of scientific data will reach such a high level that FDA's inertia threshold will be exceeded and the agency's machinery will lurch forward. (c) An effort to produce a set of harmonized international guidelines, currently being spearheaded by the EEC, might receive FDA endorsement in the name of cooperation and international commercial competitiveness.

Having reviewed some of the key points of current guidelines, we now turn to recent efforts to develop a uniform set of international reproductive and developmental guidelines. This effort originated in the EEC and has been supported by the International Federation of Teratology Societies (27), the World Health Organization, the International Clearinghouse for Birth Defects Monitoring Systems, and related entities. In this proposal the three-segment approach standard in reproductive toxicology throughout the world is maintained, but the overlap among the segments, especially between segments 1 and 2, is largely eliminated. This is done to make each segment a test of a subset of events of reproduction and development, rather than blurring the boundaries as happens now.

The critical information for the present discussion is contained within those portions of the international proposed guideline dealing with developmental neurotoxicity. No evaluation for developmental neurotoxicity is included in segment 1, since the F_1 generation is not allowed to come to term. Developmental neurotoxicity is included in both segments 2 and 3, however. The inclusion of reflex tests is inexplicably discouraged in the international guideline with the unsubstantiated statement that such tests are no better than information provided by offspring body weight. However, there are no persuasive data demonstrating that CNS injury is tightly associated with body mass. The document then states that if such tests are performed, they should include more than a single assessment age, which appears sound given that single assessment points run a serious risk of missing effects. This is shown by the following example: Pivoting represents a behavior that emerges as a stage of locomotor ontogeny that might loosely be regarded as analogous to human crawling. The analogy is only to provide a feel for this behavior and should not be pressed. Nevertheless, pivoting is similar to crawling in that it is a form of ambulation that emerges at a specific stage in early development and disappears later in a predictable way with the emergence of more adult modes of ambulation. Pivoting, as its name implies, is characterized by turning movements. In performing these movements the rat turns its head in one direction and the forelimbs gradually follow. During the turn, the hindlimbs move little and are usually splayed. In Sprague-Dawley rats, pivoting emerges around day 7 and disappears around day 11 or 12. In between, pivoting first increases then decreases in frequency across successive days. Simultaneously, full quadrupedal ambulation begins to emerge around day 9 and increases steadily thereafter. Phenytoin, as noted in the section on anticonvulsants, is a potent developmental neurotoxin, and one of its many effects is that it increases pivoting. If, however, one picked only one day to test for phenytoin's effects, say day 7 or 8, the effect would be entirely missed because the effect of phenytoin on this behavior does not emerge until day 9 (57,63). It takes only one exception to disprove an hypothesis, and this illustration shows how the single day assessment hypothesis is flawed.

The proposed international guideline's discouragement of developmental assessments of behavior without adequate justification is a weakness of the proposal. However, the proposal also contains several strengths. For example, the proposed guidelines are laudable in espousing a more diversified range of behavioral assessments than the current EEC guidelines. The proposed guideline would include the functional categories of "sensory function, reflexes, motor activity, and learning/memory" (27, p. 16). These evaluations should be performed on at least one male and one female per litter according to the description and it further states that a single pair may be used for all such tests.

It should be noted that segment 2 in the international proposed guideline is what might be termed a Japanese-style segment 2, rather than being an FDA/EEC segment 2 design. The Japanese segment 2 study has exposure during organogenesis as do all segment 2 studies, but in the Japanese study twice the usual number of pregnant dams are prepared and half of each group is evaluated by near-term fetal autopsy for teratogenic effects and the remaining half are retained for delivery and postnatal evaluation for mortality, morbidity, general development, and behavior.

While the international proposal clearly has some deficiencies, it is a positive step on two counts: (a) it specifies that developmental neurotoxicity be routinely included in all segment 2 and 3 studies, as opposed to the current situation where they are required in Japan, negotiable in the EEC, and entirely absent in the

United States; and (b) the breadth of behavioral categories listed is greater than the current Japanese or EEC language, which should promote a more comprehensive and thorough examination of the nervous system for toxic effects.

The principal weaknesses of the international proposal are that absolutely no guidance is provided on what kinds of methodology are suitable; what criteria must be met by the applicant for establishing the validity, reliability, and sensitivity of the tests chosen; any critical features that must be included; or even how often and at what ages tests are most appropriately conducted. For example, should motor activity testing be performed at one age only, or are multiple test ages preferable? Is a 30 sec test of activity adequate, or does it have to be long enough to demonstrate exploration and habituation? Or does it have to be long enough to demonstrate ultradian or even circadian rhythms? What does "learning/memory" mean? Do the authorities mean that one is to treat these as one and the same, or does it imply separate tests of learning and of memory? On tests of learning, would the authorities accept one-trial acquisition paradigms or must a learning curve be demonstrated? These are just the most basic concepts left unmentioned. Ultimately there are more issues and more complex factors within these issues that must be considered. Vagueness in the guidelines might promote creative and thoughtful approaches from some applicants, but may permit minimalist or even unsound approaches by others. The best types of guidelines go far enough to provide a scientifically sound structure, but not so far as to stifle innovation and scientific judgment. This ideal in guidelines is not met in the international proposal.

Environmental, Industrial, and Agricultural Chemicals

It is a curious situation that whereas Japan and Europe have been progressive in the implementation of developmental neurotoxicity safety assessments for pharmaceuticals, they have been immobile when it comes to environmental, industrial, and agricultural agents, the exact opposite of what one might predict. The US Environmental Protection Agency (EPA) issued neurotoxicity test guidelines in 1985, revised these guidelines, and added new ones for developmental neurotoxicity in the spring of 1991 (18). By contrast, the equivalent regulatory agencies in Japan and Europe have done nothing, nor is there any evidence of imminent action.

It is clear that in the United States this reflects a difference in the responsible government entities, and it is probable that the same is true in other countries. It is ironic that the US EPA would develop adult and developmental neurotoxicity guidelines before the FDA, given that the EPA is a much younger and less developed regulatory agency, given EPA's slow pace caused by its episodic political upheavals, and given the comparative legal constraints that have confronted the EPA compared to the FDA.

The EPA created its developmental neurotoxicity study guideline *de novo* and it resembles no other in the world. A point of interest is that prior to these revisions, most of the adult neurotoxicity guidelines existed only for toxic substances, and with only a very limited number of specialized assays for the neurotoxicity of pesticides. With the advent of the 1991 guidelines, a whole new spectrum of neurotoxicity assays has been brought into the pesticides regulatory arena.

The entire EPA 1991 developmental neurotoxicity guideline will not be reviewed here, and the reader is referred to the full text for details (18). This discussion will focus on several of its unique features.

First among these is the guideline's inclusion of an ontological framework and specification of procedural requirements. These provide a level of guidance lacking in the proposed international guidelines described for pharmaceutical agents above. Also in contrast to the international proposal, in the EPA guideline, different pairs of males and females from each litter are assigned to different tasks. The merit of this approach is that it reduces the risks of carry-over effects from one task to another, which is especially important in the EPA guideline because the approach is longitudinal, i.e., each pair is tested at different ages on the same tests. By contrast, the international proposed guideline places all its eggs in one basket (i.e., one pair per litter). If the test pair is unaffected then the effect will be missed. Both guidelines rely on testing a number of litters per group to circumvent this problem. For the EPA this is 20 litters per group, and for the international proposal it is 16 litters per group for the postnatal phase of segment 2 and 20 litters per group for segment 3.

It is worth considering the potential gain to be realized if more offspring per litter were tested on more of the tests in both the EPA and international proposed guidelines. This may not be appropriate for motor activity, a test known to be affected by experience under certain specific conditions (although the issue should be addressed empirically), but it may be useful on other tests since there is no evidence of crossover effects between, for example, startle reflex testing and tests for learning and memory. Therefore, pairs for these tests could undergo both procedures and group sizes would be doubled.

The EPA guideline requires that the dams be evaluated using an FOB (functional observations battery), a sort of gross neurological examination. The offspring are also to be evaluated using the FOB, and in addition physical landmarks of male and female maturation, motor activity, acoustic startle reflex habituation, and tests of learning *and* memory [emphasis added]. Thus, the number of functions being evaluated exceeds those in current Japanese and EEC guidelines, a step in the right direction. This still may not be adequate, however (see below).

Unlike the international proposal, the EPA guideline is truly developmental in concept. The EPA specifies that motor activity testing be done on days 13, 17, 21, and 60 ± 2 days of age; startle on 22 and 60 ± 2 days of age; and learning and memory tests on 21 to 24 and 60 ± 2 days of age. While questions have been raised about the specific days selected, especially the propriety of testing motor activity on day 13, conceptually the EPA guideline is on the right track.

Another improvement in the EPA guideline compared to the international proposal is the inclusion in the EPA guideline of key criteria for each test, such as demonstrating an habituation curve when testing for motor activity, a learning curve when testing for learning, and inclusion of nonassociative controls when testing for memory. The EPA also makes at least a partial attempt to list some learning and memory tests it regards as being within the universe of acceptable tests. This is something the international proposal should consider, although one would hope a more diversified list of possibilities would be sampled than was provided by EPA.

There is one realm in which the EPA approach is seriously flawed. Whereas the international proposal recognizes the central role of developmental neuro-toxicity for understanding a test agent's effects on all of development and therefore includes it routinely in all segment 2 and 3 studies, the EPA has excluded it from its developmental toxicity guideline and made it a separate stand-alone guideline. The problem is that the EPA approach will effectively sideline devel-opmental neurotoxicity from the mainstream of toxicity evaluation. It also sends an implicit message that brain development is not central to development, an odd proposition to say the least. Rather, brain development is treated as a spe-cialized area needed only when specific indicators point in its direction. It is difficult to comprehend how any scientist of any stripe could accept this impli-cation. Thus, at its core the EPA approach presents a problem by segregating developmental neurotoxicity from overall development. One can only be wary of an approach that separates nervous system from non-nervous system devel-opment in such an artificial manner. It is certainly within the agency's power to require both developmental toxicity and developmental neurotoxicity assess-ments in tandem, but this does not appear likely. The sidelining of developmental neurotoxicity within EPA guidelines should be a source of serious concern to all those interested in human health protection, and especially those concerned with child development.

A NEUROPSYCHOLOGICAL BASIS FOR NEUROTOXICITY: A CONCEPTUAL PROPOSAL

Early in this chapter the idea was raised that functional neurotoxicology lacks an adequate conceptual framework, and this situation might be redressed by drawing upon neuropsychological theory. This idea is not entirely new. As noted previously, Hartman (26) has already suggested that neuropsychology forms a logical basis for approaching problems presented by people exposed to possible neurotoxins. Hartman (26) presents a solid case for why the database and test batteries already developed for human neuropsychological assessment of brain injuries represents the best validated approach to evaluating effects from exposures having unknown effects on the brain. The validity of neuropsychological instru-ments is based on the development of tests which differentiate individuals with focal lesions arising from head trauma, strokes, tumors, surgical procedures, and certain neurodegenerative and metabolic diseases. Because the site of brain injury has been documented in the patients used to validate each task, the overall test batteries have strong structure-function foundations that can be used to advantage in both adult and developmental human neurotoxicological assessments. By analogy, this same approach can be extrapolated to neurobehavioral assessments for neurotoxicological effects in experimental animals.

The concept of using a neuropsychological approach as the basis for assessments in animals has also been proposed previously. Kolb (31) has discussed the application of a neuropsychological structure-function approach to nonhuman primates and rodents with various focal brain lesions induced experimentally. Kolb and his associates, most notably Whishaw, have had a long-standing interest in brain injury and development, especially early versus adult lesions and the degree to which lesioned animals exhibit recovery of function. Their work is instructive to developmental neurotoxicology because it could serve as a basis for the development of a more comprehensive and systematic assessment of developmental neurotoxic effects and provide the field with a more sound conceptual footing.

Kolb and Whishaw's recovery of function work began, in part, with investigations of the Kennard effect, i.e., the notion that early brain injury leads to greater recovery of function than later damage to the same brain region (31). Kolb and Whishaw (32) demonstrated that rats with anterior cortex removal early in life (day 7) had better recovery for some behaviors compared to rats lesioned as adults. This was seen most notably in learned behaviors. For other behaviors, the early lesioned rats showed less recovery than adults, particularly for species-typical behaviors. The latter included such things as "feeding, swimming, hoarding, or defensive burying." In this and their later work, these authors and others challenge the hypothesis that the Kennard effect represents a general characteristic of recovery from early versus later brain lesions [see review by Kolb (30)]. They note instead that some lesions lead to better recovery of some functions in young organisms, while other behaviors seem not to be recovered or are even worse in those with early lesions (30,31). This observation is striking because it matches a similar conclusion reached in behavioral teratology (70). The puzzle, then as now, is to understand why some functions are more recoverable than others. Kolb offers some evidence to suggest that the amount of dendritic arborization following insult is one key factor in predicting later recovery (31). He suggests that better recovery in day 1 and 10 hemidecorticated rats occurs because of increased neuronal arborization in the remaining cortex, whereas day 1 frontal lesions lead to decreased neuronal arborization in remaining cortex and severe long-term deficits.

Apart from the basis for recovery or the ages and sites leading to different degrees of recovery, there is another aspect of Kolb and Whishaw's approach that is instructive. Kolb (30) has reviewed the functions of the frontal cortex in rats and other species and provides a comparative analysis. One of the things he does is to review the data in terms of categories of function that are affected by cortical lesions in different species. This approach is at the heart of establishing brain-behavior relationships and dovetails with the concept of testing functional domains. The concept of testing functional domains was proposed for developmental neurotoxicity a number of years ago (29,58,64).

Categories of function identified by Kolb include motor function, response inhibition, temporal ordering, spatial function, social interactions, spontaneity, olfaction, habituation, associative learning, and language function. Most of the categories are, in turn, further divided into symptom types (ref. 30, Table II). He provides comparable categorization, symptoms and supporting references for effects of the same lesion in monkeys and rats (ref. 30, Tables III and IV). For purposes of the present discussion, the focus will be on Kolb's Table IV for effects in rats.

For motor symptoms, the effects in rats are loss of distal effector movements, poor complex chains of movements, and restricted tongue mobility. Examples include impaired digit and forelimb movement, reduced tongue extension, and impaired claw cutting. Examples of chains of movements include impaired adult-style swimming, especially loss of forelimb inhibition. Swimming patterns, especially with an emphasis on ontological progression to the adult style, has been used extensively in developmental neurotoxicity (68,69), but tongue and digit movements have not.

The next category is response inhibition. Kolb (30) notes that lesioned rats have impairments in reversal learning and on response extinction following conditioning. Such tests have been used in developmental neurotoxicology, although extinction paradigms are underrepresented.

The next category is serial ordering. Kolb (30) described disruptions in patients with cortical lesions of planned behaviors and an inability to distinguish stimuli order (i.e., which of several stimuli were presented last). Rats with frontal cortex lesions have impaired delayed-response acquisition, delayed alternation, and spontaneous alternation performance. These tests have only recently received attention in neurotoxicology (see references on learning and memory tasks in the EPA, 1991 developmental neurotoxicology guideline; and ref. 57 for spontaneous alternation). Species-specific behaviors such as food hoarding, pup retrieval, and nest building are also serially ordered behaviors. Rats with frontal cortex lesions are reported to be impaired on such behavioral sequences. These behaviors have been used in the experimental FAS literature (see review by Meyer and Riley, ref. 38) but not elsewhere in developmental neurotoxicology.

The next category, spatial orientation, Kolb (30) divides into three subcategories: taxis, mapping, and praxis. Taxis represents a task with a cue to the location of the goal or reinforcement. Examples would be cued discrimination tests or the visible platform version of the Morris maze. Rats with frontal cortex lesions are unimpaired on such tasks. Mapping tasks use a point in space as a cue to obtaining the reinforcement with reliance on distal cues. The Morris hidden platform and Olton radial arm mazes are examples of such tests. Frontal cortex lesioned rats are impaired on both of these tasks. Praxis represents a case where the route to the goal is the critical characteristic and the responses learned are a series of fixed sequential movements. Complex mazes are examples of these. Kolb (30) gives the Lashley III maze as a case in point, but the Biel and Cincinnati water mazes are also examples (65). Kolb (30) notes a study in which rats with frontal cortex lesions exhibit evidence of impaired Lashley III maze performance.

In sum, taxis tasks have been used to various extents in developmental neurotoxicology, mapping tasks only rarely, and praxis tasks extensively. Obviously, a more systematic and balanced approach is suggested by the lesion findings for future application to developmental neurotoxicology.

Another category of effects is social behaviors. Although complex to test, Kolb (30) cites work from his own lab that rats with frontal cortex lesions show increased shock-induced aggression after lesioning at some frontal sites and no effects after lesioning at other frontal sites. Testing for social behaviors has been suggested in developmental neurotoxicology (29) but largely ignored in practice.

Effects on spontaneity in humans with frontal cortex lesions are revealed by reduced conversational speech and lower frequencies of spontaneous facial expressions. In animals, Kolb (30) suggests that vocalizations in guinea pigs are similarly reduced after frontal cortex lesions, but no data exist for rats. Developmental vocalization frequency has been suggested as a method for use in developmental neurotoxicology (1,3).

Olfactory discrimination is impaired in humans and rats with frontal cortex lesions according to data reviewed by Kolb (30). Various olfactory tasks have been utilized in developmental neurotoxicology (4,59). These tests appear to merit continued use.

Habituation deficits are one of the most prominent features of animals with frontal cortex lesions (30). Hole-poking, locomotor activity, and startle habituation are examples of behaviors showing this phenomenon, and all these have been widely used in developmental neurotoxicology. The lesion research literature supports their continued use.

Associative learning is found to be impaired in humans and monkeys with frontal cortex lesions (30). Associative tests of learning overlap extensively with the tests of learning noted above (reversal learning, mazes, discrimination). However, in this category Kolb emphasizes tasks in which the stimuli and responses are arbitrarily associated, as between colors and locations, as opposed to those responses that bear a natural relationship to the stimuli, e.g., swimming mazes. In developmental neurotoxicology the data are with operant conditioning tasks, which would be one type of task that would fit into this category. Operant conditioning tests have had mixed success at best in developmental neurotoxicology (7).

Finally, other behaviors affected by frontal cortex lesions in rats have included effects on activity, homeostasis (appetite, body weight, temperature regulation, etc.), and contralateral neglect. The latter is not relevant to developmental neurotoxicology as far as is known, since developmental neurotoxins are not known to induce unilateral focal lesions. Activity and homeostatic measures, however, have been widely used in developmental neurotoxicology and their continued use appears well-founded.

The categories developed by Kolb (30) are only for cortical lesions, yet the list is extensive (see Table 2). Acoustic startle represents a complex brain stem-mediated response that is usually included in developmental neurotoxicology evaluations as a method for detecting damage to these regions and to the inferior colliculi in the midbrain (44). Morris and Olton mazes are known to be sensitive to hippocampal damage. Thus, some of the tasks already in use in developmental neurotoxicology are appropriate for the detection of damage to other brain regions. If these are combined with Kolb's categories, then the series appearing in Table 2 is obtained. The categories given in Table 2 represent a more rational approach to the evaluation of developmental neurotoxicity than what is currently being done and should be given serious consideration. Whishaw et al. (73) have also written about even more extensive ways of examining rats for dysfunction based on an even wider variety of lesions, and the reader is referred to that discussion for a more complete presentation of their approach.

One of the most interesting findings of Kolb and Whishaw is that species-specific behaviors show a marked difference in effects from lesion injuries than

TABLE 2. *Neuropsychologically based neurobehavioral test categories for animal assessment in neurotoxicology*

1. Motor	5. Social
Distal effector movements	Aggression
Chains of movements	Play
2. Response Inhibition	6. Spontaneity
Reversal learning	Vocalization
Extinction	7. Olfaction
3. Serial Ordering	Olfactory discrimination
Learned Behaviors	Olfactory orientation
Delayed response learning	8. Habituation
Delayed alternation learning	Nose-poking
Spontaneous alternation	Locomotion
Species-typical behaviors	Startle
Food hoarding	9. Other
Pup retrieval	Activity
Nest building	Homeostasis
Defensive burying	10. Reactivity
4. Spatial Orientation	Startle
Taxis (signal cue)	
Cued discrimination	
Visible platform Morris maze	
Mapping (spatial cue)	
Morris hidden platform maze	
Olton radial-arm maze	
Praxis (route cues; complex mazes)	
Lashley III maze	
Biel maze	
Cincinnati maze	
Hebb-Williams maze	

do learned behaviors. If nothing else, developmental neurotoxicology should explore this realm more thoroughly to test the concept that pre- or neonatal exposure to neurotoxic agents may also have differential effects on these two broad categories of function.

CONCLUSIONS

This chapter has reviewed the organizing concepts in developmental neurotoxicology. These concepts were illustrated for several developmental neurotoxins for which effects in both humans and animals are well-established. While these generalities are not hard-and-fast rules, they provide the basic structure around which the effects of neurotoxins on the developing nervous system are characterized. It is clear from the broad nature of these concepts that much remains to be learned about the details of how such agents act. Nevertheless, the very existence of the cases of human developmental neurotoxicity reviewed here has already changed the way drugs and chemicals are being tested for human health effects around the world.

The latter represents the second topic of this chapter, i.e., the strategies and regulations used to detect developmental neurotoxicity for the purpose of preventing such effects in human beings. While considerable progress has been made in this realm, it is evident from the review that there are serious conceptual flaws and large gaps in coverage in the existing regulatory guidelines of every nation that has such standards, to say nothing of those that do not.

Third, and finally, this chapter presents a neuropsychological approach for organizing the approach to animal neurobehavioral evaluation. This approach has already been espoused in brain lesion research, and herein it is suggested that this model be employed in both adult and developmental neurotoxicology. The strength of the approach is that it is based on brain structure-function relationships developed from human cases of brain injury and animal studies of induced brain lesions, as opposed to the present state in neurotoxicology, in which compounds are evaluated without regard to any consistent conceptual framework. This is not to imply that this approach is not without difficulties, since brain damage occurring at different sites can often induce similar functional effects. Moreover, the human/animal neuropsychological analogy has not been completely developed, so its adaptation to neurotoxicology will include all the problems attendant to the incorporation of an incompletely understood system. Nevertheless, the neuropsychological approach represents a sound starting point for developing a basis for cross-species extrapolation, a goal that is often discussed and seldom approached in neurotoxicity. In this regard, see the discussion by Kolb and Whishaw (33) on comparative issues in brain-behavior relationships. Ultimately, the success of neurotoxicology as a predictive science will rest upon the successful development of a more comprehensive approach. If neuropsychological theory can move the field closer to this goal, then it will be well worth it.

Finally, to place developmental neurotoxicity in perspective, let us consider the larger problem of developmental disorders. If developmental disorders represent the universe of child development problems that may occur, then some fraction of these may ultimately be attributable to exposure to developmental neurotoxins. Thus, it is instructive to know the breadth of the universe of developmental disorders. In a recent report from the National Center for Health Statistics, Zill and Schoenborn (74) report results of a nationally representative sample of 17,110 US children 17 years of age and under. The survey was conducted in 1988 for the National Health Interview Survey of Child Health. The survey found that the prevalence of developmental delays in US children was 4% or 2.5 million children. For learning disabilities, the prevalence was 6.5% or 3.4 million children. Finally, for emotional or behavioral problems lasting 3 months or more or requiring treatment, the prevalence was 13.4% or 7 million children. Excluding those under 3 and combining the three categories, this amounted to 19.5% or 10.2 million children who have or have had a significant developmental behavioral problem. By contrast, population estimates of chronic childhood diseases are 3.5 million with bronchitis, 3.2 million with asthma, 2.2 million with dermatitis, 1.8 million with orthopedic impairments, and 1.1 million with heart murmurs. The authors conclude, "Clearly, psychological disorders rank among the most prevalent health conditions of modern childhood" (74). These are substantial numbers, amounting to 1 in 5 of all US children. This aggregate reveals little about the multifactorial causes for these disorders, but nevertheless reveals the potential magnitude of the problem and the scale on which benefits may be realized if preventive research in developmental neurotoxicology is successful. Given the size of the problem, even modest improvements in the prevalence of developmental disorders would result in a substantial impact on the lives of thousands of children. Furthermore, when prevention of such disorders is measured in terms of lifetime impact, it is evident that the alleviation of early onset dysfunctions has benefits which accrue over the life span. No problem of this magnitude should go unattended and it is to this end that developmental neurotoxicity research need to progress.

ACKNOWLEDGMENT

The author wishes to express his sincere appreciation to Drs. E. Mollnow and K. D. Acuff-Smith for their invaluable suggestions on an earlier version of this manuscript.

REFERENCES

1. Adams J. Ultrasonic vocalizations as diagnostic tools in studies of developmental toxicity: an investigation of the effects of hypervitaminosis A. *Neurobehav Toxicol Teratol* 1982;4:299–304.
2. Adams J. High incidence of intellectual deficits in 5 year old children exposed to isotretinoin "in utero." *Teratology* 1990;41:614.

3. Adams J, Miller DR, Nelson CJ. Ultrasonic vocalizations as diagnostic tools in studies of developmental toxicity: an investigation of the effects of prenatal treatment with methylmercuric chloride. *Neurobehav Toxicol Teratol* 1983;5:29–34.

4. Adams J, Buelke-Sam J, Kimmel CA, Nelson CJ, Reiter LW, Sobotka TJ, Tilson HA, Nelson BK. Collaborative behavioral teratology study: protocol design and testing procedures. *Neurobehav Toxicol Teratol* 1985;7:579–586.

5. Adams J, Vorhees CV, Middaugh LD. Developmental neurotoxicity of anticonvulsants: human and animal evidence on phenytoin. *Neurotoxicol Teratol* 1990;12:203–214.

6. Boyd TA, Ernhart CB, Greene TH, Sokol RJ, Martier S. Prenatal alcohol exposure and sustained attention in the preschool years. *Neurotoxicol Teratol* 1991;13:49–55.

7. Buelke-Sam J, Kimmel CA, Adams J, Nelson CJ, Vorhees CV, Wright DC, St Omer V, Korol BA, Butcher RE, Geyer MA, Holson JF, Kutscher CL, Wayner MJ. Collaborative behavioral teratology study: results. *Neurobehav Toxicol Teratol* 1985;7:591–624.

8. Burbacher TM, Rodier PM, Weiss B. Methylmercury developmental neurotoxicity: a comparison of effects in humans and animals. *Neurotoxicol Teratol* 1990;12:191–202.

9. Butcher RE. Behavioral testing as a method for assessing risk. *Environ Health Perspect* 1976;18: 75–78.

10. Cattabeni F, Abbracchio MP. Behavioral teratology: an inappropriate term for some uninterpretable effects. *Trends Pharmacol Sci* 1988;9:101–104.

11. Choi BH. The effects of methylmercury on the developing brain. *Prog Neurobiol* 1989;32:447–470.

12. Clarren SK, Astley SJ, Bowden DM. Physical anomalies and developmental delays in nonhuman primate infants exposed to weekly doses of ethanol during gestation. *Teratology* 1988;37:561–569.

13. D'Amato RJ, Lipman ZP, Snyder SH. Selectivity of the Parkinsonian neurotoxin MPTP: toxic metabolite MPP$^+$ binds to neuromelanin. *Science* 1986;231:987–989.

14. Davis JM, Otto DA, Weil DW, Grant LD. The comparative developmental neurotoxicity of lead in humans and animals. *Neurotoxicol Teratol* 1990;12:215–229.

15. Dietrich KN, Succop PA, Berger OG, Hammond PB, Bornschein RL. Lead exposure and the cognitive development of urban preschool children: the Cincinnati Lead Study cohort at age 4 years. *Neurotoxicol Teratol* 1991;13:203–211.

16. Driscoll CD, Streissguth AP, Riley EP. Prenatal alcohol exposure: comparability of effects in humans and animal models. *Neurotoxicol Teratol* 1990;12:231–237.

17. Elmazar MMA, Sullivan FM. Effect of prenatal phenytoin administration on postnatal development of the rat: a behavioral teratology study. *Teratology* 1981;24:115–124.

18. Environmental Protection Agency. Pesticide assessment guidelines, Subdivision F, Hazard Evaluation: Human and Domestic Animals, Addendum 10, Neurotoxicity, Series 81, 82, and 83. US Environmental Protection Agency, Health Effects Division, Office of Pesticide Programs. Washington, DC.

19. Fox DA, Lowndes HE, Bierkamper GG. Electrophysiological techniques in neurotoxicology. In: CL Mitchell, ed. *Nervous system toxicology.* New York: Raven Press, 1982;299–335.

20. Geelen JAG. Hypervitaminosis A induced teratogenesis. *CRC Crit Rev Toxicol* 1979;6:351–376.

21. Goodlett CR, Thomas JD, West JR. Long-term deficits in cerebellar growth and rotarod performance of rats following "binge-like" alcohol exposure during the neonatal brain growth spurt. *Neurotoxicol Teratol* 1991;13:69–74.

22. Greene T, Ernhart CB, Martier S, Sokol R, Ager J. Prenatal alcohol exposure and language development. *Alcoholism: Clin Exp Res* 1990;14:937–945.

23. Greene T, Ernhart CB, Ager J, Sokol R, Martier S, Boyd T. Prenatal alcohol exposure and cognitive development in the preschool years. *Neurotoxicol Teratol* 1991;13:57–68.

24. Hanson JW, Smith DW. The fetal hydantoin syndrome. *J Pediatr* 1975;87:285–290.

25. Hanson JW, Myrianthopoulos NC, Sedgwick-Harvey MA, Smith DW. Risks to the offspring of women treated with hydantoin anticonvulsants, with emphasis on the fetal hydantoin syndrome. *J Pediatr* 1976;89:662–668.

26. Hartman DE. *Neuropsychological toxicology: identification and assessment of human neurotoxic syndromes.* New York: Pergamon Press, 1988.

27. International Federation of Teratology Societies. Draft guideline on "Detection of toxicity in reproducing animals" for pharmaceutical products, Draft 9, March 15, 1991. International Federation of Teratology Societies.

28. Kalter H. *Teratology of the central nervous system.* Chicago: University of Chicago Press, 1968.
29. Kimmel CA, Buelke-Sam J, Adams J. Collaborative behavioral teratology study: implications, current applications and future directions. *Neurobehav Toxicol Teratol* 1985;7:669–773.
30. Kolb B. Functions of the frontal cortex of the rat: a comparative review. *Brain Res Rev* 1984;8: 65–98.
31. Kolb B. Brain development, plasticity, and behavior. *Am Psychol* 1989;44:1203–1212.
32. Kolb B, Whishaw IQ. Neonatal frontal lesions in the rat: sparing of learned but not species-typical behavior in the presence of reduced brain weight and cortical thickness. *J Comp Physiol Psychol* 1981;95:863–879.
33. Kolb B, Whishaw IQ. Problems and principles underlying interspecies comparisons. In: TE, Robinson ed. *Behavioral approaches to brain research.* New York: Oxford University Press, 1983;237–310.
34. Koob GF, Bloom FE. Cellular and molecular mechanisms of drug dependence. *Science* 1988;242: 715–723.
35. Lammer EJ, Chen DT, Hoar RM, Agnish ND, Benke PJ, Braun JT, Curry CJ, Fernhoff PM, Grix AW, Lott IT, Richard JM, Sun SC. Retinoic acid embryopathy. *N Engl J Med* 1985;313: 837–841.
36. Lammer EJ, Adams J, Battaglia B, Holmes LB. Study design of phase 2 of the longitudinal study of infants exposed to isotretinoin in utero. *Teratology* 1990;41:614.
37. Laughlin NK. Animal models of behavioral effects of early lead exposure. In: Riley EP, Vorhees CV, eds. *Handbook of behavioral teratology.* New York: Plenum Press, 1986;291–319.
38. Meyer LS, Riley EP. Behavioral teratology of alcohol. In: Riley EP, Vorhees CV, eds. *Handbook of behavioral teratology.* New York: Plenum Press, 1986;101–140.
39. Minnema DJ, Michaelson IA, Cooper GP. Calcium efflux and neurotransmitter release from rat hippocampal synaptosomes exposed to lead. *Toxicol Appl Pharmacol* 1988;92:351–357.
40. National Research Council. *Biologic markers in reproductive toxicology.* National Academy of Sciences. National Academy Press, Washington, DC.
41. Nebert DW, Gonzalez FJ. P450 genes and evolutionary genetics. *Hosp Pract* 1987; March:63–74.
42. Needleman HL, Schell A, Bellinger D, Leviton A, Allred EN. The long-term effects of exposure to low doses of lead in childhood: an 11-year follow-up report. *N Engl J Med* 1990;322:83–88.
43. Nolen GA. The effects of prenatal retinoic acid on the viability and behavior of the offspring. *Neurobehav Toxicol Teratol* 1986;8:643–654.
44. Parham K, Willott JF. Effects of inferior colliculus lesions on the acoustic startle response. *Behav Neurosci* 1990;104:831–840.
45. Purves D, Lichtman JW. *Principles of neural development.* Sunderland, Massachusetts: Sinauer Associates, 1985.
46. Rice DC, Gilbert SG. Sensitive periods for lead-induced behavioral impairment (nonspatial discrimination reversal) in monkeys. *Toxicol Appl Pharmacol* 1990a;102:101–109.
47. Rice DC, Gilbert SG. Lack of sensitive period for lead-induced behavioral impairment on a spatial delayed alternation task in monkeys. *Toxicol Appl Pharmacol* 1990b;103:364–373.
48. Risau W, Wolburg H. Development of the blood-brain barrier. *Trends Neurosci* 1990;13:714–778.
49. Rodier PM. Chronology of neuron development: animal studies and their clinical implications. *Dev Med Child Neurol* 1980;22:525–545.
50. Schull WJ, Norton S, Jensh RP. Ionizing radiation and the developing brain. *Neurotoxicol Teratol* 1990;12:249–260.
51. Streissguth AP, Aase JM, Clarren SK, Randels SP, LaDue RA, Smith DF. Fetal alcohol syndrome in adolescents and adults. *JAMA* 1991;265:1961–1967.
52. Streissguth AP, Landesman-Dwyer S, Martin JC, Smith DW. Teratogenic effects of alcohol in humans and laboratory animals. *Science* 1980;209:353–361.
53. Streissguth AP, Sampson PD, Barr HM. Neurobehavioral dose-response effects of prenatal alcohol exposure in humans from infancy to adulthood. *Ann N Y Acad Sci* 1989;562:145–158.
54. Tilson HA, Jacobson JL, Rogan WJ. Polychlorinated biphenyls and the developing nervous system: cross-species comparisons. *Neurotoxicol Teratol* 1990;12:239–248.
55. VanOverloop D, Schnell RR, Holmes LB. The effects of prenatal exposure to phenytoin and other anticonvulsants on intellectual function at 4 to 8 years. *Neurotoxicol Teratol* 1991;13 (in press).

56. Vorhees CV. Some behavioral effects of maternal hypervitaminosis A in rats. *Teratology* 1974;10: 269–273.
57. Vorhees CV. Fetal anticonvulsant syndrome in rats: dose- and period-response relationships of prenatal diphenylhydantoin, trimethadione, and phenobarbital exposure on the structural and functional development of the offspring. *J Pharmacol Exp Ther* 1983;227:274–287.
58. Vorhees CV. Comparison of the Collaborative Behavioral Teratology Study and Cincinnati Behavioral Teratology test batteries. *Neurobehav Toxicol Teratol* 1985a;7:625–633.
59. Vorhees CV. Behavioral effects of prenatal methylmercury in rats: a parallel trial to the Collaborative Behavioral Teratology Study. *Neurobehav Toxicol Teratol* 1985b;7:717–725.
60. Vorhees CV. Principles of behavioral teratology. In: Riley EP, Vorhees CV, eds. *Handbook of behavioral teratology.* New York: Plenum Press, 1986a;23–48.
61. Vorhees CV. Comparison and critique of government regulations for behavioral teratology. In: Riley EP, Vorhees CV, eds. *Handbook of behavioral teratology.* New York: Plenum Press, 1986b;49–66.
62. Vorhees CV. Retinoic acid embryopathy. *N Engl J Med* 1986c;315:262–263.
63. Vorhees CV. Fetal hydantoin syndrome in rats: dose-effect relationships of prenatal phenytoin on postnatal development and behavior. *Teratology* 1987a;35:287–303.
64. Vorhees CV. Reliability, sensitivity, and validity of behavioral indices of neurotoxicity. *Neurotoxicol Teratol* 1987b;9:445–464.
65. Vorhees CV. Maze learning in rats: a comparison of performance in two water mazes in progeny prenatally exposed to different doses of phenytoin. *Neurotoxicol Teratol* 1987c;9:235–241.
66. Vorhees CV. Behavioral teratology: what's not in a name: a reply to Cattabeni and Abbracchio. *Neurotoxicol Teratol* 1989;11:325–327.
67. Vorhees CV, Brunner RL, McDaniel CR, Butcher RE. The relationship of gestational age to vitamin A induced postnatal dysfunction. *Teratology* 1978;17:271–276.
68. Vorhees CV, Brunner RL, Butcher RE. Psychotropic drugs as behavioral teratogens. *Science* 1979a;205:1220–1225.
69. Vorhees CV, Butcher RE, Brunner RL, Sobotka TJ. A developmental test battery for neurobehavioral toxicity in rats: a preliminary analysis using monosodium glutamate, calcium carrageenan and hydroxyurea. *Toxicol Appl Pharmacol* 1979b;50:267–282.
70. Vorhees CV, Mollnow E. Behavioral teratogenesis: long-term influences on behavior from early exposure to environmental agents. In: Osofsky JD, ed. *Handbook of infant development,* 2nd ed. New York: John Wiley, 1987;700–744.
71. Vorhees CV, Weisenburger WP, Acuff-Smith KD, Minck DR. An analysis of factors influencing complex water maze learning in rats: effects of task complexity, path order and escape assistance on performance following prenatal exposure to phenytoin. *Neurotoxicol Teratol* 1991;13:213–222.
72. Weisenburger WP, Minck DR, Acuff KD, Vorhees CV. Dose-response effects of prenatal phenytoin exposure in rats: effects on early locomotion, maze learning, and memory as a function of phenytoin-induced circling behavior. *Neurotoxicol Teratol* 1990;12:145–152.
73. Whishaw IQ, Kolb B, Sutherland RJ. The analysis of behavior in the laboratory rat. In: Robinson TE, ed. *Behavioral approaches to brain research.* New York: Oxford University Press, 1983;141–211.
74. Zill N, Schoenborn CA. Developmental, learning, and emotional problems: health of our nation's children, United States, 1988. Advance Data, Vital and Health Statistics, National Center for Health Statistics, No. 190, pp. 1–18, 1990.

Neurotoxicology, edited by Hugh Tilson and
Clifford Mitchell. Published by Raven Press, Ltd.,
New York 1992.

13

Risk Assessment for Neurotoxicants

David Gaylor and William Slikker, Jr.

*National Center for Toxicological Research, US Food and Drug Administration,
Jefferson, Arkansas 72079*

Risk assessment is a process that is generally considered to consist of four parts: hazard identification, dose-response description, human exposure assessment, and quantitative risk characterization (estimation). Hazard identification is the process of establishing the potential for toxic effects in humans as a result of exposure to a substance. This process may utilize data collected from humans and/or experimental animals. Dose-response relationships are generated from epidemiological data or from bioassays of laboratory animals in order to estimate the relationship between the risk (proportion of individuals developing a toxic effect or disease) and dose of a toxic substance, for various conditions of exposure. Exposure assessment is the process of determining the nature and extent of human exposure from toxic substances present in the environment. This process involves determining the distribution of toxic substances in the environment and the resulting dose and duration of exposure to humans by inhalation, ingestion, and/or dermal absorption. Further, pharmacokinetic and/or pharmacodynamic data may be generated to estimate target tissue exposure. The final step of the risk assessment process involves the integration of the available data to estimate risk for various exposure conditions. This final step generally requires the use of assumptions to extrapolate the results both from the information available for one set of exposure conditions to other exposure conditions and from results obtained from laboratory animals to humans.

The discussion in this chapter is focused on quantitative risk estimation. Thus, the emphasis is on the development of dose-response relationships to provide numerical estimates of risk as a function of dose. Some discussion is provided on the difficult problem of extrapolating results obtained from experimental dose levels used to produce toxic effects in animals to generally much lower environmental exposure levels that may produce subtle effects in humans. The uncertainties involved in extrapolating results from animals to humans is beyond the scope of this chapter. Risk assessment generally is a very uncertain process, but the regulation of toxic substances usually cannot be postponed until precise estimates of risk are obtained. The purpose of this chapter is not to justify the application of the risk assessment process for neurotoxic effects, but to discuss some procedures which may be applicable to the estimation of neurotoxic risk from exposure to potential neurotoxicants.

REFERENCE DOSE

It is generally assumed that for all biological effects, other than genotoxic carcinogens, a threshold dose for a toxic substance exists below which a given toxic effect will not occur in any individual. The assumption of the existence of threshold doses is also based on observations of no discernible differences in biological effects between individuals exposed to low dose levels and unexposed individuals.

The problem inherent with an observation of no discernible effects at low doses is that it is impossible to determine whether the risk is actually zero (i.e., the dose is below a threshold dose) or whether the statistical resolving power of a study is inadequate to detect small risks. Every study has a statistical limit of detection which depends on the number of individuals or animals involved. For example, it would be relatively unusual to conduct an experiment on a neurotoxicant with as many as 100 animals per dose. If no deleterious effects were observed in 100 animals at a particular dose, one might feel relatively comfortable that this dose level is below the threshold dose. However, we can only be 95% confident that the true risk is less than 0.03. That is, if 3% of the animals in a population actually develop a toxic effect at this dose, there is a 5% chance that a group of 100 animals would not show any effect. The observation of no toxic effects in an extremely large sample of 1,000 animals only indicates with 95% confidence that the true risk is less than 0.003, etc. Hence, threshold doses cannot be established from animal bioassay data. In general, thresholds cannot be demonstrated and are therefore assumed.

The assumption of the existence of threshold doses for neurotoxicants has led to a regulatory process of dividing the no adverse effect level (NOAEL) for a toxic substance by a safety factor (SF) or uncertainty factor (UF) to establish an allowable (reference) dose (7). The NOAEL is taken as the largest dose for which a toxic effect is not observed. Often this dose is based on the lack of obtaining a statistically significant effect. However, the statistical resolving power associated with such results generally is not stated. The NOAEL is not necessarily below the threshold dose. For most experiments, a risk of 1% would go undetected and even a dose with a risk as high as 10% could be designated as the NOAEL. To obtain an allowable (reference) dose for a neurotoxicant, a NOAEL based on animal data is generally divided by a safety factor of 10 to allow for the possible increased sensitivity of humans compared to animals and divided by another safety factor of 10 to account for variation in sensitivities among individuals (7). That is, the dose equal to the NOAEL/100 is assumed to pose no or negligible risk. This procedure currently serves as the basis for the regulation of most neurotoxicants. If the risk at the NOAEL were 10% and if the risk at low doses were proportional to dose, then the risk at the NOAEL/100 would be 0.001 (1 in 1,000), assuming that humans were no more sensitive than animals to the neurotoxicant. The NOAEL/Safety Factor approach does not provide any estimate of the risk. In some instances, the resulting risk could be unknowingly unacceptable. The purpose of this chapter is to discuss alternative procedures where estimates of risk are provided as a function of dose.

Even though a neurotoxic effect may require a threshold dose to be initiated in an individual, a threshold dose may not exist for a population of individuals. If a substance augments a neurotoxic process which is already present in some individuals who are not exposed to the substance, then the addition of any dose of the toxic substance will produce an increase in risk to the population. This assumes that the administration of a substance results in an increase in the amount of neurotoxic agent at the target tissue site. Even if a threshold dose exists, the presence of the endpoint used to determine a neurotoxic effect in individuals not exposed to the substance indicates that existing exogenous or endogenous factors have already surpassed the threshold dose of the active agent in these individuals. Hence, the addition of any dose, no matter how small, which augments a neurotoxic process which is already occurring will produce an additional risk. Crump et al. (4) also show that the dose-response in such cases will be approximately linear in the vicinity of the background risk. Since many toxic effects appear in unexposed individuals, it is difficult to argue that a threshold dose occurs for a population, below which no individual will exhibit a toxic effect, unless it can be shown that the substance administered acts through an entirely different process than any existing processes which may be producing the same neurotoxic effect. Thus, there is no guarantee that the dose corresponding to the NOAEL/100 will be below a population threshold dose.

Some improvement and consistency in establishing reference doses could be achieved by replacing the NOAEL by a precisely defined quantity which can be estimated from bioassay data (1). For example, the effective dose corresponding to an increased incidence of a neurotoxic effect in 10% of the individuals tested (ED10) could be estimated. A lower 95% confidence limit (LED10) can be calculated to account for experimental variability and uncertainty. The LED10 would be akin to the lowest observed adverse level (LOAEL) and would require an additional safety factor of 10. Hence, the reference dose would be established by the LED10/1,000. Use of the LED10, in place of the LOAEL, has the advantage that it has a precise definition, uses all of the bioassay data to obtain the LED10, and encourages better experiments to obtain larger values of the LED10 (6). In many cases, an adequate estimate of the dose which causes an incidence of neurotoxic effects in 1% of the experimental animals (ED01) can be obtained with bioassay data. The lower confidence limit (LED01) is akin to a NOAEL. Hence, the reference dose could be obtained from the LED01/100. In addition to quantal responses, Crump (3) also gives a procedure for determining the confidence limit for the dose corresponding to a given percentage change from the control average for a continuous measurement.

RISK ESTIMATION

A dose-response relationship must be obtained in order to calculate estimates of the risk of a toxic effect as a function of dose. Risk is defined as the probability (from 0 to 1) that a toxic effect in an individual will develop. If the estimate of risk is for a group (population) of individuals, then the risk may be expressed as the proportion or percentage of individuals developing a toxic effect for a given set of exposure conditions.

Obviously, the most relevant estimate of risk for exposure to a neurotoxicant would be obtained from humans exposed to the same conditions. This information is seldom available since accurate exposure conditions are seldom available for humans. Confounding factors, such as nutritional or health status, alcohol or drug consumption, and exposure to other, possibly unknown, neurotoxicants, make it difficult to estimate the effects of a neurotoxicant in humans. When human dose-response information is available, the data often arise from occupational groups or from a clinical setting where exposures are higher than those existing in the usual environment. Thus, risk estimates for environmental exposure levels would require extrapolation to lower exposures.

Since adequate human dose-response information is not usually available, risk estimates generally must be based on animal bioassay data. Such experiments usually employ a few dose levels with less than a dozen animals per dose. Hence, in order to elicit measurable neurotoxic effects in a small number of animals, the incidence must be relatively high. This generally requires the use of relatively high doses that are above environmental exposure levels. This produces two major problems in risk assessment: high-to-low dose extrapolation and animal-to-human extrapolation. The problems involved in extrapolation of results in animals to humans are beyond the scope of this chapter. The remainder of this section is devoted to discussing dose-response models and risk estimation.

The *first* step is to establish a mathematical dose-response relationship between a measure of a biological effect, or biomarker, and the dose of the substance administered. If possible, the dose of the active agent at the target tissue site should be used. The *second* step is to determine the distribution (variability) of individual measurements of biological effects or their biomarkers around the average value estimated by the dose-response curve. The *third* step is to establish an adverse, or abnormal, level of a biological effect or biomarker in an unexposed population. The *fourth* and final step is to combine the information from the first three steps to estimate the risk (proportion of individuals exceeding an adverse or abnormal level of a biological effect or biomarker) as a function of dose.

Dose-Response

Since exposure to many neurotoxicants is unavoidable, it is desirable to establish a relationship between the incidence (probability) of a neurotoxic effect and exposure (dose). This allows an informed judgment to be made regarding the effects of exposure to a neurotoxicant as a function of dose. First, a mathematical model is needed that describes the relationship between a biological effect, or biomarker, and the dose of a neurotoxicant. With the limited number of doses employed in most bioassays, it is difficult to ascertain the shape of a dose-response curve. In the absence of knowledge of the biological information which might suggest the shape of a dose-response curve, a general polynomial function might be used to simply provide a mathematical model which describes the data in the experimental dose range

$$y = b_0 + b_1 d + b_2 d^2 + \cdots \qquad [1]$$

where y represents a neurotoxic measure (e.g., serotonin level in a specific region of the brain), d represents the dose, and the b's are parameters to be estimated from the data by statistical curve-fitting procedures. For reasons to be discussed in the next section, it may be desirable to use the logarithm of the response, $\ln y$. Whereas the polynomial model may be adequate to describe the dose-response function in the experimental dose range, it provides no assurance of the validity of the estimation of the response at lower doses experienced by humans.

It is desirable to utilize a dose-response model which has a plausible biological basis to provide greater validity for low-dose estimates. Slikker and Gaylor (8) suggest that the relationship between the levels of a neurochemical biomarker and the administered dose of a neurotoxicant may be subject to saturable processes. The metabolic biotransformation of a parent neurotoxic compound to its active metabolite or the interaction of the proximate toxicant with target cell membranes or receptor sites is likely to contain a saturable step due to the presence of finite amounts of enzyme or receptor sites.

In many cases, a saturation model may be used to describe the relationship between the estimated average level of a neurochemical measure $\bar{y}(d)$, expressed as percent of the control average level at a dose (d) of a neurotoxicant

$$\bar{y}(d) = \frac{1 + b_0 d}{1 + b_1 d} \times 100\%. \qquad [2]$$

At dose $d = 0$, the level is 100% of the normal average. The constants b_0 and b_1 are estimated empirically by statistical curve-fitting procedures from the data. Here, a nonlinear regression procedure is required. The ratio, $(b_0/b_1) \times 100\%$, provides an estimate of the minimum percent of the control mean that is obtained at large doses and $(b_0 - b_1) \times 100\%$ is the percent change (slope) of the normal average per unit of dose for low doses approaching zero. The saturation curve, Eq 2, is illustrated in Fig. 1.

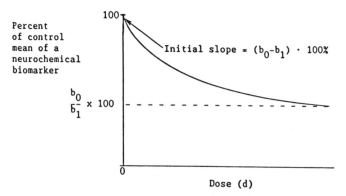

FIG. 1. General saturation dose-response curve.

Distribution of Measurements

The second step in the risk estimation process is to determine the distribution (variation) of individual measurements of biological effects, or their biomarkers, around the average value estimated by the dose-response curve. For example, measurements of the concentrations of the neurotransmitter serotonin (5-HT), its metabolite, 5-hydroxyindoleacetic acid (5-HIAA), and the numbers of degenerated axons per unit area in various regions of the brain were observed by Gaylor and Slikker (5) to be compatible with log-normal distributions. Wyzga (9) also utilized the log-normal distribution to describe blood lead levels in humans. That is, the logarithm of the measurement tends to be adequately described by a normal distribution. In such cases, the proportion of measurements expected to fall within various limits can be readily estimated.

A log-normal distribution is described by the average ($\bar{y}_{\ln y}$) and standard deviation ($s_{\ln y}$) of the logarithms of the measurements about $\bar{y}_{\ln y}$. The average of the logarithms of measurements is also the geometric mean of the measurements. Standard statistical normal deviates (Z_p) can be used to estimate the proportion (p) of individual measurements expected to fall within various limits. A proportion p of measurements are estimated to fall within

$$\bar{y}_{\ln y} \pm Z_p s_{\ln y} \qquad [3]$$

where $\bar{y}_{\ln y}$ is estimated by the dose-response curve and $s_{\ln y}$ is the standard deviation of ln y about the dose-response curve. The value of Z_p can be obtained from standard normal tables, (e.g., ref. 2). For example, 95% of individual measurements are estimated to fall within the limits given by $\bar{y}_{\ln y} \pm 1.96\ s_{\ln y}$. Or, 99% of the measurements are estimated to be above

$$\bar{y}_{\ln y} - 2.33 s_{\ln y}$$

and 99.9% larger than

$$\bar{y}_{\ln y} - 3.09 s_{\ln y}.$$

As seen in the next section, this value will be used to establish an abnormal level of a measurement for untreated control animals.

Adverse / Abnormal Level

The third step in the risk estimation process is to establish an adverse or abnormal level of a biological effect or biomarker. For example, silver stain neurohistology is used to identify dead or dying neurons. A maximum number of dead and/or dying neurons could be selected, which is considered to indicate the boundary between safe and adverse levels that should not be exceeded. From a model that relates the number of dead neurons to dose of neurotoxicant and from a measure of the variability in the number of dead neurons among animals exposed to a given dose, it is possible to estimate the risk (proportion of animals) of exceeding an adverse or safe number of dead neurons.

Suppose the average number of degenerating or dead axons in the caudate of the brain of untreated rats is $2 \times 10^3/\text{mm}^2$. Further, suppose the upper level of the normal range, as defined by three standard deviations above the average, is $4 \times 10^3/\text{mm}^2$. If the number of degenerated axons then exceeded $4 \times 10^3/\text{mm}^2$ due to exposure to a neurotoxicant, this would be considered an abnormal level and could be considered an adverse effect. Risk could be defined as the proportion of animals that exceeded 4.0×10^3 degenerated axons per mm^2.

In the absence of a well-established adverse effect level, an abnormal level may be used which is based on the distribution of levels observed for unexposed control animals. For example, a level may be considered abnormal if it is lower than the level at the 0.1 percentile or above the 99.9 percentile of the background levels observed in an unexposed control population of animals. That is, a level of an endogenous neurochemical may be considered abnormal if that level is in the range expected in fewer than 1 per 1,000 individuals in an unexposed control population. In general, measurements are not available on 1,000 or more individuals. Thus, the 0.1 percentile, or 99.9 percentile, cannot be observed directly. However, as discussed in the previous section, if the measurements are log-normally distributed, the 0.1 percentile of ln y is estimated by

$$\ln y_{.001} = \bar{y}_{\ln y} - 3.09 s_{\ln y} \qquad [4]$$

where $\bar{y}_{\ln y}$ is the average of the logarithms of the neurotoxic measure of the untreated control animals and $s_{\ln y}$ is the standard deviation of the ln y.

If there is concern about large values of a neurotoxic measure, then levels above the 99.9 percentile may be considered abnormal and may be undesirable. The 99.9 percentile for the logarithm of the neurotoxic measure is estimated by

$$\ln y_{.999} = \bar{y}_{\ln y} + 3.09 s_{\ln y}. \qquad [5]$$

Other percentiles could be chosen for abnormal levels. In this case, 3.09 would be replaced by the standard normal deviate, Z_p, for the pth percentile.

Quantitative Risk Estimation

To reiterate, quantitative risk estimation for a neurotoxic measure requires four steps: (a) determination of a dose-response model, (b) determination of the distribution of measurements about the model, (c) determination of an adverse, or abnormal, level of the neurotoxic measure, and (d) estimation of the probability that a measure is beyond the abnormal level as a function of dose.

In general, let the dose-response curve be represented by

$$\bar{y}(d) = f(d). \qquad [6]$$

Often, this function may be a saturation curve as described by Eq 2. The average value, $\bar{y}(d)$, of the neurotoxic measure at dose $= d$ is estimated by a function of dose that is obtained by fitting this function to dose-response data by an appropriate statistical curve-fitting technique.

Since Gaylor and Slikker (5) noted that several neurotoxic measures are log-normally distributed, it is likely that the logarithms of the neurotoxic measure will be normally distributed around the geometric mean. The distribution of the measures around the dose-response function needs to be examined in each specific case. In some cases the original measures may be normally distributed, and in some cases a transformation other than the logarithm may be more appropriate. In general, and for purposes of illustration here, it is assumed that the neurotoxic measure is approximately log-normally distributed.

As discussed in the previous section, in the absence of an established adverse level of a neurotoxic measure, the pth percentile will be considered an abnormal level. For a log-normally distributed measure, the pth percentile is defined by

$$\ln y_p = \bar{y}_{\ln y} + Z_p s_{\ln y} \qquad [7]$$

where $\bar{y}_{\ln y}$ is the average of the logarithm of the neurotoxic values for the control animals, $s_{\ln y}$ is the standard deviation for $\ln y$, and Z_p is the standard normal deviate for the pth percentile. For example, the 0.1 percentile is

$$\ln y_{.001} = \bar{y}_{\ln y} - 3.09 s_{\ln y}.$$

The probability that a log-normally distributed neurotoxic measure, $\ln y(d)$, is less than $\ln y_p$ at a dose $= d$ is

$$P[\ln y(d) \le \ln y_p]$$

which is equal to the probability that the standard normal deviate, Z, is less than

$$P\left[Z < \frac{\ln y_p - \ln \bar{y}(d)}{s_{\ln y}(d)}\right] \qquad [8]$$

where $s_{\ln y}(d)$ is the standard deviation at the dose $= d$. Gaylor and Slikker (5) noted that $s_{\ln y}(d)$ generally is nearly constant across dose levels for each case investigated. That is, the percent error (coefficient of variation) often remains constant across dose levels. In such cases, this improves the precision of the estimate of $s_{\ln y}(d)$ since the pooled variance among ln y within dose groups can be used. If a constant coefficient of variation is not achieved, the calculation of $s_{\ln y}(d)$ may be complex; discussion of this case is beyond the scope of this chapter.

If there is concern about large values of a neurotoxic measure, then the probability that a log-normally distributed neurotoxic measure, ln $y(d)$, is greater than the abnormal level is the probability that the standard normal deviate, Z, is greater than

$$P\left[Z > \frac{\ln y_p - \ln \bar{y}(d)}{s_{\ln y}(d)}\right]. \qquad [9]$$

A dose (d) corresponding to a selected level of risk, presumably very low, can be calculated from the above process. This dose could be divided by an additional uncertainty (safety) factor, if deemed desirable, to allow for potentially greater sensitivity of humans or for possible experimental shortcomings such as less-than-lifetime exposure. Preferably, the latter type of limitation could be corrected by a different experimental design.

Example

The procedure described in the previous section is illustrated here with data on the neurotransmitter serotonin (5-HT) measured in the hippocampus of male rats 30 days after exposure to methylenedioxymethamphetamine (MDMA). MDMA was administered orally on 4 consecutive days. Dose is expressed in terms of mg of MDMA per kg of body weight per day. The average levels of 5-HT are given in Table 1.

As noted by Slikker and Gaylor (8), these data are described by a saturation curve of the type given by Eq 2 and illustrated in Fig. 1. The Statistical Analysis System (SAS) (SAS Institute, Inc., Box 8000, Cary, North Carolina) procedure

TABLE 1. *Average levels, expressed as percent of the controls, of 5-HT in the hippocampus of male rats 30 days after oral dosing on 4 consecutive days with MDMA*

Dose mg/kg/day	Percent of control level
0	100
5	88
10	79
80[a]	46
160	53
160[a]	54

[a] Animals sacrificed 14 days after dosing.

NONLIN was used to fit the saturation curve to the average levels of MDMA expressed as percent of the control level, $\bar{y}(d)$, giving

$$\bar{y}(d) = \frac{1 + .031d}{1 + .068d} \times 100\%.$$

The initial slope is $(.031 - .068) \times 100\% = -3.7\%$. That is, at low doses there is 3.7% reduction in the level of 5-HT in the rat hippocampus per mg/kg/day of MDMA. At large doses, saturation limits the minimum average level of 5-HT to $(.031/.068) \times 100\% = 46\%$ of the average level of 5-HT observed in unexposed controls.

The levels of 5-HT for individual rats are approximately log-normally distributed. That is, the values of ln (5-HT) are approximately normally distributed and the standard deviation is $s_{\ln y} = 0.318$. By definition, for the controls, $\bar{y} = 100\%$ and $\bar{y}_{\ln y} = \ln(1.0) = 0$. If the 0.1 percentile is used for the abnormal level

$$\ln y_{.001} = -3.09(.318) = -0.983$$

and

$$_e \ln y_{.001} = {}_e -0.983 = 0.374.$$

Thus, the 0.1 percentile is 37.4% of the mean control level of 5-HT in the rat hippocampus. This is less than the average minimum level of 5-HT of 46% of the mean control level estimated at large doses of MDMA, but levels in individual animals will be even lower than 37.4% of the mean control level when exposed to MDMA.

The estimate of the mean value of 5-HT at a dose of 1 mg/kg/day of MDMA is

$$\bar{y}(1) = \frac{1 + .031}{1 + .068} \times 100\% = 96.5\%$$

of the mean control level. The estimate of the risk (probability of an animal with a 5-HT level below the abnormal level of 37.4% of the control level) at a dose of 1 mg/kg/day is the probability that the standard normal deviate, Z, is less than

$$Z = \frac{\ln y_{.001} - \ln \bar{y}(1)}{s_{\ln y}} = \frac{\ln\left(\dfrac{y.001}{\bar{y}(1)}\right)}{s_{\ln y}}$$

$$= \frac{\ln\left(\dfrac{37.4}{96.5}\right)}{0.318} = -\frac{.948}{.318} = -2.98$$

which is 0.0014. The additional risk above the risk for the controls is (0.0014 − 0.001) = 0.0004, i.e., 4 in 10,000. Similarly, the additional risk at any other dose level could be estimated.

The dose corresponding to any given level of additional risk can be estimated. For example, suppose it is of interest to estimate the dose at which the additional risk is 0.0001, i.e., 1 in 10,000. Since the risk for the control animals is by definition equal to 0.001, we need to find the dose for which the total risk is 0.0011. The standard normal deviate corresponding to a probability of 0.0011 is −3.07. The dose corresponding to this level of risk is the solution to the equation

$$Z = \frac{\ln\left(\dfrac{y.001}{\bar{y}(d)}\right)}{s_{\ln y}} = \frac{\ln\left(\dfrac{37.4}{\bar{y}(d)}\right)}{0.318} = -3.07$$

which gives $\bar{y}(d) = 99.3\%$. That is, the estimated dose has a mean value of 99.3% of the control mean. Thus, the estimated dose is the solution of

$$99.3 = \frac{1 + .031\,d}{1 + .068\,d} \times 100\%$$

which is $d = 0.2$ mg/kg/day.

For these data, if the lowest dose of 5 mg/kg/day is taken to be a no observed adverse effect level, the reference dose is $5/100 = 0.05$ mg/kg/day. This dose is considerably more conservative than the dose of 0.2 mg/kg/day estimated to have an additional risk of 0.0001.

SUMMARY

It is generally assumed that neurotoxicants have threshold doses below which no adverse effects occur. Usually a no observed adverse effect level is established from dose-response bioassay data. For current regulatory procedures, a reference dose is calculated by dividing the NOAEL by a combination of safety factors, usually 100, to account for extrapolation from animals to humans and to allow for sensitive humans in the population. It is assumed that the risk at the reference dose is zero or negligible. However, this process does not provide any estimate of the risk at the reference dose.

If a neurotoxic effect occurs in unexposed individuals, exogenous or endogenous factors may be adequate to exceed the threshold dose. Thus, for a toxic substance which augments the processes involved in the production of neurotoxicity, the population threshold has already been surpassed. In this case, a risk exists for any additional dose, including the reference dose. Thus, a procedure is needed to estimate risk as a function of dose in order to effectively regulate such neurotoxicants.

A four-step procedure is described which provides an estimate of the risk (probability) of an adverse or abnormal neurotoxic effect as a function of the dose of the toxicant. The first step establishes a mathematical dose-response relationship between a measure of a biological effect or biomarker (e.g., level of serotonin in the hippocampus) and the dose of the neurotoxicant administered. The second step requires definition of the distribution (variability) of individual measurements of the biological effect or biomarker around the average value estimated from the dose-response relationship. It is noted that the logarithms of biological effects or biomarkers frequently are distributed normally. This facilitates the estimation of probabilities. The third step requires establishing an adverse or abnormal level of a biological effect or biomarker in an unexposed population. In the absence of an established adverse level, it is suggested that an estimate of the 0.1 percentile or 99.9 percentile in an unexposed population represents an abnormal level that could be used as a reference level for estimating risks. The fourth step combines the information from the first three steps to give an estimate of the risk (probability) of individuals reaching the adverse or abnormal level of a biological effect or biomarker as a function of dose. Risk estimates may be improved if the dose of toxicant in the target tissue(s) can be measured.

The validity of a risk estimate is improved if the mathematical relationship used to describe the dose-response curve is based on suspected or plausible biological mechanisms. For example, the metabolic transformation of a parent compound to its active metabolite or the interaction of the proximate toxicant with target cells, membranes, or receptor sites are events which are likely to contain saturable processes due to definitive amounts of enzyme or receptor sites. In such cases the dose-response relationship would be expected to follow a saturable curve.

The regulation of neurotoxicants by establishing a reference dose based on the NOAEL does not make efficient use of dose-response data and does not encourage better experimentation. The calculation of a numerical estimate of risk as a function of dose provides a quantitative basis for regulatory decisions.

REFERENCES

1. Barnes DG, Dourson ML. Reference dose (RfD): description and use in health risk assessments. *Regul Toxicol Pharmacol* 1988;8:471–486.
2. Beyer WH. *Handbook of tables for probability and statistics,* 2nd ed. Cleveland, OH: The Chemical Rubber Co., 1968.
3. Crump KS. A new method for determining allowable daily intakes. *Fundam Appl Toxicol* 1984;4: 854–871.
4. Crump KS, Hoel DG, Langley CH, Peto R. Fundamental carcinogenic processes and their implications for low dose risk assessment. *Cancer Res* 1976;36:2973–2979.
5. Gaylor DW, Slikker W, Jr. Risk assessment for neurotoxic effects. *Neurotoxicol* 1990;11: 211–218.

6. Kimmel CA, Gaylor DW. Issues in qualitative and quantitative risk analysis for developmental toxicology. *Risk Analysis* 1988;8:15–20.
7. Lehman AJ, Fitzhugh OG. 100-fold margin of safety. *Q Bull Assoc Food Drug Officials* 1954;18: 33–35.
8. Slikker W, Jr, Gaylor DW. Biologically-based dose-response model for neurotoxicity risk assessment. *Korean J Toxicol* 1990; 6:205–213.
9. Wyzga RE. Towards quantitative risk assessment for neurotoxicity. *Neurotoxicol* 1990;11: 199–208.

Neurotoxicology, edited by Hugh Tilson and
Clifford Mitchell. Published by Raven Press, Ltd.,
New York 1992.

14

Qualitative and Quantitative Issues in Assessment of Neurotoxic Effects

William F. Sette* and Robert C. MacPhail†

*Health Effects Division, Office of Pesticide Programs, US Environmental Protection
Agency, Washington, DC 20460; †Neurotoxicology Division, Health Effects
Research Laboratory, Office of Research and Development, US Environmental
Protection Agency, Research Triangle Park, North Carolina 27751*

The purpose of this chapter is to review several issues that have arisen in the assessment of neurotoxicity data in the Office of Pesticides and Toxic Substances (OPTS) of the Environmental Protection Agency (EPA). While the actions of OPTS represent only a part of the activities among US federal agencies (1, 2) to test and assess neurotoxic effects, and this chapter reflects only our own experiences, we would suggest that these are issues generally relevant to this area for risk assessments conducted under different statutes and for a variety of purposes. Some of these issues are common to many types of target organ toxicity, while others have features particular to the evaluation of effects on the nervous system. Some are issues that have long been debated, while others are now arising in the wake of activities being undertaken to change existing policies.

Risk assessment has been defined as consisting of essentially four steps: hazard identification, dose response assessment, exposure assessment, and risk characterization (3). The risk assessment is then used as part of the basis for risk management, which then takes into consideration the economic benefits and costs of potential regulation, technical factors, and other factors that help shape regulatory options, and legal and political considerations to reach a regulatory conclusion.

Briefly, we will discuss the evolution of the EPA testing guidelines because they form a significant focus for the hazard identification and characterization of neurotoxic and behavioral effects. Also examined in this chapter are general differences in the adversity of different types of neurotoxicity data, including the interpretation of behavioral data in animals exhibiting other types of target organ toxicity. We distinguish between different uses of the term threshold in relation to risk assessment; review the major approaches used in quantitative assessment of dose-response data, including a survey of some of the different specific tech-

The research described in this article has been reviewed by the Health Effects Research Laboratory,
U.S. Environmental Protection Agency, and approved for publication. Approval does not signify
that the contents necessarily reflect the views and policies of the Agency nor does mention of trade
names or commercial products constitute endorsement or recommendation for use.

niques proposed for neurotoxicity data; and describe some implications of the dose-response assessment in risk characterization.

The past decade has been witness to a substantial increase in concern about and interest in neurotoxicology and regulation. These developments have stemmed in large part from a number of reports of behavioral and neurological disorders in humans exposed to a wide range of environmental and industrial chemicals (4). More recently, concern has been expressed about the potential role of environmental factors in the increased incidence of neurodegenerative diseases, including amyotropic lateral sclerosis and Parkinson's disease (5, 6). Simultaneously, there has been increasing recognition of how little is actually known about the neurotoxic potential of most chemicals to which humans are exposed (7). As a consequence, broad support has been expressed by national and international expert committees, the American public, and the federal government for improvements in the way chemicals are evaluated and regulated based on neurotoxic potential. The EPA has initiated a number of activities over the last decade in an effort to address some of these concerns with respect to industrial chemicals and pesticides. At the same time, however, the EPA and other federal regulatory agencies have been criticized for either providing too broad (industry's point of view) or too narrow (environmental organizations' point of view) a regulatory scope.

In the last 30 years, regulation of neurotoxicants has chiefly come from the recommendations of threshold limit values (TLVs) and related exposure limits by the American Council of Governmental Industrial Hygienists (ACGIH) adopted by the Occupational Safety and Health Administration (OSHA) for chemicals in the workplace, and the derivation of exposure limits derived and adopted by regulatory agencies, chiefly as has been done for pesticides and drinking water contaminants (2). The values derived by the ACGIH were adopted in two mass actions as the basis of OSHA standards: the first in 1974 and the second in January 1989 (1, 8). The TLVs come from the conclusions of deliberative bodies and are based on different kinds of data. They do not follow an overt or rigid formula, and may provide larger or smaller uncertainty factors for different situations; e.g., in some cases, levels are set just below those associated with complaints of workers. In contrast, most other exposure limits are defined by the regulatory agencies themselves, rely largely on animal data, and employ operationally defined formulae for selecting uncertainty factors for determining a reference dose (9, 10). To be fair, the levels defined by ACGIH with adoption by OSHA become regulatory doses, as defined by Barnes and Dourson, and are acknowledged as different from the more consistently defined reference doses, which are devoid of risk management considerations. The selection by OSHA of limits recommended by ACGIH, in some cases in preference to National Institute for Occupational Safety and Health (NIOSH) recommended limits, has been a subject of some debate (1).

Explicit testing for regulatory purposes of the neurotoxic potential of most types of chemicals is a relatively recent development. Among US federal agencies, only the EPA and the Food and Drug Administration (FDA) require toxicity testing by manufacturers, and only the EPA has published explicit guidelines for neurotoxicity testing. Food additives and drugs are tested by standard systemic toxicity studies which include clinical signs and some histopathology of the nervous system as well as other target organs. Drugs that are designed for neuropharmacological activity are evaluated by the manufacturer in a variety of tests. Data on these drugs are reviewed separately by the FDA (1, 2).

In the early 1980s, the EPA developed and published the first set of explicit test guidelines for neurotoxicity testing. The first neurotoxicity guideline published was the hen-test requirement issued by the EPA's Office of Pesticide Programs in 1982 (11). This requirement specified clinical and morphological assessments of hens for determining whether organophosphates and related compounds produced delayed neurotoxicity. In addition, the pesticide data requirements included a conditional requirement for a mammalian neurotoxicity study, but no specific guideline was proposed. The only other explicit references to the detection and assessment of neurotoxic effects of chemicals were some specification of nervous system histopathology and the cageside behavioral observations made in acute, subchronic, and chronic systemic toxicity studies. Those guidelines called for periodic clinical examinations and cageside observations, including changes in "(E) autonomic and central nervous system; (F) somatomotor activity; (G) behavior pattern"; and, for acute studies, "(H) tremors, convulsions, salivation, lethargy, sleep, and coma." While the need to determine the behavioral and neurological consequences of chemical exposures is unequivocally stated, no firm guidance was provided regarding what signs to look for, how the evaluations should be carried out, or how detailed reporting needed to be. Absent, too, was the need to demonstrate the proficiency of laboratory personnel in being able to recognize either subtle or gross signs of neurotoxicity. As a result, a commonly reported conclusion in these toxicity testing studies was simply that no clinical signs of intoxication were seen.

For subchronic studies, some histopathological examination of the nervous system was also required, including three levels of brain and peripheral nerve, and conditional on signs or target organ involvement, three levels of spinal cord. For chronic studies, spinal cord sections were required. Last, determination of cholinesterase activity in subchronic and chronic studies was specified as a measurement that "may be considered necessary for an adequate evaluation" (11). As with the clinical evaluations, however, the quality and rigor with which these were made fell far short of advancing standards in neurotoxicology.

In 1985, the Office of Toxic Substances of EPA promulgated for the first time a series of testing guidelines for explicit neurotoxicological evaluations (12). Many of the guidelines were developed as a result of recommendations made by a number of national and international expert committees (13). Several of the guidelines were primarily enhancements/refinements to existing toxicological studies described above. These included a functional observational battery, motor activity, neuropathology, and an assay for neurotoxic esterase to accompany the tests of delayed neurotoxicity in hens. Other tests were proposed for inclusion as necessary as supplements to the core of tests listed above. One of these was a test of schedule-controlled operant behavior, and was included because of concern over the possible adverse effects of chemicals on acquired (learned) behavior and performance. Another test, nerve conduction velocity, was included as a functional assessment of the peripheral nervous system, which has been shown frequently to be a target site for neurotoxic chemicals. More recently, a guideline for a developmental neurotoxicity study to evaluate the neurotoxic potential of exposure during development has been prepared, and along with revisions to the other testing guidelines, was recently published by OPTS for evaluating the neurotoxic potential of both industrial and agricultural chemicals (14). A test guideline for sensory evoked potentials has also been written and is about to be proposed for public comment.

Testing performed according to these guidelines will serve as the primary basis for decision-making in these EPA offices regarding the neurotoxic potential of chemicals and the need to regulate exposures in order to protect public health. While there appears to be general consensus in the scientific community that these studies will remedy many of the past deficiencies in the hazard identification of neurotoxic substances, differences of opinion are common regarding the interpretation of data for risk assessment purposes. To a large extent these differences concern the validity of the test data. As the next section explains, however, the concept of validity is multidimensional.

ADVERSITY OF FUNCTIONAL
AND MORPHOLOGICAL ENDPOINTS

In an earlier paper, a framework for analysis of neurotoxicity data in terms of the different kinds of validity involved was described (15). That framework consisted of consideration of the extent to which effects could be said to be a consequence of exposure (content validity), the extent to which effects of different types, i.e., functional and morphological, were correlated with one another (concurrent validity), the extent to which effects in animals predicted effects in humans (predictive validity), and the extent to which effects could be said to be adverse or neurotoxic (construct validity). We now propose to further distinguish among different types of endpoint data with respect to their construct validity as adverse effects, or, in the case of behavioral endpoints, as neurotoxic effects as well.

The National Academy of Sciences (NAS) (16) has defined adverse as "impairment of functional capacity or a decrement of the ability to compensate for additional stress, a detectable decrement in the ability to maintain homeostasis, or enhancement of susceptibility to the deleterious effects of environmental influences." This definition is notable for its breadth and exclusive reliance on functional criteria. Clearly, too, these operational criteria reflect continuous phenomena, so that some judgment is required to assess the degree of change as well as its impact to support a conclusion, i.e., not all changes may be regarded as adverse.

In addition, the construct of adversity also involves other subjective elements (17). The interpretation of an exposure-related change in an endpoint may involve both the knowledge of the meaning of that endpoint with respect to the function or structure of the nervous system, as well as some judgment of the acceptability of such a change following that exposure. The acceptability of a change, in turn, may be related to psychological perception of its nature or the exposure situation. Slovic has analyzed the views of the public in terms of their perception of the risks of many phenomena and related them to a number of factors, such as whether they are voluntary (18). For example, a risk one may accept voluntarily in a hazardous occupation may be totally unacceptable for children eating breakfast. In addition, a 10% increase in reaction time may be relatively inconsequential at home, but fatal on the freeway. Decisions about the acceptability of risks may also be influenced by the way the problem is stated. Tversky and Kahneman have shown that reversals in preference can be obtained by changing the statement of the problem for questions relating to the loss of human lives, as well as other problems (19).

The principal types of endpoints in neurotoxicity studies are morphological, neurophysiological (including both electrophysiological and neurochemical measures), and behavioral measures. They differ somewhat in the way in which historically or operationally one may consider their adversity.

Most neuroanatomical changes will be regarded as adverse. This is consistent with the traditional notion in toxicology that toxicity is largely synonymous with histopathology, and the inference that such changes are generally permanent and irreversible, particularly within the central nervous system. It may be that because neuropathology data provide permanent physical evidence of effects and are less apparently affected by other variables that may affect functional measures, they are viewed as more substantive. Of course, there is a great deal of reserve capacity in the nervous system so that the consequences of a particular lesion may not be apparent (20), and peripheral effects are well known to be reversible (21). Nevertheless, most histopathology of target organs in general and the nervous system in particular are regarded as adverse.

Whether neurophysiological changes will be regarded as adverse will depend largely on their correlation with changes in anatomy or behavior. Changes in electrophysiological parameters, for example, may be related to increases in seizure susceptibility or to distal axonopathy (22). Similarly, inhibition of acetylcholinesterase (AChE) in the blood can be related to prolonged neuronal firing and the resultant signs of autonomic stimulation, e.g., diarrhea and pupillary constriction. But whether many neurophysiological changes, as commonly measured, should be regarded as adverse may be more difficult to assess for several reasons.

The sequence of physiological changes following a specific measured change and preceding the ultimate adverse consequence may not be known. Neurotoxic esterase (NTE) inhibition in brain and spinal cord is generally well correlated with the central-peripheral distal axonopathy caused by certain organophosphates, but most of the intervening steps are not understood. In contrast, the sequence of changes relating cholinesterase inhibition and autonomic stimulation is well known.

The complexity of interpretation of neurochemical changes measured in the blood (23) points to another more general problem: the available and readily obtained measures may be removed from the target. Blood changes are a step removed from the neural locus of action. While blood or urine levels may be imprecise estimators of the critical target organ concentration, they are often as good as it gets. For even the best studied neurotoxicants, like lead, they may be the best available and only practical indices of exposure and effect on which risk assessments can be based (24). However, inhibition of cholinesterase in the blood also reflects a measure of binding potency. Thus, they are more specifically related to the putative chain of events than that dictated by the pharmacokinetics governing their presence in the blood, or than simple indices of exposure where the relation of their binding to blood elements and toxicity is unknown.

Another problem is that knowledge of the extent of correlations among different measures (concurrent validity) may be limited. Many complex neurophysiological changes may be measured, and their precise or critical relation to a functional physiological consequence may be very difficult to determine. An adverse effect of a cholinesterase inhibitor may be a consequence of both the rate and duration of inhibition, for example, rather than the peak level of inhibition. Also, the correlative behavioral or morphological change may not be specific. As Bondy has noted, for example, disruption of many neurotransmitter systems may result in hyperactivity (25). This will affect interpretations of adversity of physiological effects or interpretations of neurotoxicity of behavioral effects.

Most behavioral changes may be regarded as adverse because they *are* the consequences, in terms of the definition of adverse dysfunctions given above, without reference to correlative measures. Of course, the severity and other general factors described above will be relevant to such judgments, and not all behavioral effects should be regarded as adverse. Some toxicologists have expressed concern about interpreting changes in behavior as evidence of a toxicant's effect on the nervous system. The comments heard most often are that observed changes in behavior are due either to nonspecific effects or to indirect actions on the nervous system.

Whether they are judged *neurotoxic* will depend, in part, on their correlation with anatomical, neurophysiological, or other behavioral changes. We seek the correlative changes to more clearly determine the role of the nervous system in the cause, in addition to the expression of the effect. Second, there has been a long-standing tradition of behavior analysis in neuroscience. Several decades of research have concentrated on the behavioral impact of exposure to a vast array of centrally acting drugs, of the effects of lesions of the nervous system, and their interaction. Clearly, many important correlations have been established between behavior and the nervous system. Thus, comparison to known syndromes caused by chemicals, lesions, or known physiological mechanisms, e.g., cholinesterase inhibitors, hippocampal lesions, or anoxia, form physiological or toxicological constructs that are an important means to aid in the interpretation of behavioral effects.

When interpreting changes in behavior in relation to the nervous system, one should also consider whether changes in behavior were seen only at the highest dosage, whether these were life-threatening doses, and whether there was a dose-response relationship. Hematology, clinical chemistry, and histopathological evidence of effects in other organ systems also provide important information for interpreting behavioral changes in relation to physiological mechanisms. However, it is a well-established principle in toxicology that chemicals have multiple effects. Consequently, it may prove difficult to isolate a single mechanism underlying an observed behavioral effect of a toxicant. While establishing mechanisms of action is an important goal for any science, there are few instances in neurotoxicology where mechanisms have been identified. Moreover, for a regulator charged with the protection of public health, it is probably sufficient to know that a compound produces adverse effects, and to act accordingly, regardless of the mechanisms underlying the effect.

Sickness and malaise have often been invoked to explain changes in the behavior of laboratory animals exposed to toxicants. For example, decreases in body weight have been suggested as evidence of nonspecific toxicity, which frequently occurs following exposure of rats to neurotoxic compounds. At the same time, however, different neurotoxicants produce different patterns of effects on behavior, and so it is difficult to understand how nonspecific effects on body weight could underlie these behavioral changes. In addition, we know of no effort to operationally define sickness and malaise and the signs that accompany them. In the absence of these definitions, terms such as sickness and malaise should be viewed as nonscientific concepts that lack explanatory force.

DOSE-RESPONSE ASSESSMENT

Risk assessments may range in quantitative complexity from description of an exposure level where complaints occur to those based in part on the subjective judgments of experts as defined by Wallsten et al. for dose-response functions (26) or the metaanalysis of electrophysiological studies of lead toxicity (24). But central to most of these approaches is determination of some level associated and/or not associated with the effects of concern.

The acceptable daily intake approach is usually credited to the FDA, which is also credited with origination of the 100-fold uncertainty factor for extrapolating from animal data to a human level expected to be safe (10). Dourson and Stara have reviewed the more recent history of use of safety factors in establishing acceptable daily intakes (27). Further, Barnes and Dourson have recommended several revisions in terminology, particularly, referring to uncertainty factors instead of safety factors, margins of exposure instead of margins of safety, and other changes (10). They define the margin of safety or exposure as the ratio of the estimated exposure levels to the highest exposure levels that cause no adverse effects, i.e., the NOAEL.

The US EPA uses this standardized approach to risk assessment for noncarcinogenic chemicals. The RfD (for reference dose) or RfC (reference concentration) approach typically uses data taken from a laboratory experiment divided by an uncertainty factor to arrive at an exposure level that is considered to be without significant risk (10). In other words,

$$RfD = NOAEL/UF$$

The first step in the process involves selecting a NOAEL (a non-observable adverse effect level). The NOAEL is taken from the study that is considered to be the best in terms of many experimental design considerations, definition of the most sensitive endpoint available, and establishing any dose-response relationship. If a NOAEL is not available, a LOAEL may be substituted. The uncertainty factor (UF) is included as a means of compensating for imprecise prediction of nontoxic human exposure levels on the basis of laboratory data. The uncertainty factor is generally a multiple of 10. One factor of 10 is included to compensate for individual differences in susceptibility to the toxicant. Another factor of 10 is included when extrapolating from laboratory animal data (i.e., the assumption is that humans may be 10 times more sensitive to the effects of the toxicant than is a species of laboratory animal). Another factor of 10 may be used for estimating chronic exposure levels from subchronic data. An additional factor of 3 to 10 may sometimes be included if a LOAEL is used. While not shown in the equation above, a modifying factor (MF) may also be included in the denominator. The value of the MF can range from 1 to 10, and is included to reflect the completeness of the database upon which the RfD is based.

Laboratory investigations that firmly establish relationships between exposure level and the degree (or incidence) of neurological or behavioral impairment provide fundamental contributions to the field. Such data are, of course, necessary for further studies that more precisely define the locus of neural action, delve into the mechanisms underlying the impairment, or compare related chemicals for similarity of effect. Under most circumstances, however, human exposures rarely approach those used in the laboratory with animals. We are, therefore, more interested in estimating the likelihood of impairment for humans exposed to much lower levels of the toxicant, or in estimating exposure levels that are ineffective in producing the types of impairments obtained in laboratory animal investigations.

It should be clear from the above description that the RfD, the level of exposure presumed to be without risk for the human population, can be several orders of magnitude below levels shown experimentally to be without effect. While there have been many criticisms of the RfD approach, perhaps none is more basic than the fact that the RfD is not an estimate of risk, but rather an exposure estimate associated with the absence of risk. Conversely, real-world exposures can and do exceed the RfD, and under these circumstances, little can be predicted regarding either the incidence or severity of adverse effects likely to result until exposures reach the levels tested.

There has been considerable interest recently in the use of more quantitative approaches to risk estimation in neurotoxicity. Two fundamentally different approaches have been pursued. One approach uses information on the shape of the dose-response curve to estimate levels of exposure associated with relatively small effects, for example, a 1, 5, or 10% change in a biological endpoint. Such changes are often difficult to establish experimentally but can be estimated based on extrapolation of data obtained from the measurable portion of the dose-response curve. Methods of this sort have been developed by Crump (28), Dews (29), and colleagues (30, 31).

In the approach developed by Dews, for example, effects of increasing concentrations of a solvent are determined on the performance of individual mice. Most solvents produce a stimulation of performance followed by a decrease as the concentration increases. Dews fit linear functions to the descending limb of the concentration-response function and then estimated the concentration associated with a 10% decrement in performance for each mouse. When similar calculations were carried out using a large number of mice, the distribution of values was approximately normal when concentrations were expressed as natural logarithms. As a result, by referring to the Student's t-distribution, Dews could estimate the concentration of a solvent likely to produce a 10% decrement in performance in, say, 1 in 100 or 1 in 1,000 exposed mice. The approach developed by Crump is somewhat similar. Like the approach of Dews, Crump first fits a mathematical function to data. Crump then defines a benchmark dose as the lower limit of the confidence interval associated with a small increase in risk (e.g., a 5 or 10% increase in incidence) over background. Crump then advocates applying uncertainty factors to arrive at an RfD.

The approaches of Dews and Crump are similar in that both involve fitting a mathematical function to data and providing some estimate of the variability in exposure levels associated with a relatively small effect. Both methods also place premiums on well-conducted experiments since estimates of exposure levels are generally inversely related to the magnitude of variability in the data. The methods differ, however, in terms of the data used in the calculations, and how risk estimates change with sample size (32).

An alternative approach has recently been described by Gaylor and Slikker (33, and this volume). In their approach, the variability in an endpoint between animals in an untreated control group is used to estimate extreme values, for example those greater than 3 standard deviations from the mean. A mathematical function is then fit to the control and dose group data, and the level of exposure associated with a value that exceeds 3 standard deviations of the control mean is estimated. Knowledge of the shape of the dose-response function along with the variability obtained under control conditions can then be used to estimate the likelihood of risk of an adverse, i.e., extreme, effect at each of the exposure levels.

The approach of Gaylor and Slikker is similar to that of Dews and of Crump in that it involves fitting mathematical functions to data, but differs from the other approaches by determining exposure levels associated with an adverse effect on the basis of the variability obtained under control conditions. Their approach also differs in that the greater the control variability, the larger the estimated exposure level associated with an adverse effect.

THRESHOLDS

The concept of thresholds (or their absence) for toxic effects has been a central assumption distinguishing between the two major approaches to quantitative risk assessment: derivation of reference doses through application of uncertainty factors (8) or the estimation of a virtually safe dose (34) through extrapolation of dose-response models. The use of these different approaches leads to differences in the assessment of risks, particularly at low doses, and in risk characterization.

Low-dose extrapolation techniques are used exclusively for carcinogenic risk assessment and separate it from other effects of concern, although nonthreshold models of other effects, e.g., neurotoxicity, have been suggested (35). Various dose-response models are applied to carcinogenicity data sets and estimates of tumor incidence derived for all exposures above zero. The central basis of the notion that carcinogens may not have thresholds is the mechanistic theory that a single molecular genotoxic event may trigger a carcinogenic response. It is often noted as supporting evidence, too, that if observed dose-response relations are linear in the observed range, then this is consistent with linear models assuming no threshold.

While the selection of a regulatory dose or incidence based on extrapolation of the dose-response curve is arbitrary, the most common incidence described is 10^{-6}, or one in a million. Endpoints with thresholds are assessed by the use of uncertainty factors applied to the empirical NOAEL. Most reference doses based on systemic effects use uncertainty factors of 100 to 1,000. For most chemicals where both types of effects are seen, this will lead to the prediction of risks of carcinogenicity at levels below those estimated to cause other effects, even if the data show systemic effects at lower doses. The use of low-dose extrapolation has been a source of great controversy in the regulatory arena since first used (36), and is an issue that continues to spark debate (37) within the area of carcinogenicity.

There are four different ways in which the term threshold is used. This term also has its own history and special meaning with respect to evaluation of the nervous system. We hope that by reviewing these differences and some current papers on this topic, we can help to focus discussion more clearly on the issues and the possibilities and limits of research.

Thresholds may be thought of first as a limit of detection of a system such as the visual system or an assay such as a toxicity study. The concept of thresholds in sensory systems can be traced to the origin of modern scientific psychology in the mid-19th century Germany with Weber and Fechner. Weber's law stated that the just noticeable difference between two stimuli is a constant proportion of the change in stimulus intensity. Fechner posited that there was a logarithmic relation between the change in the intensity of a stimulus and the perception of a difference (38). Signal detection theory is a modern extension of those psychophysical studies that began at that time. Signal detection methods can be applied to the analysis of assays in terms of their accuracy, independently of the relative frequencies of the outcomes and of the bias of the system (39).

The limit of detection of an assay can be thought of as its power. EPA neurotoxicity studies use ten rats/sex/dose and are required to determine a minimum or no effect level; thus, an empirical threshold will be established. Dews (40, 41) has described the relation between numbers of mice and the error variance in response rate data from schedule-controlled behavior and noted that, practically, it should be generally feasible to detect 20% changes in rate (six mice, 180 observations), or even 10% (750 observations), but it is unlikely that much lower limits could be directly measured. He argues, too, that increases in numbers of subjects will also not be sufficient and that, "in real life, we cannot detect an LD.001 even with such large groups," i.e., 300 animals. For EPA guideline studies and for explicit neurotoxicity studies, the theoretical limit given to sample size without regard to error variance is an ED.05, if you can combine the sexes. As Dews concludes, "The no-effect level is indeterminate" (31). In the most simple sense, most neurotoxicity studies will define empirical thresholds in the vicinity of an ED10 at best.

Second, a threshold can be a mathematical assumption; many common models for fitting curves to dose-response data may assume the existence of no threshold. For example, the probit model is described by the Food Safety Council as follows: "This model was originally introduced in the context of drug standardization, in which responses in the 5 to 95% range are of most interest so that it was mathematically convenient to assume no threshold for the individual tolerances" (42). In this sense, one may adopt a model for descriptive purposes without positing anything about the biological plausibility of the assumptions. It is impossible to meaningfully differentiate between different dose-response models in the observed range, i.e., they are all generally linear (30), so that assertions of the linearity of observed data should not be taken to support no threshold over threshold models or vice versa. Third, we may think of a threshold as a characteristic of the mechanism of effect of the agent in question. In the "one hit" theory of carcinogenicity, a single molecule can initiate changes in DNA whose ultimate expression is a tumor in the affected individual. Silbergeld has proposed and reviewed several mechanisms of neurotoxicity that she suggests would be expected to display different dose-response curves, and for which some would and others would not be expected to display thresholds (35). She encourages investigators to propose models that will lead to studies to validate those models. Clearly, the ultimate resolution of the shape of the critical dose-effect curve will rely on both a qualitative and quantitative understanding of the target organ dose and the mechanism of toxicity. As Burbacher et al. (43) noted for methyl mercury, interspecies extrapolation improved when target doses were compared. But in many other respects this may prove difficult for many of the practical reasons noted above.

For lead in children, there has been a steady decline in levels shown to have effects and some have suggested that no threshold is apparent (35). While this may partly be a function of the inadequacy of most measures of cumulative exposure, i.e., blood levels, as indicators of either body burden or target organ doses, in some more recent studies, better estimates of chronic exposure (dentine levels) have been available. Of course, they are still indirect measures of exposure at the neural target. But the limited data may still be "interpreted to demonstrate the lack of a threshold" for certain agents (35). And this should be a cause of concern. It would be more precise to characterize such data in a positive way, i.e., that certain low levels of exposure are associated with effects.

Fourth, thresholds may be conceived of in terms of population sensitivity; that is, if a single exposure can cause an effect, then some individual in the population may respond to such an insult, and therefore it is predicted that the observed dose-response relationship may apply for any exposure. But is this biologically plausible? It certainly seems that in a practical sense, even if such a mechanism were operating, there is a minimum level of exposure in terms of both intensity and duration for there to be a reasonable probability of a molecule's reaching the right place at the right time (44). Second, as Cornfeld has noted, just because a distribution is linear or log linear in the observed range does not mean that there are entries below the observed range, i.e., there are no 6-inch tall men (45). In short, it seems difficult to embrace a hypothesis of open-ended population sensitivity that is outside our ability to "know" in a scientific sense.

Swets reviewed the difficulty of establishing whether a sensory threshold exists for humans and concludes, "We have then, the possibility of a threshold, but it is no more than a possibility, and we must observe that, since it is practically unmeasurable it will not be a useful concept in experimental practice" (46). Dews makes a similar analysis and criticism of the possibility of ever determining the shape of the lower portion of the dose-response curve in bioassays, as noted above. Since the idea of no threshold seems experimentally untestable, it falls then to policy judgment to sustain a construct that will not be scientifically defensible and to defend any actions based on it.

"I see nobody on the road," said Alice.
"I only wish I had such eyes," the king remarked with a fretful tone. "To be able to see Nobody! And at that distance too! Why it's as much as I can do to see real people in this light."

Lewis Carroll
from *Through the Looking Glass* (47)

RISK CHARACTERIZATION

Risk characterization is the last of the four steps in the risk assessment process and consists of the description of the estimated risks of adverse health effects to those exposed. It is the integration of the dose-response and dose-effect data with the estimates of the nature, duration, and intensity of exposures and the characteristics of the exposed populations (3). Ultimately, the risk assessment approach must also satisfy the statutory language under which the material is being regulated and the policy of the regulating agency. The terms "unreasonable adverse effects" (FIFRA), "unreasonable risk of adverse effects" (TSCA), "reasonable certainty of no harm" (FFDCA), or "ample margin of safety" (Clean Water Act) are examples of such statutory language. As Barnes and Dourson have noted, the derivation of the RfD may precede and be different from the ultimate determination of regulatory doses and what may be judged to be an "ample" margin of exposure (10).

Risks may be described a number of different ways. The result of most EPA dose-response assessments is a reference dose based on a NOAEL or LOAEL. The exposure assessment may ultimately be expressed as an estimated daily intake or exposure level for one or a number of activities and populations. For example, we may be interested in the exposure to the applicator of a pesticide when mixing, loading, and applying the pesticide, to the farmworker entering the field after application, and the consumers who may be exposed to residues on the treated vegetables. For dietary consumption of commodities, EPA uses a computer program based on national food surveys by the Department of Agriculture to estimate the consumption by adults, children, and infants of different commodities. With these data, we can estimate exposures to a single commodity based on the residue data or limits (tolerance), or the total exposure to a chemical for all the registered commodities (total maximum residue concentration). Tolerances are maximum permissible residues based on efficacious use of the product and not on health or environmental effects.

The simplest expression of a risk characterization, then, may be whether exposures exceed the NOAEL or reference dose. Since the reference dose incorporates the uncertainty (safety) factors and assumes a threshold, the assumption is that exposures equal to or less than the reference dose are likely to be without adverse effect. Alternatively, we can describe the margin of exposure, the ratio of the NOAEL, and the anticipated exposure level(s). This provides for some estimate of the distance between exposures and effects, but does not prescribe what an adequate margin might be. Better perhaps are combinations of these two types of descriptors, where one looks at the margins of exposure for the reference dose as well as for the NOAEL and any other levels of interest. This may be done for each activity associated with different exposures and for groups of people exposed to different levels as a function of their dietary consumption.

Under executive order 12291 (48), for any major rule or regulation, agencies must perform a regulatory impact analysis and submit it to the Office of Management and Budget for review. This analysis is required to show that "there is adequate information concerning the need for and consequences of the proposed action; the potential benefits to society outweigh the potential costs; and of all the alternative approaches to the given regulatory objective, the proposed action will maximize net benefits to society" (49). The cost benefit analysis may include medical costs and the cost of lost production, but this will rely on the extent to which we can predict the expected number of cases.

Use of models that predict a number of cases at relevant exposure levels will allow regulatory impact analyses to estimate the number of people expected to experience an effect from particular exposures and when combined with estimates of the medical and production loss costs/case, provide quantitative data on the benefits of risk reduction. But it may be difficult to estimate in many cases the expected risk or number of expected cases at levels between the reference dose and the NOAEL. Risk assessments employing dose-response models provide estimates of the incidence of effects in a certain range, i.e., ED10, but will also not apply above the reference dose or often for many doses believed to be relevant.

Dews provides empirical estimates based on normal distribution of the probability of seeing an effect of a given size, e.g., 10% at different exposure levels. Like cancer models, such an approach does not assume a limit to sensitivity, i.e., when does the likelihood of a 10% difference become minimal? Dews argues that we should stop at a prediction of a probability of 1 in 100 (41).

For most neurotoxicity risk assessments, and probably most other risk assessments using this approach, we will only be able to say that exposures above the RfD or regulatory dose may show an effect, and perhaps that at the ED10 and NOAEL one may have an estimate of the minimal detectable incidence, i.e., with one of ten subjects we can derive an estimate of the number of cases. But ironically, the most criticized aspect of the low-dose extrapolation technique, the specific estimates of the number of cases at any dose, allows for the quantitative cost-benefit analysis sought to support regulatory action.

Another issue that arises in risk characterization is the question of sensitive subgroups in the population. Different populations may have different sensitivities to the agent in question. This may be a function of exposure, i.e., children consume more fruit per pound of their body weight than do adults, or possess different sensitivity, as in the estimated 1 in 1,000 children who developed a form of mercury poisoning, acrodynia, after exposure to certain teething powders (50). Some would suggest separate risk assessments for such groups. While information may be limited concerning these groups, it is important to consider the known characteristics of their sensitivity in describing the risks to exposed populations. It may not be socially acceptable to exclude certain groups from activities, such as fertile women in the workplace, or possible either to identify the sensitive part of the population, like the children with sensitivity to mercury, or to control their exposure, i.e., consumer products.

Some have questioned whether the selection of a margin of exposure is a risk assessment or a risk management decision (49). A report of the Office of Management and Budget observes that "The choice of an appropriate margin of safety should remain the province of responsible risk-management officials, and should not be preempted through biased risk assessments." Barnes and Dourson propose a distinction between a reference dose and a regulatory dose to acknowledge this difference. They define a regulatory dose in relation to the NOAEL and the margin of exposure that is chosen in the context of the appropriate legal language, e.g., "ample margin of safety," and a variety of factors relevant to any risk management decision, including costs, technical feasibility, and other factors. It seems that this issue lies directly at the border between risk assessment and risk management. As Ruckelhaus observed, "Nothing will erode public confidence faster than the suspicion that policy considerations have been allowed to influence the assessment of risk" (49), but it is clearly unavoidable if individual decisions are to be made in which considerations of risk management will influence the extent of the protection afforded to the public.

REFERENCES

1. US Congress, Office of Technology Assessment. *Neurotoxicity: identifying and controlling poisons of the nervous system.* OTA-BA-436. Washington: US Government Printing Office, April 1990.
2. Bass BF, Muir WR. Summary of federal actions and risk assessment methods relating to neurotoxicity. *EPA* Contract Report No. 68-02-4228. Arlington, VA: Hampshire Research Associates, Inc. 1986;63.
3. National Academy of Sciences. National Research Council. *Risk assessment in the federal government: managing the process.* Washington: National Academy Press, 1983.
4. Johnson BL, Anger WK, Durao A, Xintaras C, eds. *Advances in neurobehavioral toxicology: applications in environmental and occupational health.* Chelsea, MI: Lewis, 1990.
5. Calne DB, Eisen A, McGeer E, Spencer P. Alzheimer's disease, Parkinson's disease, and motoneurone disease: abiotropic interaction between ageing and environment? *Lancet* 1986;2:1067–1070.
6. Tanner CM, Langston JW. Do environmental toxins cause Parkinson's disease? A critical review. *Neurology* 1990;40(Suppl. 3):17–30.
7. National Academy of Sciences, National Research Council. *Toxicity testing: strategy to determine needs and priorities.* Washington: National Academy Press, 1984.
8. Sette WF, Levine TE. Behavior as a regulatory endpoint. In: Annau Z, ed. *Neurobehavioral toxicology.* Baltimore: Johns Hopkins University Press, 1986;391–403.
9. Lehman AJ, Fitzhugh OG. Quarterly report to the editor on topics of current interest: 100-fold margin of safety. *Association of Food and Drug Officials US Q Bull* 1954;18:33–35.
10. Barnes DG, Dourson M. Reference dose (RfD): description and use in health risk assessments. *Regul Toxicol Pharmacol* 1988;8:471–486.
11. US Environmental Protection Agency. Pesticide Assessment Guidelines. Subdivision F. Hazard evaluation: human and domestic animals. Springfield VA 22161: National Technical Information Service, 1982; EPA 540/9-82-025. PB 83-15319.
12. US Environmental Protection Agency, Office of Toxic Substances. Toxic Substances Control Act test guidelines: final rule. Health effects testing guidelines. Subpart G-Neurotoxicity. Federal Register. 1985;50:39458–39470.
13. Sette WF. Adoption of new guidelines and data requirements for more extensive neurotoxicity testing under FIFRA. *Toxicol Indust Health* 1989;5:181–194.
14. US Environmental Protection Agency. Pesticide assessment guidelines. Subdivision F. Hazard evaluation: human and domestic animals. Addendum 10. Neurotoxicity. Springfield, VA 22161: National Technical Information Service, 1991; EPA 540/09-91-123. PB 91-154617.
15. Sette WF. Complexity of neurotoxicological assessment. *Neurotoxicol Teratol* 1987;9:411–416.

16. National Academy of Sciences, National Research Council. *Principles for evaluating chemicals in the environment.* Washington: National Academy Press, 1975.
17. Hartung R, Durkin PR. Ranking the severity of toxic effects: potential applications to risk assessment. *Comments Toxicol* 1986;1:49–63.
18. Slovic P. Perception of risk. *Science.* 1987;236:280–285.
19. Tversky A, Kahneman D. The framing of decisions and the psychology of choice. *Science* 1981;211: 453–458.
20. Lashley KS. *Brain mechanisms and intelligence: a quantitative study of injuries to the brain.* Chicago: University of Chicago Press, 1929.
21. Morgan JP, Penowich P. Jamaica ginger paralysis: a 44 year followup. *Arch Neurol* 1978;35: 350–353.
22. World Health Organization. *Principles and methods for the assessment of neurotoxicity associated with exposure to chemicals.* Geneva: Environmental Health Criteria 60, 1986.
23. US Environmental Protection Agency. Report of the SAB/SAP Joint Study Group on Cholinesterase. May 1990. EPA-SAB-EC-90-14.
24. Davis M, Svendsgaard D. Nerve conduction velocity and lead: a critical review and meta-analysis. In: Johnson BL, Anger WK, Durao A, Xintaras C, eds. *Advances in neurobehavioral toxicology: applications in environmental and occupational health.* Chelsea, MI: Lewis, 1990;353–376.
25. Bondy SC. Especial considerations for neurotoxicological research. *CRC Crit Rev Toxicol* 1985;14: 381–402.
26. Wallsten TS, Forsyth BH, Bedescu DV. Stability and coherence of health experts' upper and lower subjective probabilities about dose-response functions. *Org Behav Human Perf* 1983;31: 277–302.
27. Dourson ML, Stara JF. Regulatory history and experimental support of uncertainty (safety) factors. *Reg Toxicol Pharmacol* 1983;3:224–238.
28. Crump KS. A new method for determining allowable daily intakes. *Fundam Appl Toxicol* 1984;4: 854–871.
29. Dews PB. Estimation of low risks. *Pharmacologist* 1980;22:159.
30. Glowa JR, Dews PB. Behavioral toxicology of volatile organic solvents. IV. Comparison of the behavioral effects of acetone, methyl ethyl ketone, ethyl acetate, carbon disulfide, and toluene on the responding of mice. *J Am Coll Toxicol* 1987;6:461–469.
31. Glowa JR, DeWeese J, Natale ME, Holland JJ. Behavioral toxicology of volatile organic solvents. I. Methods: acute effects. *J Am Coll Toxicol* 1983;2:175–185.
32. Glowa JR. Dose-effect approaches to risk assessment. *Neurosci Biobehav Rev* 1991;15:153–158.
33. Gaylor DW, Slikker W. Risk assessment for neurotoxic effects. *Neurotoxicol Teratol* 1990;11: 211–218.
34. US Environmental Protection Agency. Guidelines for carcinogen risk assessment. Federal Register September 24, 1986;51(185):33992–34003.
35. Silbergeld EK. Developing formal risk assessment methods for neurotoxicants: an evaluation of the state of the Art. In: Johnson BL, Anger WK, Durao A, Xintaras C, eds. *Advances in neurobehavioral toxicology: applications in environmental and occupational health.* Chelsea, MI: Lewis, 1990;133–148.
36. Hutt PB. A history of risk assessment. From the transcript of an address. *International Life Sciences Institute.* Washington DC: Brookings Institute, November 22, 1989.
37. Abelson PH. Testing for carcinogens with rodents. *Science* 1990;249:1357.
38. Stebbins WC. *Animal psychophysics: the design and conduct of sensory experiments.* New York: Plenum Press, 1970.
39. Swets JA. Measuring the accuracy of diagnostic systems. *Science* 1988;240:1285–1293.
40. Dews PB. Epistemology of screening for behavioral toxicity. *Environ Health Perspect* 1978;26: 37–42.
41. Dews PB. Some general problems of neurobehavioral toxicology. In: Annau Z, ed. *Neurobehavioral toxicology.* Baltimore: Johns Hopkins University Press, 1986;424–434.
42. Food Safety Council. Quantitative Risk Assessment. In: proposed system for food safety assessment. Final report of the scientific committee of the food safety council. June 1980. Washington: Food Safety Council; 137–160.
43. Burbacher TM, Rodier PM, Weiss B. Methylmercury developmental neurotoxicity: a comparison of effects in humans and animals. *Neurotoxicol Teratol* 1990;12:191–202.

44. Dinman BD. Non-concept of NO-threshold": chemicals in the environment. *Science* 1972;175: 495–497.
45. Cornfeld J. Carcinogenic risk assessment. *Science* 1977;198:693–699.
46. Swets JA. Is there a sensory threshold? *Science* 1961;134:168–177.
47. Burger EJ. Health as a surrogate for the environment. In: Risk, ed. *Daedalus* 1990;Fall:133–154.
48. Reagan, R. Regulatory impact analyses. Executive Order 12291, Federal Register. February 19, 1981;46:33,13193.
49. US Office of Management and Budget, Office of Information and Regulatory Affairs. Regulatory program of the United States Government, 1990.
50. Warkany J, Hubbard DM. Acrodynia and mercury. *J Pediatr* 1953;42:365–386.

Neurotoxicology, edited by Hugh Tilson
and Clifford Mitchell. Raven Press, Ltd.,
New York © 1992.

15

Assessment of Neurotoxicity in Humans

W. Kent Anger

*Center for Research on Occupational and Environmental Toxicology,
The Oregon Health Sciences University, Portland, Oregon 97201*

Workplace and environmental exposures to chemicals have grown concomitantly with industrialization. As a result of this growth, there have been efforts to control chemical exposures based on available evidence of adverse effects on human health. Research to identify and characterize such health effects has contributed to many successful efforts to reduce exposure through voluntary self-monitoring in industry and regulation by federal and state governments (1). This chapter reviews human subject research aimed at identifying and characterizing the effects of chemicals on one critical organ system, the nervous system, with objective behavioral tests. This research serves as one basis for decisions on reducing exposure concentrations at the workplace.

Research directed at studying chemicals found in the workplace or in the environment takes varied forms. Clinical assessment of working populations (2,3) provided the early data on neurotoxic chemicals. Private and public health officials, however, recognized the need to more thoroughly evaluate chemical exposures. Since human exposures at the workplace did not afford sufficient control of the independent variable (chemical exposure), experimental convenience, or the potential for destructive tests (e.g., pathology), scientists began to study the effects of chemicals in animals (4). Animals are necessarily surrogates for the subject of concern regarding workplace exposures. Despite the conservative tendencies of evolution, animals differ sufficiently from humans that many questions about the neurotoxic effects of chemicals in humans cannot be answered by animal research. Thus, in the 1960s, experimental and quasi-experimental (epidemiological) research to assess the effects of chemicals on the nervous system was undertaken in human subjects (5,6).

DISTINCTIONS BETWEEN LABORATORY
AND WORKSITE RESEARCH

Human Behavioral Neurotoxicology (HBN) employs two types of research to assess the effects of toxicants on the nervous system: (a) experimental *labor tory research* in which volunteer human subjects are purposefully exposed to industrial chemicals; and, (b) quasi-experimental *worksite research* in which behavioral or neurophysiological tests are given to volunteer subjects who have been exposed to industrial chemicals where they work. The latter is termed "quasi-experimental" because, as in all epidemiological research, the independent variable is not manipulated, only characterized (7). Both types of research have been used to study the same chemicals with similar or related test methods, but these types of research have very different limitations and answer very different questions.

Human laboratory experiments typically involve short-duration exposures (2–6 hours) to industrial chemicals by inhalation, either through a mask or in a controlled environmental chamber. Most human laboratory studies expose the subjects only once to one concentration of a chemical, although as many as 6–10 subject exposures have been employed in some human laboratory studies (8,9). Therefore, these studies are classified as short-duration or acute exposure studies. Chemical concentrations are generally measured before, during, and after the exposures, and behavioral or electrophysiological tests are used to measure the effects of these chemicals on the nervous system (6).

Worksite research typically involves employees who are exposed for extended periods of time (months or years) to chemicals by inhalation, dermal contact, and/or ingestion in uncontrolled and undocumented patterns typically unique to the individual employees and their jobs. Generally, exposure measurements of airborne concentrations and, less frequently, in biological samples are taken once in a subset of job stations or workers, and behavioral tests are administered once to volunteer subjects at the worksite (5). The procedures for conducting human laboratory (6,10,11,12) and worksite research (5) have been described recently.

Chemicals Studied in Laboratory and Worksite Research

Twelve industrial solvents plus solvent combinations have been extensively studied in human laboratory experiments (13–26). These solvents are listed in the left column of Table 1, along with lead, carbon monoxide, and organophosphates which have also been studied in human laboratory research (6). Worksite research has investigated a larger number of industrial chemicals. Most prevalent in recent years is that on mixed solvent exposures. Several metals and individual solvents have also been studied extensively. Individual chemicals which have been the subject of worksite research are listed in the right column of Table 1. At least 29 different chemicals, plus several chemical combinations or mixtures, have been the focus of worksite studies (27). Those chemicals studied in both laboratory and worksite research are listed in bold type. Based on the total number of published studies, toluene is the chemical studied most extensively in laboratory research (6). Lead, mercury, and carbon disulfide are the most extensively studied chemicals in worksite research (27).

TABLE 1. *Chemicals studied in laboratory and worksite research*

Laboratory research	Worksite research
Acetone	Acrylates
Acetone and MEK	Amines
Carbon monoxide	Carbon disulfide
Carbon tetrachloride	**Carbon monoxide**
Fluorocarbon 113	Chromium
Halon 1301	Diazinon
Lead	Ethylene oxide
Methyl ethyl ketone (MEK)	Formaldehyde
Methyl chloride	Gasoline
Methyl chloroform	Jet Fuel
Methylene chloride	**Lead**
Methyl isobutyl ketone (MIBK)	Lead Stearate
MIBK and Toluene	Manganese
Organophosphates	Menthol
Propylene glycol dinitrate	Mercury
Styrene	Methyl bromide
Tetrachloroethylene	**Methyl chloride**
Toluene	**Methylene chloride**
Toluene and Ethanol	Organophosphates
Toluene and MEK	Pentaborane
Toluene and Xylene	Polybrominated biphenyls
Trichloroethylene	Silver
Trichloroethylene and Ethanol	Solvents (multiple)
White Spirits	**Styrene**
Xylene	**Tetrachloroethylene**
	Toluene
	1,1,1-Trichloroethane
	Trichloroethylene
	Trimethyltin, Methyl chloride
	White Spirits (solvents)

Limitations in Worksite and Laboratory Research

Table 1 demonstrates that there are only nine chemicals which have been studied after both short-duration (laboratory) and extended-duration (worksite) exposures. One reason for this lack of overlap is that laboratory research is limited to extensively studied chemicals that do not produce irreversible effects (due to concern for subject safety), while workplace studies are constrained by availability of subjects with relevant exposures and companies willing to allow scientists to conduct worker testing and exposure measurements in their workplace. Thus, different factors control decisions on selection of chemicals for study in laboratory and worksite research.

In recent years, ethical considerations have placed limits on the concentrations of chemicals (primarily industrial solvents) studied in laboratory experiments. Typically, the limit or maximum is the United States Permissible Exposure Limit (PEL) or, in some European countries, the Maximum Allowable Concentration (MAC) (12,28). Consequently, because of ethical and regulatory considerations, human laboratory research with solvents has not produced a definitive database with dose–response curves ranging from no effect to those concentrations which produce marked CNS depression (i.e., narcosis). Research is restricted to the lower end of the dose–response curve, where critical effects measures analogous to the measures used in traditional toxicology (e.g., ED-50) are not available.

Different problems beset worksite research. Worksite research often involves poorly characterized exposures to changing chemicals and concentrations of uncertain duration that cannot clearly be related to individual workers (i.e., study subjects). Similarly, lack of ongoing exposure and biological burden measurements to characterize exposures, and related burdens, make it difficult to relate worksite research at higher concentrations to laboratory research at lower concentrations.

The problems of restricted dose–response range and the functional importance of subclinical effects detected in human laboratory research were major topics at an international workshop held in 1970 in Prague, Czechoslovakia. From this workshop (29) and in subsequent publications (30), two general experimental approaches were identified to deal with the restrictions on developing dose-response relationships in human experimentation: (a) development of increasingly sensitive tests and, (b) use of reference substances with biologic effects similar to those of the chemical under study (29). In recent years, the increased use of tests of cognitive functions (e.g., alertness, attention, arousal, and memory) rather than tests of sensory or motor functions (12,28,31) suggests that the laboratory research scientists are adopting the approach of using more sensitive tests as well as reference compounds (32).

Worksite research poses similar concerns. Increased standardization of test methods and the use of reference substances was advocated by Laties (33) for all behavioral toxicology research. Reference substances have not been used at all in field studies. Methodological concern in worksite research has focused on the development of standardized tests and, particularly, screening batteries to identify subclinical effects of chemical exposures (11,34).

SELECTING TESTS FOR RESEARCH

A central problem in pursuing laboratory or worksite research is selecting the most appropriate tests. Criteria for selecting tests for laboratory and worksite research are difficult to establish. Valciukas (35) identified several factors to consider in the selection of tests, including:

- Reliability
- Validity
- Sensitivity
- Specificity

Reliability is essential for any test, and this is usually established by administering the test twice to a given population (test/retest reliability) or dividing a given test into two equivalent forms (split/half or alternate forms reliability). A reliability correlation of 0.90 or higher is desirable (36). However, the reliability correlations for many tests used in Human Behavioral Neurotoxicology are not published or are below 0.90. Letz (34) is one of the few to publish test/retest reliability correlations for tests in his battery (the Neurobehavioral Evaluation System). The test/retest correlations were calculated from results in laboratory studies for 10 tests and ranged from 0.51–0.91. The majority of the correlations were above 0.80 (36). Variability inherent in a population can greatly influence reliability scores and correlations need to be determined in a population very similar to the population of interest (36).

Validity is not as simple to address. There are several types of validity. The main type identified in behavioral research is construct validity, or, "does the test measure a particular psychological construct?" (36). This is a valuable asset when characterizing the nervous system deficits produced by a chemical. However, it is doubtful that construct validity is necessary in selecting screening tests. Criterion validity is the more relevant form of validity. In Behavioral Neurotoxicology, this may be equated with test sensitivity or with the ability of a test to reflect nervous system changes produced by chemicals. This form of validity still presents problems. Even when the chemical to which the subject population has been or will be exposed is established, it must be recognized that the effects of most chemicals are not well characterized. Moreover, there is always a potential for unanticipated interactions from multiple chemical exposures. Thus, important tests may be omitted from a study because a particular functional effect has not been previously identified for the chemical(s) under study. Clearly, the degree to which sensitivity to chemicals is a sufficient basis for selecting valid tests is arguable.

The specificity of a test is crucial for a test intended to characterize an identified deficit, but it is less important for a screening test. Specificity is considered on a case-by-case basis. While the four criteria discussed above are widely acknowledged as important, they are rarely addressed in the published literature. As a result, the extent to which they are weighed in the test selection process is unclear.

Other factors which should be considered (35) include:

• adequacy of a test's research base
• degree of challenge to the subject
• credibility to the subject and to the profession
• requirements for test administrators
• ruggedness and portability of test equipment (for worksite research)
• potential for clinical distinction (for worksite research)

These factors are frequently considered in selecting tests for both worksite and laboratory research. However, financial factors often play the most important role in test selection. The most frequently cited factors are the time required to administer the test and the cost of test equipment. The former affects workplace productivity; employers are often reluctant to grant more than an hour per worker for testing unless well-acknowledged problems are driving the study. For laboratory research, as well, subject costs are directly related to the time required to participate in the study. Ideally, published data on the chemical under study serves as a guide to test selection. Studies of compounds related by structure or mechanism of action can also be useful in selecting tests.

Potential confounding factors can also play a role in the test selection decision. Sex, education, and age are well established confounding factors that should be considered in any study. The degree to which these factors affect test performance varies (37). These factors can be handled by equalizing (balancing) them across test subjects, by matching, or by experimental design and statistics. Perhaps the issue most carefully considered is the power of a given test to discriminate group differences with small numbers of subjects. This highly practical factor should always be weighed when determining the number of subjects.

Frequently Employed and Sensitive Tests

Behavioral tests used in cross-sectional research to study adults in worksite settings have been recently catalogued. Two hundred and fifty unique behavioral tests in 185 published worksite studies through 1989 have been identified (27). Table 2 lists general functional areas evaluated (column 1) in behavioral worksite research and the most frequently employed test of each function (column 3) in the 185 studies reviewed in the 1990 publication (27). Column 2 lists the number of studies in which the function has been assessed; column 4 lists the number of studies in which each specific test has been administered (the functions listed are somewhat arbitrarily assigned; other categorization schemes could be developed by other authors).

Table 2 identifies **frequently used** tests in worksite research. However, a measure of criterion validity can also be drawn from the same review (27). Due to the relatively small number of research reports, it is necessary to look to the entire research database (i.e., all chemicals studied in laboratory and worksite research) to identify tests sensitive to workplace chemicals. Tests sensitive to chemicals studied in worksite research are listed in Table 3, ranked by a ratio. The S/S ratio represents the number of studies in which the test was used to the number of times the results were significant in 185 studies reviewed by Anger (ref 27; Table 3). The tests in Table 3 can be considered among the **most sensitive** in worksite research and can thus provide a basis for selecting tests for future research. The caveat must be added, however, that these results may not be predictive of test sensitivity in future research, due to the small number of studies conducted with a low number of chemicals.

A similar compilation is not available for laboratory research, but tests used frequently in laboratory research that are regularly sensitive to the effects of several solvents include: Dual Task (32), Spokes test (38), Identical Number (38), Time Discrimination (15), Digit Span (39), Compound Reaction Time (40), Simple Reaction Time (31), Tapping (9), and the Steadiness test (26).

TABLE 2. *Most frequently administered tests in worksite research*

Function	Number[a]	Test	Number[a]
Cognitive			
Memory	214	Digit Span	49
Intelligence	127	Picture Completion	26
Spatial Relations	82	Block Design	67
Coding	75	Digit Symbol	62
Vigilance	66	Bourdon-Wiersma	19
Acquisition	51	Paired Associates	29
Vocabulary	34	Synonyms	12
Calculation	22	Arithmetic	13
Concept Shifting	20	Trail Making	16
Categorization	19	Figure Classification	9
Distractibility	15	Embedded Figures	14
Motor			
Coordination/Speed	77	Santa Ana	29
Coordination	62	Tapping	25
Response Speed	55	Simple Reaction Time	52
Speed/Decision	32	Choice Reaction Time	30
Steadiness	19	Stylus in hole	5
Grip Strength	11	Dynamometer	9
Sensory			
Vision-flicker	14	Flicker Fusion/Frequency variants	13
Vision-color	12	Farnsworth variants	6
Vision-pattern	11		
Equilibrium	4		
Affect and Personality			
Personality	36	Eysenck Maudsley Personality Inventory	9
Affect	14	Neurobehavioral Evaluation System Mood Test	4

[a] Number of tests of this function (column 2), or the number of times the identified test of the function has been used (columns 4 and 6) [Drawn from ref. 27] [Modified from ref. 61].
[b] The Similarities test has been used in 25 studies.

TABLE 3. *Most sensitive tests used in worksite research[a]*

Cognitive		Motor	
Test	S/S[b]	Test	S/S[b]
Rey Memory Tests	0.81	Stylus-in-hole	0.80
Raven Progressive Matrices	0.80	Mira Test	0.69
Sternberg Memory tests	0.80	Santa Ana	0.62
Embedded Figures	0.64	Symmetry drawing	0.57
Bourdon–Wiersma	0.63	Choice reaction time	0.55
Arithmetic	0.54	Simple reaction time	0.52
Benton Memory tests	0.50	Tapping	0.48
Similarities	0.48	Grooved pegboard	0.44
Picture Completion	0.46	Michigan Eye–Hand Coordination	0.44
Digit Span	0.43		
Digit–Symbol/Symbol–Digit	0.42		
Block Design	0.40		

[a] Certain tests were excluded from Table 3. Questionnaires that are uniformly sensitive in both laboratory and field research are not included in Table 3 because none are used consistently from study to study. Autonomic and electrophysiologic tests are used infrequently in either laboratory or worksite research and are thus not included in Table 3. Since tests used in less than five studies have an insufficient history on which to base decisions, such tests are not included in Table 3.
[b] Number of studies in which test detected a *significant* change divided by the number of *studies* in which the test was used.

The tests identified above as sensitive in laboratory research are different from those in Table 3 identified as sensitive in worksite research. This difference reflects, in part, the different influences in the respective research fields. Experimental psychologists, drawing on their rich century-old tradition, employ tests typical of their field in laboratory research, while neuropsychologists, drawing on their equally long tradition with its roots in intelligence testing, employ tests typical of their field in worksite research. Some crossover does exist. As noted above, the Digit Span test from the Weschler Adult Intelligence Scale (WAIS) has been among the most sensitive cognitive tests in laboratory research (and in worksite research; see Table 3). Conversely, the Sternberg and Bourdon–Wiersma tests, both drawn from experimental psychology research, are among the most sensitive cognitive tests used in field research. Most of the motor tests used in both field and laboratory research are experimental psychology tests.

FUNCTIONS AFFECTED BY SHORT- AND LONG-DURATION EXPOSURES

Some of the most sensitive worksite tests in Table 3 are, surprisingly, tests of general intelligence (Raven Progressive Matrices) or classic tests of intelligence from the WAIS (Similarities, Picture Completion, Digit Span, Digit–Symbol or its variant Symbol–Digit, and Block Design). Such tests have not been frequently employed in laboratory research, nor have they been sensitive to the chemicals studied when they have been used (41,42). This suggests that long-term exposures affect some very significant and broad intellectual functions. By contrast, in laboratory research, simple laboratory tests of monitoring and attention have been sensitive to short-duration exposures (6). This may be due to test selection or to a fundamental difference in the types of functions affected by short-duration and long-duration exposures.

While many tests have been used in both laboratory and worksite research, only in a small number of cases have the same tests been used to assess subjects exposed to the same chemical for both short- and long-duration exposure periods. This is an important question that needs study since chronic exposures encountered at the workplace could potentially affect workers adversely before worksite research discovered the problem. Laboratory research could provide an efficient and safe means of predicting effects of extended-duration chemical exposures (28). This would require a much greater understanding of the predictive value of short-duration exposure effects for long-duration exposure effects. According to this hypothesis, the established value of laboratory research for identifying effects related to narcosis, irritation and safety for regulatory (PELs, MACs) and recommended (REL, TLV®) workplace standards (43,44) would be significantly increased.

Table 4 lists the three tests in which significant adverse effects have been identified in *both* laboratory and worksite research on human subjects exposed to the same chemical: Digit Span, Simple Reaction Time, and Choice Reaction

TABLE 4. *Tests used in both laboratory and worksite research and the chemicals studied*

Test Focus and Test Name	Chemicals in laboratory studies	Chemicals in worksite studies
Cognitive		
Digit Span	Halothane	Anesthetics
	Nitrous oxide	Carbon disulfide
	Toluene	Ethylene oxide
	Trichloroethylene	Formaldehyde
		Gasoline
		Lead
		Mercury
		Organophosphates
		Pesticides
		Solvents
		Tetrachloroethylene
		Toluene
Motor		
Simple Reaction Time	Acetone	Anesthetics
	Methyl chloroform	Carbon disulfide
	Methyl chloride	Carbon monoxide
	Methyl ethyl ketone	Ethylene oxide
	Methyl isobutyl ketone	Jet fuel
	Nitrous oxide	Lead
	Styrene	Manganese
	Toluene	Mercury
	Trichloroethylene	Organophosphates
	White spirit	Solvents
		Styrene
		Tetrochloroethylene
Choice Reaction Time	Acetone	**Toluene**
	Halon 1301	Anesthetics
	Methyl chloroform	Carbon disulfide
	Methyl ethyl ketone	Carbon monoxide
	Methyl isobutyl ketone	Formaldehyde
	Nitrous oxide	Lead
	Styrene	Mercury
	Toluene	Methyl chloride
	Xylene	Organophosphates
		Pesticides
		Solvents
		Styrene
		Tetrachloroethylene
		Trichloroethylene

Time tests. Also listed in Table 4 are all other chemicals studied in research which employed these tests, regardless of outcome. Chemicals in **bold** type are those used in both laboratory and worksite research for a given test. The two chemicals that have been studied with the same tests are toluene and styrene. The results of the studies on these two chemicals thus provide a basis for assessing the relationships between short- and long-duration exposures. The results are summarized in Table 5 where *only* statistically significant deficits in performance are listed from worksite and laboratory studies of toluene and styrene.

TABLE 5. *Significant findings in worksite and laboratory research on the same chemicals in which one test was significant in both types of research*

Worksite (extended-duration) Exposures			Laboratory (short-duration) Exposures		
Test	Function	Reference	Test	Function	Reference
			Toluene		
Digit Symbol	Coding	45	Spokes	Attention	38
Embedded Figures	Distractibility	91	Digit Span	Memory	22
Benton Visual Retention	Memory	45	Pattern Recognition	Memory	22
Block Design	Spatial Relations	91	Visual Vigilance	Vigilance	13,94,95
Grooved Pegboard	Coord.,* speed	46,47	Identical Number	Vigilance	38
Simple Reaction Time	Speed, coord.	24	Dual Task	Vigilance+	40
			One-hole	Coordination	22
			Pegboard	Coordination	49
			Simple Reaction Time	Speed, coord.	22
			Compound Reaction Time	Speed, coord.	40
			Choice Reaction Time	Speed, coord., decision	38
			Landolt Ring	Vision-acuity	49
			Color Discrimination	Vision-color	49
			Styrene		
Logical Memory	Memory	52	Crawford	Coord.	58
Word Memory	Memory	52	Flanagan Coordination	Coord.	58
Bourdon-Wiersma	Vigilance	52,92	Wire Spiral	Coord.	40
Simple Reaction Time	Speed, coord.	50,53,54,55,56	Simple Reaction Time	Speed, coord.	40
Choice Reaction Time	Speed, coord., decision	51,52,57	Compound Reaction Time	Speed, coord.	20
Kuhnburg Figure Matching	Vision-pattern	92,93	Choice Reaction Time	Speed, coord., decision	20
Rorschach Inkblot	Personality	92,93	Vestibular Response	Equilibrium	96

* Coord. = coordination.

Toluene

As documented in Table 5, only one test, Simple Reaction Time, has detected significant adverse effects in both laboratory (22) and worksite research (24) on toluene. Deficits have also been reported on other tests involving both speed and coordination following extended toluene exposures on the grooved pegboard (46,47) and after short-duration exposures on the Choice and Compound Reaction Time (40,48), One Hole (22) and Pegboard tests (49). This supports the findings with the Simple Reaction Time test of the same function. There is also significant overlap in the functions studied. Memory deficits have been seen on the Benton Visual Retention test following extended toluene exposures (45) and on the Digit Span and Pattern Recognition tests following short-duration exposures (22).

Styrene

Tests of both Simple Reaction Time and Choice Reaction Time, as documented in Table 5, detected statistically significant performance changes following both extended-duration (50–57) and short-duration exposures (20, 40) to styrene. Other tests involving speed and/or coordination have also been sensitive to styrene exposures. These are Compound (visual and auditory) Reaction Time (40), the Crawford and Flanagan (58) and Wire Spiral tests of manual dexterity (40) in studies of short-duration styrene exposures.

Considering that 185 worksite studies (27) and over 70 laboratory studies have been conducted, the lack of comparable results is discouraging. An examination of the results from the studies of common chemicals using common tests does not begin to provide the answer to the question of the relation between short- and extended-duration exposure effects. The comparable test data are too sketchy to draw significant conclusions, although there is no real support in these data for the hypothesis that nervous system effects of short-duration exposures are related to those produced by long-duration exposures in humans. Rather, the lack of use of similar tests points to the need to conduct parallel research in a systematic fashion if this question is to be answered. Tables 4 and 5 provide a direction for beginning that process. More laboratory research is needed on toluene and styrene to expand existing knowledge. The short-duration exposure studies on several solvents employed tests that could be applied in worksite studies of people exposed to the same solvents for extended periods of time. Worksite studies on individuals exposed to acetone, methyl ethyl ketone, and methylene chloride could provide new comparisons, since a relatively large laboratory research base exists on these chemicals (6). Of course, further research at the worksite and in the laboratory on all five of these chemicals is needed to provide a sufficient base for reliable conclusions, to provide expansion in the functions tested, and to replicate the sparse findings in Table 5.

It must be repeated that the limitations noted above militate against the systematic choice of chemicals for research. Rather, availability of a subject population is the controlling factor in worksite research, while laboratory research is limited to relatively safe chemicals and very low concentrations. However, when possible, research which would contribute to the comparison of short- and extended-duration exposures should be pursued because it will eventually provide the basis for an evaluation of the impact of the exposure duration variable on nervous system changes produced by neurotoxic chemicals.

BEHAVIORAL TEST BATTERIES

Two significant methodological changes occurred in Human Behavioral Neurotoxicology research during the 1980s. One change was a huge growth in the diversity of tests used in field research (27); the second change was the development of standardized behavioral testing batteries used primarily in field research. This strong trend toward the use of test batteries in worksite research may bring a degree of standardization to this area of research in the early 1990s and may have an impact on laboratory research as well.

The first behavioral test battery for use in neurotoxicology research was developed during the 1970s at Finland's Institute of Occupational Health (FIOH) (59). This battery has had a significant impact on worksite neurotoxicology (11), although its developer has recently proposed the use of a more focused battery for screening (60) based on results with the more extensive FIOH battery. Since that time, several batteries have been implemented in worksite settings. Some of these batteries (61) are no longer used (London School of Hygiene Battery, [47]; Test Battery for Investigating Functional Disorders, [62]; Microtox Test System or MTS, [63]), have a limited application base (Hänninen's FIOH battery; Swedish Performance Evaluation System or SPES, [64]; [65, 66], or are too new to evaluate (most computer-implemented batteries, [67, 68], [60, 69]).

There are five batteries available for administration in the English language which can be considered suitable candidates for broad adoption in worksite research. These are in current use, have seen variously broad application in occupational settings, and appear suitable for screening situations. They are Williamson's (70) unnamed battery based on Information Theory; the NCTB (11); the NES (34); CNS-B (71), and POET (72). The tests in each of these five batteries are listed in Table 6.

The Neurobehavioral Core Test Battery (NCTB) was developed at a 1983 meeting in Cincinnati sponsored by the World Health Organization (WHO) and the U.S. National Institute for Occupational Safety and Health (NIOSH) (73). Experts conducting worksite research in the 1970s and early 1980s were convened to consider the possibility that a single test battery could be recommended to identify neurotoxic effects in working populations. The concept developed at that meeting was a core set of tests to be used routinely, recommending selection

TABLE 6. *Behavioral test batteries used in behavioral neurotoxicology research*

Australian Battery[a]	World Health Organization (WHO) Neurobehavioral Core Test Battery (NCTB)[b]
Critical Flicker Fusion (CFF)	Santa Ana
Vigilance	Aiming
Simple Reaction Time	Simple Reaction Time
Visual Pursuit	Digit Symbol
Hand Steadiness	Benton Visual Retention Test
Sensory Store Memory	Digit Span
Sternberg Memory Test	Profile of Mood States (POMS)
Paired Associates (STM)	
Paired Associates (LTM)	

Neurobehavioral Evaluation System (NES)[c]

Psychomotor	Memory and Learning	Cognitive
Symbol-Digit[d]	Digit Span[d]	Vocabulary
Hand-Eye Coordination	Paired-Associate Learning	Horizontal Addition
Simple Reaction Time[d]	Paired-Associate Recall	Switching Attention
Continuous Performance Test	Visual Retention	Grammatical Reasoning
Finger Tapping	Pattern Memory	Color Word
	Memory Scanning	
	Serial Digit Learning	

Perceptual Ability	Affect
Pattern Comparison	Mood Test

California Neuropsychological Screening (CNS/B I and II) Battery[e]	Pittsburgh Occupational Exposures Test (POET) Battery[g]
Information (WMS)[h]	Information (WMS)[h]
Orientation (WMS)	Visual Reproduction (WMS)
Mental Control (WMS)	Associate Learning (Verbal) (WMS)
Memory Passages (WMS)	Digit Span (WAIS)[f]
Visual Reproduction (WMS)	Block Design (WAIS)
Associate Learning (WMS)	Digit Symbol (WAIS)
Digit Span (WAIS)[f]	Similarities (WAIS)
Block Design (WAIS)	Picture Completion (WAIS)
Arithmetic (WAIS)	Symbol Digit
Digit Symbol (WAIS)	Symbol Digit Recall (delayed 30 min.)
Trail Making A & B	Trail Making
Dynamometer	Grooved Pegboard
Finger Tapping Test	Incidental Memory
Purdue Pegboard	Recurring Words
Benton Visual Retention Test	Boston Embedded Figures
Neurotoxic Anxiety Scale	Mental Rotation
Visual Information Processing	

[a] Source: references 70 and 83.
[b] Source: reference 11.
[c] Source: references 34 and 78.
[d] Variant of WHO core test.
[e] Source: reference 71.
[f] WAIS = Wechsler Adult Intelligence Scale.
[g] Source: reference 72.
[h] WMS = Wechsler Memory Scale.

of additional tests depending on the chemical exposure and the setting where testing will be carried out (11). Individually, the tests in the WHO-recommended NCTB (Table 6) are among the most widely employed in past worksite research on persons exposed to neurotoxic chemicals (27). The NCTB has been used, as a 7-test battery, primarily in control studies to develop a normative data base (in the U.S., numbering 900 subjects from ages 16–65) (37).

When the NCTB was developed, there was concern over its feasibility for neurotoxicity assessments in populations which differed from those found in the Western European and derivative cultures where the tests were developed and validated, and where their sensitivity to chemicals was determined. To solve this problem of the feasibility of its use in diverse cultures, a Cross-Cultural Assessment (CCA) was developed under WHO auspices aimed at administering the battery in a minimum of eight countries with diverse cultures (73). By mid-1991, 2300 subjects (1303 males and 1028 females) between the ages of 16 and 65 years in ten countries were administered the NCTB. These subjects were primarily working adults who were *not* exposed to neurotoxic chemicals where they worked and were tested in field settings. The countries (and primary subject occupations) contributing data to the CCA were Hungary (agricultural workers), Poland (power plant workers), Austria (household workers), France (auto manufacturers), Italy (cleaners), the Netherlands (city residents), People's Republic of China (embroidery workers), USA (postal workers), Canada (utility workers), and Nicaragua (cattle farmers). These countries represent a variety of cultures, although the European and U.S. base remains dominant. Subjects in most countries were paid or received time off from work to participate in the CCA, and test administrators had a set of administration instructions (an Operational Guide) and received common training on how to administer the tests.

Exemplary results from the CCA are shown in Figures 1 and 2. The results demonstrate that mean scores on some NCTB tests (Simple Reaction Time, Benton) are very similar from country to country, while there is more variability between countries on the Digit Span, Digit Symbol, Santa Ana, and Aiming test scores. The data shown in the figures are representative of all age ranges. Data from male Nicaraguan male subjects is substantially inferior to that in other countries on all but the Santa Ana test. This could reflect cultural differences in this country, but a more likely explanation lies in the lack of education in those subjects. About 74% of the Nicaraguan subjects had 3 or less years of education, while virtually all subjects in other countries had 8 or more years of education. Performance on the NCTB tests correlated very highly with years of education (0.33–0.69) in the Nicaraguan subjects (37).

The variables of age and sex had a significantly smaller effect on performance in the CCA, as shown in Figure 3. Age- and sex-related differences were greatest in the Digit–Symbol test, a fact consistent with results in the neuropsychological literature in different types of populations (74,75). Figure 3 demonstrates the limited impact these variables had on the Digit–Symbol test. Clearly, the NCTB produced consistent results in many countries, although it is equally clear that

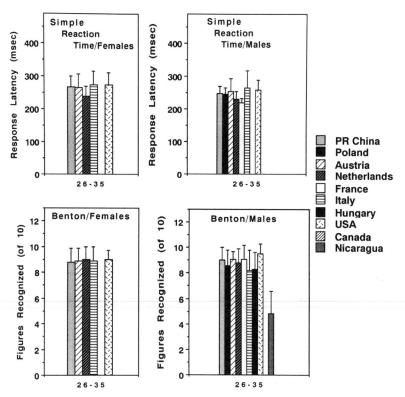

FIG. 1. Response latency in msec on the Simple Reaction Time and number of figures recognized in the Benton Visual Retention Test from all countries supplying data on 10 or more male or female subjects in the 26–35 age range. Error bars reflect +1 s.d.

control data from one country cannot be used as a comparison basis for data collected in exposed subjects in another country. Very likely the same can be said of within-country research (37,76).

The Neurobehavioral Evaluation System (NES) is the behavioral battery with the broadest and deepest application in human behavioral neurotoxicology (34,77,78). The tests are listed in Table 6. The NES is implemented on a computer, which provides a highly efficient means of administering tests to literate subjects (who must read the instructions) and recording data from them. The sensitivity of computer-implemented behavioral tests to chemicals is under study, as most such tests derive from tests administered by an individual (who speaks the instructions and essentially verifies or judges that they were understood). The computer-implemented variants of these tests thus, strictly speaking, do not have demonstrated sensitivity to chemicals (criterion validity) in their modified form. Some results of studies with chemical exposures are encouraging, but others

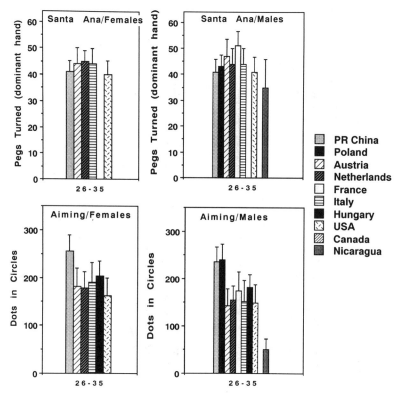

FIG. 2. Number of pegs turned on the Santa Ana Dexterity Test and the number of dots placed in circles on the Aiming Test for all countries supplying data on 10 or more male or female subjects in the 26–35 age range. Error bars reflect +1 s.d.

appear mixed from the standpoint of demonstrating the sensitivity of the tests to neurotoxic chemical exposures (34). However, it is likely that lower concentration exposures are now being studied, due to reduced exposure concentrations in industry and in laboratory studies where the sensitivity of this battery must be evaluated.

The NES is distributed widely and has been used extensively in international settings where it has been translated into several languages. This battery has been used in several studies of workplace chemicals. Three NES tests (Simple Reaction Time, Symbol-digit, and Serial Digit Learning) are now being used in the U.S. National Health and Nutrition Evaluation Study (NHANES) III, which will create a huge representative database (>5,000 people) for these tests. Many other NES tests also have a large representative database (34).

The CNS-B (71) and Pittsburgh Occupational Exposures Test (POET) battery

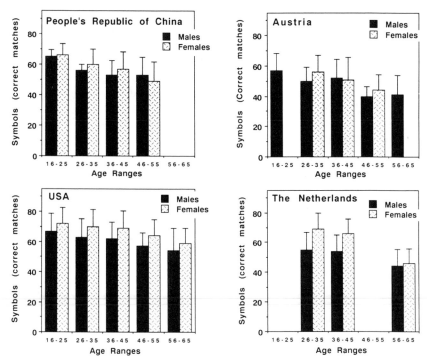

FIG. 3. Number of symbols (+s.d.) correctly matched on the Digit Symbol Test by male and female subjects from the People's Republic of China, the Netherlands, the United States, and Austria.

(72) have been developed from the neuropsychological tradition with its strength in individual analysis and diagnosis (the tests in these similar batteries are listed in Table 6). Neuropsychological testing was developed to assess intellectual abilities and has been applied to testing injury victims and to certain neurological diseases where such test results have successfully localized CNS damage (79). This addresses a significant issue in worksite behavioral neurotoxicology evaluations where it is important to identify people in need of medical attention or compensation. However, these neuropsychological batteries and some of their component tests have not been extensively used in community or occupational settings with normal adults. Thus, the normative base may not yet be sufficient or appropriate for analysis of individual subjects in the relevant settings. Further, neuropsychological tests have only been used infrequently to assess chemically-exposed individuals (80–82), so their predictive utility and potential for characterizing chemical effects is not yet established for neurotoxic chemical exposures. The value of these batteries for screening for nervous system effects would appear to be established by their use of tests proven sensitive to chemical exposures

(e.g., tests in the NCTB; [27,80]). It is not clear that they have greater potential than other established batteries such as the NCTB or that the goal of individual assessment will be achieved. Further research is needed to answer these questions.

The [Information Theory] Battery of Williamson et al. (70,83) has been used in Australia with working populations, and the individual tests have been used with volunteers, in most cases, with college students in laboratory research in the U.S. and Europe. Its strength is its base in a theoretical framework. While Information Theory is somewhat dated, it has stood the test of time without extensive fundamental modification, although refinements continue to be made (84). Williamson (83,85,86) has studied workers exposed to lead, mercury, and oxygen deficits (abalone divers) and found different explanations for their deficits based on the different results on the various tests. If Information Theory proves to be substantially correct, this battery clearly has the potential for very important contributions to the field of characterizing major categories of nervous system effects. However, a larger base of empirical data is needed for this battery before it can be recommended over batteries with tests of proven sensitivity to neurotoxic chemicals.

SELECTING A BEHAVIORAL TEST BATTERY

The "best" test battery depends on the basis for selections. The NCTB provides a recommended core set of tests, and suggests the addition of other tests when possible and when indicated by the chemical or identified symptoms under study (11). The NCTB is the most obvious choice as a suitable battery due to its use of tests of proven sensitivity to neurotoxic chemicals. The CNSB and POET can be recommended for the same reason, but the utility of the additional tests selected to provide a more complete neuropsychological assessment for neurotoxic chemical exposures remains unproven. The NES provides an array of tests and a reliable and efficient mode for test administration. The NES is also a good battery due to its use of tests with demonstrated sensitivity, although the fact that they are modifications of those tests is a concern that must be acknowledged at present. Another advantage of the NES and NCTB is that both have been applied in a range of countries and cultures. Their instructions have been translated into other languages (34,37,76). However, evidence of their utility in nonindustrialized countries is limited at present. It must be remembered that sensitivity to a handful of neurotoxic chemicals (now studied) as a basis for selecting tests does not guarantee sensitivity to other, as yet unstudied chemicals. Further, there is a lack of knowledge about mechanisms of effect or structure-activity relationships for most neurotoxic chemicals, so the selection of tests sensitive to structurally related chemicals is not a useful basis for choosing tests. Although there are clear limitations to the use of sensitivity to neurotoxic chemicals as the criterion for selecting tests, it is the best available criterion at the present time.

In a laboratory setting, other strategies can be employed more readily for test battery selection. Some laboratory studies have used ethanol as a positive control, particularly for solvents, to demonstrate test sensitivity (13,32) of untried tests in pilot experiments. The lack of knowledge of the relation of ethanol's effects to the effects of other chemicals, however, limits this approach. Williamson (70), in the development of her battery, employed Information Theory as a basis for selecting tests. This theory has stood the test of time, but it's explanatory value is limited to a subset of cognitive functions (84), and Williamson has reported only on lead, mercury, and presumed oxygen lack (83,85,86) to demonstrate the value of the battery for identifying and characterizing (by the theory) the effects of occupational chemicals.

Selection of tests or a test battery thus poses a challenge for most researchers, although, based on current knowledge, some suitable choices exist. However, the fragility of the bases for selection must be acknowledged, and researchers must be vigilant to new knowledge in the neurosciences, cognitive theory, and neurotoxicology that could lead to better test selection strategies.

INTERNATIONAL RECOMMENDATIONS

Development of methods for research on the nervous system in humans has been the focus of four triennial meetings titled the International Symposia on Neurobehavioral Methods in Occupational and Environmental Health. In the second meeting of this Symposium, there were strong supporters and detractors of the WHO-recommended NCTB centered in Europe where this meeting was held (87). In 1988, at the third convening of this Symposium, the scientists assembled in a workshop to discuss test batteries recommended, "without dissent," inclusion of the NCTB as a core "in all clinical and field studies of neurobehavioral function" (88). In 1991, the fourth such meeting also convened workshops on computerized and noncomputerized (i.e., human-administered) batteries. The latter group continued to support the use of the NCTB as a core, although it also acknowledged the need to maintain flexibility in developing new, supplementary tests. The most important needs identified were the evaluation of potential confounding factors (e.g., education, sex, age) (89) and, in the workshop on computerized batteries, the identification of a core battery for computerized tests (90). As noted above, Iregren and Letz (90,90a) have suggested three tests for such a battery. The use of standardized methods also needs attention by scientists conducting laboratory research (33,34,64).

The focus brought by these meetings continues to be on the development of standardized test batteries for worksite research and perhaps for laboratory research (although, as pointed out by Dick and Johnson in 1986 [6], the driving force in laboratory research appears to be the need for maximally sensitive tests by using strategies such as varying test difficulty). The development of standardized batteries has been the most exciting area of Human Behavioral Neurotoxicology research in recent years due to the wide-scale adoption of the NES in industrialized countries (34) and widespread use of the core tests of the NCTB in many studies (27).

ACKNOWLEDGMENTS

Substantial contributions to the laboratory research sections of this chapter by Dr. Robert B. Dick (Division of Biomedical and Behavioral Sciences, National Institute for Occupational Safety and Health, Cincinnati, Ohio, U.S.A.) are acknowledged with appreciation. Wendy Carlton and Joyce Mikelson provided skillful word processing of the manuscript.

REFERENCES

1. Stewart RD. Solvent seminar keynote address. In: Xintaras C, Johnson BL, deGroot I, eds. *Behavioral Toxicology.* Cincinnati: NIOSH (DHHS) Pub. No. 74-126, 1974;35–40.
2. Hamilton A. Industrial poisons encountered in the manufacture of explosives. *J Am Med Assoc* 1917;LXVII:1445-1447.
3. Vigliani EC. Carbon disulphide poisoning in viscose rayon factories. *Br J Ind Med* 1954;11:235–244.
4. Irish DD, Adams EM, Spencer HC, Rowe VK. The response attending exposure of laboratory animals to vapors of methyl bromide. *J Ind Hyg Toxicol* 1940;22:218-230.
5. Anger WK. Workplace exposures. In: Annau Z, ed. *Neurobehavioral Toxicology.* Baltimore: Johns Hopkins University Press, 1986;331-347.
6. Dick RB, Johnson BL. Human experimental studies in neurobehavioral toxicology. In: Annau Z, ed. *Neurobehavioral toxicology.* Baltimore: Johns Hopkins University Press, 1986;348-87.
7. Elmes DG, Kantowitz BH, Roediger III, HL. *Research Methods in Psychology,* 2nd ed. New York: West Publishing Company, 1985.
8. Stewart RD, Hake CL, Wu A, et al. *Methyl chloride: Development of a biological standard for the industrial worker by breath analysis.* NTIS Report No. PB81-167686, 1977.
9. Savolainen K, Riihimaki V, Seppäläinen AM, Linnoila M. Effects of short-term m-xylene exposure and physical exercise on the CNS. *Int Arch Occup Environ Health* 1980;45:105–121.
10. Gamberale F. Use of behavioral performance tests in the assessment of solvent toxicity. *Scand J Work Environ Health* 1985;11(suppl 1):65–74.
11. Johnson BL, Baker EL, El Batawi M, Gilioli R, Hänninen H, Seppäläinen AM, Xintaras C, eds. *Prevention of Neurotoxic Illness in Working Populations.* New York: John Wiley and Sons, 1987.
12. Gamberale F. Critical issues in the study of the acute effects of solvent exposure. *Neurotox Teratol* 1989;11:565-570.
13. Dick RB, Setzer JV, Wait R, Hayden MB, Taylor BJ, Tolos B, Putz-Anderson V. Effects of acute exposure of toluene and methyl ethyl ketone on psychomotor performance. *Int Arch Occup Environ Health* 1984;54:91–109.
14. Putz VR, Johnson BL, Setzer JV. A comparative study of the effects of carbon monoxide and methylene chloride on human performance. *J Environ Pathol Toxicol* 1979;2:97–112.
15. Putz-Anderson V, Setzer JV, Croxton JS, Phipps FC. Methyl chloride and diazepam effects on performance. *Scand J Work Environ Health* 1981;7:8–13.
16. Matsushita T, Goshima E, Miyakaki H, Maeda K, Takeuchi Y, Inoue T. Experimental studies for determining the MAC value of acetone. 2. Biological reactions in the "six-day exposure" to acetone. *Sangyo Igaku* 1969;11(10):507–15.
17. Stewart RD, Gay HH, Erley DS, Hake CL, Peterson JE. Human Exposure to Carbon Tetrachloride vapor—Relationship of expired air concentration to exposure and toxicity. *J Occup Med* 1961;3:586–590.

18. Reinhardt CF, McLaughlin ME, Maxfield ME, Mullin LS, Smith PE. Human exposures to fluorocarbon 113 (1,1,2-trichloro-1,2,2-trifluoroethane). *Am Ind Hyg Assoc J* 1971;32:143–52.
19. Pierson DL, Cintron NM, Pool SL. *Halon 1301 human inhalation study.* Houston, TX, National Aeronautics and Space Administration, Report JSC 23845, August 1989.
20. Gamberale F, Hultengren M. Exposure to styrene II. Psychological functions. *Work Environ Health* 1974;11:86–93.
21. Gamberale F, Annwall G, Olson BA. Exposure to trichloroethylene III. Psychological functions. *Scand J Work Environ Health* 1976;4:220–224.
22. Echeverria D, Fine A, Langolf G, Schwork A, Sampaio C. Acute neurobehavioral effects of toluene. *Br J Ind Med* 1989;46:483–495.
23. Hjelm EW, Hagberg M, Iregren A, Lof A. Exposure to methyl isobutyl ketone: toxicokinetics and occurence of irritative and LNS symptons in man. *Int Arch Occup Environ Health* 1990;62: 19–26.
24. Iregren A. Effects on psychological test performance of workers exposed to a single solvent (toluene)—A comparison with effects of exposure to a mixture of organic solvents. *Neurobehav Toxicol Teratol* 1982;4:695–701.
25. Olson BA, Gamberale F, Iregren A. Coexposure to toluene and p-xylene in man: Central Nervous System functions. *Br J Ind Med* 1985;42:117–122.
26. Ferguson RK, Vernon RJ. Trichloroethylene in combination with CNS drugs. *Arch Environ Health* 1970;20(4):462–67.
27. Anger WK. Worksite behavioral research: Results, sensitive methods, test batteries and the transition from laboratory data to human health. *Neurotoxicology* 1990;11:629–720.
28. Dick RB. Short duration exposures to organic solvents: The relationship between neurobehavioral test results and other indicators. *Neurotox Teratol* 1988;10:39–50.
29. Horvath M. *Adverse effects of environmental chemicals and psychotropic drugs-quantitative interpretation of functional tests. Vol. 1.* New York: Elsevier, 1973.
30. Horvath M. *Adverse effects of environmental chemicals and psychotropic drugs-neurophysiological and behavioral tests, Vol. 2.* New York: Elsevier, 1976.
31. Gamberale F, Annwall G, Hultengren M. Exposure to methylene chloride II. Psychological functions. *Scand J Work Environ Health* 1975;1:95–103.
32. Dick RB, Setzer JV, Taylor BJ, Shukla R. Neurobehavioral effects of short duration exposures to acetone and methyl ethyl ketone. *Br J Ind Med* 1989;46:111–121.
33. Laties VG. On the use of reference substances in behavioural toxicology. In: Horvath EM, ed. *Adverse Effects of Environmental Chemicals and Psychotrophic Drugs.* New York: Elsevier, 1973;83–88.
34. Letz R. The Neurobehavioral Evaluation System: An international effort. In: BL Johnson, WK Anger, A Durao, and C Xintaras, eds. *Advances in Neurobehavioral Toxicology: Applications in Environmental and Occupational Health.* Chelsea: Lewis Publishers, Inc., 1990;189–201.
35. Valciukas JA. *Foundations of environmental and occupational neurotoxicology.* New York: Van Nostrand Reinhold, 1991.
36. Walsh WB. *Tests and Measurements, 4th ed.* Englewood Cliffs: Prentice Hall, 1989.
37. Anger WK, Cassitto MG, Liang Y-x, et al. *Comparison of performance from three continents on the WHO-recommended Neurobehavioral Core Test Battery (NCTB).* Submitted.
38. Gamberale F, Hultengren M. Methylchloroform exposure II. Psychophysiological functions. *Work Environ Health* 1973;10:82–92.
39. Cook TL, Smith M, Starkweather JA, Winter PM, Eger EI. Behavioral effects of rare and subanesthetic halothane and nitrous oxide in man. *Anesthesiology* 1978;49:419–424.
40. Oltramare M, Desbaumes E, Imhoff C, Michiels W. *Toxicology of styrene monomer: Experimental and clinical research in man.* Geneva: Medicine & Hygiene, 1974;84–100.
41. Bruce DL, Bach MJ. *Effects of trace concentrations of anesthetic gases on behavioral performance of operating room personnel.* DHEW-NIOSH Report No. 76-169. Cincinnati: NIOSH Publication Office, 1976.
42. Garfield JM, Garfield FB, Sampson J. Effects of nitrous oxide on decision-strategy and sustained attention. *Psychopharmacologia (Berl)* 1975;42:5–10.
43. National Institute for Occupational Safety and Health (NIOSH). *NIOSH pocket guide to chemical hazards.* Cincinnati: National Institute for Occupational Safety and Health (DHHS), June, 1990.
44. American Conference of Governmental Industrial Hygienists (ACGIH). Threshold Limit Values for Chemical Substances and Physical Agents in the Workroom Environment, 1990–1991. Cincinnati: ACGIH Publication Office.

45. Ørbæk P, Nise G. Neurasthenic complaints and psychometric function of toluene-exposed rotogravure printers. *Am J Ind Med* 1989;16:67–77.
46. Cherry N, Johnston JD, Venables H, Waldron HA, Buck L, Mackay J. The effects of toluene and alcohol on psychomotor performance. *Ergonomics* 1983;26(11):1081–87.
47. Cherry N, Venables H, Waldron HA. Description of the tests in the London School of Hygiene Test Battery. *Scand J Work Environ Health* 1984;10(suppl 1):18–19.
48. Gamberale F, Hultengren M. Toluene exposure II. Psychophysiological functions. *Work Environ Health* 1972;9:131–139.
49. Baelum J, Andersen I, Lundqvist G, Molhave L, Pedersen O, Væth M, Wyon D. Response of solvent-exposed printers and unexposed controls to 6-hour exposure. *Scand J Work Environ Health* 1985;11:271–280.
50. Kjellberg A, Wigaeus E, Engstöm J, Åstrand I, Ljungquist E. Långtidseffekter av styrenexposition vid en plastbåtsindustri. *Arbete och Halsa* 1979;18:1–23 (abstract reviewed).
51. Mutti A, Mazzucchi A, Rustichelli P, Frigeri G, Arfini G, Franchini I. Exposure-effect and exposure-response relationships between occupational exposure to styrene and neuropsychological functions. *Am J Ind Med* 1984;5:275–286.
52. Mutti A, Mazzucchi A, Frigeri G, Falzoi M, Argini G, Franchini I. Neuropsychological investigation on styrene exposed workers. In: Gilioli R, Cassitto MG, Foá V, eds. *Advances in the Biosciences. Neurobehavioral Methods in Occupational Health.* New York: Pergamon Press, 1983;46:271–281.
53. Mutti A, Arfini G, Ferrari M, Muzzucchi A, Posteraro L, Franchini I. Chronic styrene exposure and brain dysfunction. A longitudinal study. In: *Neurobehavioural Methods in Occupational and Environmental Health.* Copenhagen, World Health Organization (Environmental Health Document 3), 1985;136–140.
54. Götell P, Axelson O, Lindelöf B. Field studies on human styrene exposure. *Work Environ Health* 1972;9:76–83.
55. Cherry N, Waldron HA, Wells GG, Wilkinson RT, Wilson HK, Jones S. An investigation of the acute behavioural effects of styrene on factory workers. *Br J Ind Med* 1980;37:234–240.
56. Gamberale F, Lisper HO, Anshelm-Olson B. The effect of styrene vapour on the reaction time of workers in the plastic boat industry. In: Horváth M, ed. *Adverse Effects of Environmental Chemicals and Psychotropic Drugs.* New York: Elsevier Scientific Publishing Co., 1976;2:135–148.
57. Mackay CJ, Kelman GR. Choice reaction time in workers exposed to styrene vapour. *Hum Toxicol* 1986;5:85–89.
58. Stewart RD, Dodd HC, Baretta ED, Schaffer AW. Human exposure to styrene vapor. *Arch Environ Health* 1968;16(5):656–62.
59. Hänninen H, Lindström K. Neurobehavioral Test Battery of the Institute of Occupational Health. Helsinki: Institute of Occupational Health, 1979, 1989 (revised).
60. Hänninen H. The neuropsychological screening test battery: Validation and current uses in Finland. In: Johnson BL, Anger WK, Durao A, Xintaras C, eds. *Advances in Neurobehavioral Toxicology: Applications in Environmental and Occupational Health.* Chelsea: Lewis Publishers, Inc., 1990;257–262.
61. Anger WK, Johnson BL. Human behavioral neurotoxicology: Workplace and community assessments. In: Rom W, ed. *Environmental and Occupational Medicine, 2nd ed.* New York: Little, Brown and Company, in press.
62. Hogstedt C, Hane M, Axelson O. Diagnostic and health care aspects of workers exposed to solvents. In: Zenz C, ed. *Developments in occupational medicine.* Chicago: Year Book Medical Publishers, 1980;249–258.
63. Eckerman DA, Carroll JB, Foree D, et al. An approach to brief field testing for neurotoxicity. *Neurobehav Toxicol Teratol* 1985;7:387–393.
64. Gamberale F, Iregren A, Kjellberg A. Computerized performance testing in neurotoxicology: Who, what, how, and where to? The SPES Example. In: Russell RW, Flattau PE, Pope AM, eds. *Behavioral Measures of Neurotoxicity.* Washington D.C.: National Academy Press, 1990;359–394.
65. Angotzi G, Cassitto MG, Camerino D, et al. Rapporti tra esposizione a mercurioe candizioni di salute in un gruppo di lavoratori addetti alla distillazione di mercurio in uno stabilimento della provincia di Siena. (Correlations between exposure to mercury and health in a group of workers at a mercury distillation plant in the province of Siena). *Med Lav* 1980;71:463–480.

66. Schneider H, Seeber A. Psychodiagnosis in the assessment of neurotoxic effects of chemical substances. *Z Psychol* 1979;187:178–205.

67. Kennedy RS, Wilkes RL, Dunlap WP, Kuntz LA. Development of an automated performance test system for environmental and behavioral toxicology studies. *Percept Mot Skills* 1987;65: 947–962.

68. Englund CE, Reeves DL, Shingledecker CA, Thorne DR, Wilson KP, Hegge FW. *Unified triservice cognitive performance assessment battery (UTC-PAB) I. Design and specification of the battery.* Naval Health Research Center Report No. 87-10. San Diego: Naval Health Research Center, 1987.

69. Almirall-Hernández P, Mayor-Rios J, del Castillo-Martin N, Rodriguez-Notario R, Román-Hernández J. Manual de recomendaciones para la evaluación psicológica en trabajadores expuestos a sustancias neurotóxicas. Havana, Ministerio de Salud Publica, Instituto del Trabajo, Departmento de Psicologiá, 1987.

70. Williamson AM. The development of a neurobehavioral test battery for use in hazard evaluations in occupational settings. *Neurotoxicol Teratol* 1990;12:509–514.

71. Bowler RM, Thaler CD, Becker CE. California Neuropsychological Screening Battery (CNS/B I & II). *J Clin Psychol* 1986;42:946–955.

72. Ryan CM, Morrow LA, Bromet EF, Parkinson DK. Assessment of neurological dysfunction in the workplace: Normative data from the Pittsburgh Occupational Exposures Test battery. *J Clin Exp Neuropsychol* 1987;9:665–679.

73. Anger WK, Cassitto MG. Individual-administered human behavioral test batteries to identify neurotoxic chemicals. Submitted.

74. Nolan BH, Swihart AA, Pirozzolo FJ. The neuropsychology of normal aging and dementia: An introduction. In: Wedding D, Horton Jr. AM, Webster J, eds. *The Neuropsychology Handbook: Behavioral and Clinical Perspectives.* New York: Springer Publishing Co., 1986;410–446.

75. Albert MS, Heaton RK. Intelligence Testing. In: Albert MS, Moss MB, eds. *Geriatric Neuropsychology,* 1988;13–32.

76. Cassitto MG, Camerino D, Hänninen H, Anger WK. International Collaboration to Evaluate the WHO Neurobehavioral Core Test Battery. In: Johnson BL, Anger WK, Durao A, Xintaras C, eds. *Advances in neurobehavioral toxicology: Applications in environmental and occupational health.* Chelsea: Lewis Publishers, Inc., 1990;203–223.

77. Baker EL, Letz R, Fidler A. A computer-administered neurobehavioral evaluation system for occupational and environmental epidemiology. Rationale, methodology, and pilot study results. *J Occup Med* 1985;27:206–212.

78. Letz R, Baker EL. Computer-administered neurobehavioral testing in occupational health. *Sem Occup Med* 1986;1:197–203.

79. Lezak MD. *Neuropsychological assessment,* 2nd ed. New York: Oxford University Press, 1983.

80. Bowler RM, Mergler D, Huel G, Harrison R, Cone J. Neuropsychological impairment among former microelectronics workers. *Neurotoxicology* 1991;12:87–104.

81. Parkinson DK, Ryan C, Bromet EJ, Connell MM. A psychiatric epidemiologic study of occupational lead exposure. *Am J Epidemiology* 1986;123:261–269.

82. White RF, Feldman RG. Neuropsychological assessment of toxic encephalopathy. *Amer J Ind Med* 1987;395–398.

83. Williamson AM, Teo RKC, Sanderson J. Occupational mercury exposure and its consequences for behavior. *Int Arch Occup Environ Health* 1982;50:273–286.

84. Simon HA, Kaplan CA. Foundations of cognitive science. In: Posner MI, ed. *Foundations of cognitive science.* Boston: Massachusetts Institute of Technology Press, 1989;1–47.

85. Williamson AM, Teo RKC. Neurobehavioral effects of occupational exposure to lead. *Br J Ind Med* 1986;43:374–380.

86. Williamson AM, Clarke B, Edmonds C. Neurobehavioral effects of professional abalone diving. *Br J Ind Med* 1987;44:459–466.

87. Grandjean P. *Neurobehavioral Methods in Occupational and Environmental Health: Symposium Report.* Copenhagen: World Health Organization Regional Office for Europe, 1985.

88. Eckerman DA. What should clinical and field-testing batteries for neurobehavioral screening encompass? In: BL Johnson, WK Anger, A Durao, and C Xintaras, eds. *Advances in Neurobehavioral Toxicology: Applications in Environmental and Occupational Health.* Chelsea: Lewis Publishers, Inc., 1990;477–480.

89. Anger WK, Cassitto MG. Report of workshop IIa. WHO (NCTB) and other neurobehavioral test batteries. Submitted.
90. Iregren A, Letz R. Computerized testing in neurobehavioral toxicology. *Arbete och Halsa,* in press.
90a. Iregren A, Letz R. Summary of workshop on "Computerized Test Batteries," submitted.
91. Hänninen H, Antti-Poika M, Savolainen P. Psychological performance, toluene exposure and alcohol consumption in rotogravure printers. *Int Arch Occup Environ Health* 1987;59:475–483.
92. Lindström K, Harkonen H, Hernberg S. Disturbances in psychological functions of workers occupationally exposed to styrene. *Scand J Work Environ Health* 1976;3:129–139.
93. Lindström K, Harkonen H. The effects of styrene exposure on psychological performance. In: Klimikova-Deutschova E, Lukas E, eds. *Proceedings of the Second International Industrial and Environmental Neurology Congress.* Prague: Universita Karlova, 1976;74–77.
94. Horvath M, Frantik E, Krekule P. Diazepam impairs alertness and potentiates the similar effect of toluene. *Act Nerv Super (Praha)* 1981;23(3):177–79.
95. Stewart RD, Hake CL, Forster HV, Lebrun AJ, Peterson JE, Wu A. *Toluene: Development of a biological standard for the industrial worker by breath analysis.* Cincinnati: DHEW-NIOSH Contract Report No. 99-72-84, 1975.
96. Odkvist LM, Larsby B, Tham R, et al. Vestibulo-oculomotor disturbances in humans exposed to styrene. *Acta Otolaryngol (Stockh)* 1982;94:487–93.
97. Harkonen H, Lindström K, Seppäläinen AM, Asp S, Hernberg S. Exposure-response relationship between styrene exposure and central nervous functions. *Scand J Work Environ Health* 1978;4:53–59.
98. Repko JD, Jones PD, Garcia LS, Schneider EJ, Roseman E, Corum CR. *Behavioral and Neurological Effects of Methyl Chloride.* USDHEW (NIOSH) Publication No. 77-125. Cincinnati: NIOSH Publication Office, 1976.

Subject Index